MW01248811

PERINATAL STEM CELLS

PERINATAL STEM CELLS

Edited by

Curtis L. Cetrulo, M.D.
Kyle J. Cetrulo
Curtis L. Cetrulo, Jr., M.D.

A JOHN WILEY & SONS, INC., PUBLICATION

Published by John Wiley & Sons, Inc., Hoboken, New Jersey
Published simultaneously in Canada

Library of Congress Cataloging-in-Publication Data

Perinatal stem cells / edited by Curtis L. Cetrulo, Kyle J. Cetrulo, Curtis L. Cetrulo Jr.
p. ; cm.
Includes bibliographical references and index.
ISBN 978-0-470-42084-3 (cloth)
1. Stem cells. 2. Placenta. 3. Amniotic liquid. I. Cetrulo, Curtis L. II. Cetrulo, Kyle J. III. Cetrulo, Curtis L., 1975-
[DNLM: 1. Stem Cells. 2. Placenta—cytology. 3. Umbilical Cord—cytology. QU 325 P445 2009]
QH588.S83P47 2009
616'.02774—dc22

2009020888

Printed in the United States of America

10 9 8 7 6 5 4 3 2 1

CONTENTS

PREFACE

The cover of the February 9, 2009 issue of *Time** magazine proclaimed that the "revolution in stem cells could save your life." They were right to do so, for advances in stem cell research have already saved many lives. The authors of the article, however, in their exclusive examination of embryonic stem cells and induced pluripotent stem cells, failed to acknowledge the multiple sources of potentially valuable, life-sustaining stem cells, including cells from bone marrow, umbilical cord blood, adult peripheral blood, amniotic fluid, the placenta, the lining of the amniotic sac, and the Wharton's jelly of the human umbilical cord.

With this book, we wish to educate the public, both scientists and lay people alike, about the incredible potential of perinatal stem cells, the stem cells generated as a result of pregnancy that fall on the continuum between embryonic stem cells and adult stem cells. We know, for instance, that stem cells from umbilical cord blood are medicinally valuable as a treatment of hematological disorders, and we suspect that stem cells from the Wharton's jelly may prove to be the source used for co-transplantation with hematopoetic stem cells to improve engraftment and aid the treatment of hypoxic ischemic encephalopathy or even cancer.

When American scientists decided to try to put a man on the moon, they did not explore a single option for doing so, but instead, threw their collective effort behind the many avenues for success in seeking a new paradigm. So too now must the current scientific community cast their net wide to rein in the multiple solutions available from the full variety of stem cells proven to be life-sustaining. And we don't need to go to the moon to get them—they are right here, readily accessible through non-invasive, non-controversial means. The countdown begins...

*February 9, 2009, V173, No.5, p38–43.

CONTRIBUTORS

Selim Aractingi, MD, PhD, Developmental Physiopathology Laboratory EA4053, UPMC University of Paris 6, Paris, France

Karen K. Balien, MD, Division of Hematology/Oncology, Department of Medicine, Massachusetts General Hospital Boston, MA

Monica Betancur, MSc, Tufts Medical Center, Molecular Oncology Research Institute Division of Hematology and Oncology, Boston, MA

Laurent Boissel, PhD, Tufts Medical Center, Molecular Oncology Research Institute Division of Hematology and Oncology, Boston, MA

Christian Breymann, MD, Feto Maternal Haematology Unit, Obstetric Research, University Hospital Zurich, Switzerland

Curtis L. Cetrulo, Jr., MD, Division of Plastic and Reconstructive Surgery, Department of Surgery, Keck School of Medicine, University of Southern California, Los Angeles, CA

Julio C. Davila, PhD Pfizer Inc., Pfizer Global Research and Development, St. Louis, MI

Kenneth Dorko, Department of Pathology, University of Pittsburgh, Pittsburgh, PA

Ewa C.S. Ellis, PhD, Department of Pathology, University of Pittsburgh, Pittsburgh, PA

Jonas B. Galper, MD, PhD, Molecular Cardiology Research Institute, Cardiology Division, Department of Medicine, Tufts New England Medical Center, Boston, MA

Serban P. Georgescu, MD, Molecular Cardiology Research Institute, Cardiology Division, Department of Medicine, Tufts New England Medical Center, Boston, MA

Roberto Gramignoli, DSc, Department of Pathology, University of Pittsburgh, Pittsburgh, PA

Richard L. Haspel, MD, PhD, Department of Pathology, Beth Israel Deaconess Medical Center, Boston, MA

Kiarash Khosrotehrani, MD, PhD, Developmental Physiopathology Laboratory EA4053, UPMC Univ Paris 6, Paris, France

Hans Klingemann, MD, PhD, Molecular Cardiology Research Institute, Cardiology Division, Department of Medicine, Tufts New England Medical Center, Boston, MA

Dequen Kong, MD, Molecular Cardiology Research Institute, Cardiology Division, Department of Medicine, Tufts New England Medical Center, Boston, MA

Taxiarchis Kourelis, MD, Department of Pharmacology and Experimental Therapeutics, Tufts University School of Medicine, Boston, MA

Michéle Leduc, PhD, Developmental Physiopathology Laboratory EA4053, UPMC University of Paris 6, Paris, France

Barbara Lutjemeier, Department of Anatomy and Physiology, Kansas State University, Manhattan, KS

Akrivi Manola, MD, Department of Pharmacology and Experimental Therapeutics, Tufts University School of Medicine, Boston, MA

James Marchand, PhD, Tufts University, Department of Anesthesia Research and Pharmacology, Boston, MA

Fabio Marongiu, PhD, Department of Pathology, University of Pittsburgh, Pittsburgh, PA

Hanna Mikkola, MD, PhD, University of California Los Angeles, Los Angeles, CA

Toshio Miki, MD, PhD, Department of Pathology, University of Pittsburgh, Pittsburgh, PA

Jack Naggar, MD, Molecular Cardiology Research Institute, Cardiology Division, Department of Medicine, Tufts New England Medical Center, Boston, MA

Ho-Jin Park, PhD, Molecular Cardiology Research Institute, Cardiology Division, Department of Medicine, Tufts New England Medical Center, Boston, MA

Aarati R. Ranade, PhD, Department of Pathology, University of Pittsburgh, Pittsburgh, PA

Kartin E. Rhodes, University of California Los Angles, Los Angles, CA

Andreina Schoeberlein, PhD, Dept. of Obstetrics and Gynecology, and Department of Research, Inselspital, Bern University Hospital and University of Bern, Switzerland

James L. Sherley, MD, PhD, Boston Biomedical Research Institute, Watertown, MA

Margaret J. Starnes, MD, Division of Plastic and Reconstructive Surgery, Department of Surgery, Keck School of Medicine, University of Southern California, Los Angeles, CA

Stephen C. Strom, PhD, Department of Pathology, University of Pittsburgh, Pittsburgh, PA

Professor Daniel Surbek, MD, Department of Obstetrics and Gynecology, and Department of Research, Inselspital, Bern University Hospital and University of Bern, Switzerland

Rouzbeh R. Taghizadeh, PhD, Programs in Regenerative Biology and Cancer, Boston Biomedical Research Institute, Watertown, MA

Theoharis C. Theoharides, PhD, MD, Laboratory of Molecular Immunopharmacology and Drug Discovery, Department of Pharmacology and Experimental Therapeutics, and Departments of Internal Medicine and Biochemistry, Tufts University School of Medicine, Tufts Medical Center, Boston, MA

Deryl L. Troyer, DVM, PhD, Department of Anatomy and Physiology, Kansas state University, Manhatton, KS

Ming-Song Tsai, MD, Director, Dept. of OBS & GYN, Cathay General Hospital, Taipei, Taiwan, Assistant Professor, School of Medicine, Fu-Jen Catherine University, Taiwan

Anna Wagner, MD, Department of Obstetrics and Gynecology, and Department of Research, Inselspital, Bern University Hospital and University of Bern, Switzerland

Professor Mark L. Weiss, Department of Anatomy and Physiology, Kansas State University, Manhattan, KS

Yali Zhang, MD, PhD, Molecular Cardiology Research Institute, Cardiology Division, Department of Medicine, Tufts New England Medical Center, Boston, MA

INTRODUCTION: PERINATAL STEM CELLS

Curtis L. Cetrulo, MD

Professor, Obstetrics and Gynecology, Tufts University School of Medicine, Boston, MA 02111

Kyle J. Cetrulo, CEO

Auxocell Laboratories, 816 Washington Street, Brookline, MA 02446

Curtis L. Cetrulo, Jr., MD

Division of Plastic and Reconstructive Surgery, Department of Surgery, Keck School of Medicine, University of Southern California, Los Angeles, CA 90033

THE STEM CELL ERA IN MEDICINE

The study of stem cells is beginning to define a new era in medicine and biomedical research. The investigation of these remarkable cells promises potential therapies for numerous diseases and injuries. Stem cells are loosely defined as cells that are able to replicate into many different cell types, while retaining the ability to replenish their own numbers. More precisely, the essential characteristics of a stem cell include the ability to: (1) retain the potential for self-renewal and (2) sustain the population of at least one functional, differentiated cell type. Numerous sources of stem cells exist.

HUMAN EMBRYONIC STEM CELLS

Perhaps the most potentially potent and certainly the most controversial stem cell, human embryonic stem cells (hESCs) are acquired from the inner cell mass of a human blastocyst. These cells have powerful proliferative capacity and can give rise to all cell types from all three germ layers. However, despite these incredible features, therapeutic applications and even progress in basic research using hESCs has been slow to develop due to the heated ethical and political controversy surrounding the inevitable destruction of the human blastocyst that results when the inner cell mass is extracted.

Perinatal Stem Cells. Edited by C. L. Cetrulo, K. J. Cetrulo, and C. L. Cetrulo, Jr.

ADULT STEM CELLS

The study of adult stem cell types skirts the ethical quandary surrounding human embryonic stem cells, and many sources of adult stem cells have been identified, including bone marrow, peripheral blood, adipose tissue, and even periodontal tissue. While many of these stem cell populations are able to differentiate into various cell types, their use is ultimately limited by the requirement for a donor (with an accompanying invasive harvest procedure), immunologic restrictions inherent to transplantation of "non-self" cells, a low capacity for proliferation, and the fact that adult stem cells are extremely rare among differentiated functional adult cells.

INDUCED PLURIPOTENT STEM CELLS

The limitations of other stem cell types have driven innovation in the laboratory through manipulation of cell differentiation. New techniques for producing "stemness" in mature differentiated cells have emerged and include the engineering of the induced pluripotent stem cell (iPSCs). With this method a normal adult fibroblast can be induced to become phenotypically and functionally identical to a human embryonic stem cell in its pluripotency. These methods are extremely exciting and may bear invaluable and inexhaustible sources of patient-specific stem cells for many therapeutic purposes. However, these techniques are in their infancy and will require a great deal more knowledge about epigenetic reprogramming, maintenance of pluripotency, and control of stem cell differentiation and self-replication before this technology can be fully harnessed and its clinical potential realized.

PERINATAL STEM CELLS

Because of the limitations of the stem cell sources outlined above, postembryonic, perinatal sources of stem cells may yet prove to represent the most important and potentially useful source of stem cells. The term "perinatal" refers to "around the time of birth", yet technically encompasses the time period from the twentieth week of gestation through the neonatal period (the first 28 days of life). Since these tissues are discarded at the time of birth, the harvesting of stem cells from these sources represents a simple, noninvasive, and safe means for attaining powerful stem cell types.

This book highlights the characteristics of stem cell types obtained from perinatal sources. Sources of perinatal stem cells include umbilical cord blood, the Wharton's Jelly matrix of the umbilical cord, amniotic fluid, the amnion lining the amniotic cavity, postgestational maternal peripheral blood, the umbilical vein, and the placenta itself: including chorionic mesenchymal stromal cells and multipotent cells derived from the human term placenta.

In Chapter 1, Haspel and Ballen review the current state of cord blood stem cell transplantation. Cord blood stem cells represent a perinatal stem cell therapy that has been used successfully over 10,000 times in the clinical setting for decades for over 80 disorders, and exemplifies the potential of perinatal stem cells for the treatment of various diseases. In Chapter 2, Taghizadeh and Sherley describe strategies for optimization of the current use of umbilical cord blood hematopoietic stem cells through expansion of these cells.

Breymann describes the potential use of many other perinatal stem cell types for future therapies in Chapter 3, reviewing the use of amniotic, placental, chorionic, and umbilical stem cells in tissue engineering and regenerative medicine. Other potential approaches to perinatal stem cell therapy, including in utero stem cell transplantation and gene therapy using perinatal stem cells are discussed by Surbek et al. in their excellent chapter (Chapter 4).

Umbilical cord Wharton's jelly matrix cells also contain a primitive mesenchymal stromal stem cell compartment, which Boissel et al. (Chapter 5) and Weiss and co-workers (Chapter 6) discuss. Cetrulo and Starnes (Chapter 7) review the therapeutic potential of endothelial progenitor cells from perinatal sources, and in Chapter 8, Theoharides and co-workers discuss umbilical cord blood mast cells as a unique model for studying a variety of disorders.

The amniotic cavity is also a source of two different novel stem cell types. The amniotic fluid itself contains mesenchymal stem cells and the lining of the amniotic cavity (the amniotic epithelial cell layer) also contains stem cells. Tsai (Chapter 9) reviews amniotic fluid derived stem cells and Strom and co-workers (Chapter 10) describes the stem cells isolated from the amnion itself.

Recently, the unbilical vein also has been recognized as a novel source of stem cells. Park and co-workers (Chapter 11) describe the potential of human umbilical vein endothelial cells from this unique source. Mikkola and co-workers (Chapter 12) also describe characteristics of multipotent cells isolated from the human term placenta.

Luduc et al. (Chapter 13) report on fetal chimeric progeniter cells. These remarkable fetal stem cells have been isolated from the maternal peripheral blood circulation 25 years after the conclusion of the pregnancy.

In sum, a host of novel stem cell types have been identified from perinatal sources and represent opportunities for significant therapeutic advances and insight into stem cell biology, all while avoiding the ethical controversies surrounding other stem cell fields. Additionally, these cell sources represent a unique opportunity to harvest, and cryopreserve autologous primitive stem cells populations that can be used for future therapies.

1

CORD BLOOD TRANSPLANTS: PERINATAL STEM CELLS IN CLINICAL PRACTICE

Richard L. Haspel

Beth Israel Deaconess Medical Center, Department of Pathology, 330 Brookline Avenue, Yamins 309, Boston, MA 02215

Karen K. Ballen

Massachusetts General Hospital, Division of Hematology/Oncology, Department of Medicine, Boston, MA 02215

INTRODUCTION

Hematopoietic stem cells (HSCs) are currently the only perinatal stem cells routinely used for treatment of patients. This chapter reviews the collection, processing, and utility of cord blood (CB) in comparison to adult HSC sources. Also addressed are active areas of research including double CB transplants.

HEMATOPOIETIC STEM CELL TRANSPLANTS: ADULT DONOR COLLECTION

In a myeloablative hematopoietic stem cell transplant (HSCT), patients with hematologic malignancies are given high doses of chemotherapy and/or irradiation to eradicate the tumor. In the process, however, the patient's native marrow is destroyed requiring replacement by a source of HSCs. The HSCs can either be collected from the patient prior to transplant (autologous) or from a related or unrelated donor (allogeneic).

Bone marrow (BM) was the donor source for the first allogeneic HSCT performed in 1957 and is currently used in ~25% of adult transplants in the United States

Perinatal Stem Cells. Edited by C. L. Cetrulo, K. J. Cetrulo, and C. L. Cetrulo, Jr.

[Appelbaum, 2007; CIBMTR, 2007]. Typically, BM is collected from the iliac crest in an operating room with the patient under general anesthesia. Collection goals are typically 2×10^8 total nucleated cells per kilogram (recipient weight) that can lead to >1 L of product being collected. The dose of actual HSCs collected from BM, as measured by CD34+ cell content, is $\sim 2-3 \times 10^6$/kg [Bensinger, 2001; Blaise, 2000; Couban, 2002; Schmitz, 2002].

In the United States, collection of HSCs from peripheral blood has overtaken BM as the primary graft source in adult patients. Treating donors with growth factors, for example, granulocyte colony stimulating factor (GCSF) for 4 days leads to mobilization of CD34+ HSCs from the BM into the peripheral blood. These cells can then be collected by apheresis. During apheresis, donor blood is separated by centrifugation to allow selective removal of the cell layer containing the HSCs. The remaining white blood cells, red cells, platelets, and plasma are returned to the patient. In a single apheresis session, ~10–20 L of the donor's blood is processed and only one or two sessions are usually required to reach the typical minimum target dose of 2×10^6 CD34+ cells/kg. These collections yield higher numbers of CD34+ HSCs ($5-10 \times 10^6$ CD34+ cells/kg), as well as T cells compared to bone marrow [Bensinger, 2001; Blaise, 2000; Couban, 2002; Schmitz, 2002].

In general, both BM and peripheral blood stem cells (PBSC) collection are well tolerated, but there are side effects. The BM donors typically have pain from the procedure and may have nausea and vomiting from the general anesthesia. The PBSC donors can experience bone pain from the GCSF injections, as well as symptoms of hypocalcemia from the citrate anticoagulant infused during the apheresis procedure.

One of the largest studies of collection-related side effects compared 166 BM and 163 PBSC donors who were part of a randomized trial [Favre, 2003]. The rates of adverse events were similar (57% for BM vs. 65% for PBSC donors). Serious adverse events occurred in 1% of BM donors and 7% of PBSC donors. The BM donors most often had harvest-related complaints while PBSC donors complained of side effects from GCSF. The BM donors required longer hospital stays (median of 2 vs. 0 days) and had more days of restricted activity (median of 6 vs. 2 days).

This study and others confirm that while donations are usually safe, they are not risk-free [Bredeson, 2004; Favre, 2003; Heldal, 2002; Rowley, 2001]. Overall, complication rates are similar, but BM donors may take longer to recover and often require overnight hospital stays. A recent review cited an estimated 1 in 10,000 risk of dying from a BM harvest with the same or possibly less risk from a PBSC collection [Horowitz, 2005].

HEMATOPOIETIC STEM CELL TRANSPLANTS: ADULT DONOR TESTING

Extensive testing of adult allogeneic donors is performed prior to collection. Aside from evaluating the donor's overall health to ensure they can tolerate the collection procedure, the FDA (21CFR Part 1271) and the Foundation for Accreditation of Cellular Therapy (FACT standards, C6.3) require testing and screening for infectious diseases including human immunodeficiency virus (HIV), hepatitis C virus (HCV), human T-lymphotropic virus (HTLV), hepatitis B virus (HBV), and syphilis. This testing must be performed within 30 days of collection. In addition, if a related donor is not available, donors are identified through registries, for example, the National Marrow Donor Program (NMDP) and HLA typing must be confirmed.

Until recently, allogeneic transplants were matched at HLA-A, B and DR (i.e., a 6/6 match). Matching was typically low-resolution/serological at class I (A, B) and high-resolution/allele level at class II (DR). Currently, most BM and PBSC transplant are matched at high resolution for classes I and II. In addition, for unrelated donor transplants, allele-level matching is also preferred at HLA-C and DQ leading to the ideal of a 10/10 match [Kogler, 2005].

The identification of a donor and the additional testing listed above takes considerable time. Using adult stem cell sources, ~25% of the time, a patient has a first-degree relative who can provide a source of allogeneic stem cells. If there is no relative, ~80% of the time a donor is found through NMDP for Caucasian patients. Minority patients have a lower probability of finding a match (e.g., 60% chance for African–Americans). The unfortunate fact is that only 30% of patients actually make it to transplant as they can be very ill and it takes considerable time to identify, test, and prepare a donor for collection [Government Accounting Office, 2002].

HEMATOPOIETIC STEM CELL TRANSPLANTS: RECIPIENT ISSUES

In North America, the majority of allogeneic transplants are performed for hematologic malignancies, most often acute and chronic leukemia. The most common cause of death from allogeneic transplant, ~40%, is relapse [CIBMTR, 2007]. As such, disease-free survival (DFS) is an important outcome measure.

In regard to short-term morbidity and mortality, there can be significant toxicity from the chemotherapy and/or radiation. Hemorrhagic cystitis, mucositis, cardiac and renal toxicity are all concerns. In addition, it can take significant time for the donor stem cells to engraft. Typical outcome measures are time to absolute neutrophil count (ANC) >500, platelets >20,000 (short-term engraftment) and platelets >50,000 or 100,000 (long-term platelet engraftment). Time to engraftment is correlated with stem cell dose and can range for ANC >500 from <2 weeks with PBSC to over a month with CB [Bensinger, 2001; Blaise, 2000; Couban, 2002; Eapen, 2007; Laughlin, 2004; Rocha, 2001, 2004; Schmitz, 2002; Takahashi, 2004, 2007]. Although patients can be supported with red blood cell and platelet transfusions, they are at increased risk of infection and often have to be placed on antibacterial and antifungal medications. Given these early complications, transplant-related mortality (TRM), often defined as death prior to day 100 post-transplant, is another important outcome measure.

Another source of morbidity and mortality is graft-versus-host-disease (GVHD). Upon engraftment, donor immune cells may attack the recipient as foreign. Skin is most commonly involved although any organ can be affected. Aside from complications due to the GVHD itself, patients are often placed on immunosuppressive regimens that put them at greater risk for infection. The GVHD accounts for ~15% of mortality from allogeneic stem cell transplants [CIBMTR, 2007]. Due to differences in biologic mechanisms, GVHD is typically separated into symptoms occurring in the first 100 days of transplant (acute GVHD) and those occurring post-100 days (chronic GVHD).

Given the significant regimen-related toxicity, many patients are not candidates for ablative conditioning. For this reason, non-myeloablative/reduced intensity conditioning (RIC) regimens are also utilized. In this setting, the goal of the chemotherapy is not to destroy to the tumor, but to create enough immunosuppression to allow engraftment of the donor HSCs. Once the HSCs engraft, the donor immune system will attack the tumor. This phenomenon, known as the graft-versus-leukemia/lymphoma effect, was first identified in patients with chronic myelogenous leukemia (CML). In these patients, improved survival was correlated with GVHD suggesting that the donor immune system was attacking the tumor, as well as the recipient's normal tissue. Consistent with this model, further studies demonstrated that infusion of donor lymphocytes (DLI) can help treat leukemic relapses [Kolb, 1990]. The DLI is now a standard treatment for relapses posttransplant of several hematologic malignancies, but is not a possibility with cord blood transplant [Chen, 2007; Daly, 2003; Dey, 2003].

COLLECTION AND PROCESSING OF CORD BLOOD UNITS

Unlike BM and PBSC for HSCT, a major advantage of using CB is that there is no potential for harm to the donor (mother or child). As such, almost any normal pregnancy can produce a potential cord blood unit (CBU) for transplant. In addition, all required testing is done "up front" avoiding delays for transplant. One study from the University of Minnesota found that the median time of donor availability was 13.5 days for CB and 49 days for BM, an important difference to patients with aggressive leukemias [Dalle, 2004].

Cord blood units can be collected either pre- or postdelivery of the placenta. The umbilical vein is cannulated and ~100 cc of blood is collected in a bag containing citrate anticoagulant. Products are typically frozen in 10% dimethyl sulfotide (DMSO). Many centers also plasma and red cell reduce products prior to cryopreservation. This process leads to smaller volumes allowing for easier storage and reduction of cellular debris upon thawing.

The CBU collection and processing protocols are under investigational new drug (IND) status and are not standardized. Although national and international accrediting agencies, for example, FACT and AABB, provide oversight, there are still differences in collection center procedures. Recently, as a step toward product licensure, the Food and Drug Administration (FDA) has drafted a guidance document for CBUs intended for HSC transplant [FDA, 2006]. The recommendations are similar to those required for other minimally manipulated stem cell products (21CFR Part 1271) including testing and screening for HIV, HTLV, HBV, HCV, syphilis, and cytomegalovirus to determine communicable disease risk. The testing should be performed on the mother and must occur within 7 days of collection. Additional recommended testing and screening includes analysis for hemoglobinopathies and HLA and blood group typing. Although some European centers perform follow-up testing of the donors 6 months after product collection or shortly before product release to determine if there has been any significant change in the donor health, similar requirements are not in the draft guidance document.

The FDA has also proposed guidelines for validating processing procedures to confirm purity and potency. Acceptance criteria include $>= 5 \times 10^8$ TNC with $>= 85\%$ viable and $>= 1.25 \times 10^6$ CD34+ cells per CBU. In regard to cyropreservation, a validated freeze–thaw procedure must allow recovery of 70% of viable nucleated cells. As these criteria demonstrate, the total nucleated cell (TNC) and CD34+ cell dose from a CBU is approximately 10-fold lower than the dose that can be obtained from a BM collection [Eapen, 2007; Laughlin, 2004; Rocha, 2001, 2004; Takahashi, 2004, 2007].

Postthaw processing is not discussed in the FDA guidance document and there is variation between centers. Since early cord blood transplants were performed on children with non-red cell reduced CBUs, products were washed after thawing to remove the potentially toxic effects of DMSO or cellular debris. Washing, however, may also lower the cell dose [Laroche, 2005]. Chow et al. found delays in engraftment, presumably due to differences in cell dose, in washed versus nonwashed products [Chow, 2007]. Both neutrophil (ANC >500: 20 days vs. 27 days, $p < 0.02$) and platelet engraftment (>20,000: 47 vs. 54 days, $p = 0.0003$) were faster with nonwashed cord blood. Their study, as well as others, also demonstrates that unwashed products can be safely infused without significant adverse events [Chow, 2007; Hahn, 2003; Nagamura-Inoue, 2003].

Although there are no results from randomized trials, some centers infuse CBUs after thawing without a wash step.

BONE MARROW VERSUS SINGLE CORD BLOOD: PEDIATRIC

The first cord blood transplant (CBT) was performed in 1989 on a child in France with Fanconi's anemia [Gluckman, 1989]. He received a cryopreserved CBU from a female sibling. Since that time, thousands of unrelated CBTs have been performed. Due to the concern over low cell dose, the majority of early CBTs were performed in children. Observational studies comparing CB with adult hematopoietic stem cell sources have demonstrated both the advantages and disadvantages of these transplants (Tables I.1, I.2).

The largest comparison study in children looked at 785 patients ($<=$ 16 years old) with acute leukemia; 503 received cord blood and 283 received bone marrow [Eapen, 2007]. Only 7% of CBUs were a 6/6 match with the recipient (40% were a single-antigen and 53% were a two-antigen mismatch). In addition, 41% of the bone marrow transplant (BMT) patients were allele-level matched at both class I and class II. The CBUs were obtained at a median of 10–16 months while BM was obtained at a median of 20–23 months. As will be the case for all the cord blood studies reviewed below, CB recipients received an approximately 10-fold lower TNC dose.

Time to engraftment was delayed in recipients of cord blood. Recovery of ANC >500 and platelets >20,000 occurred at a median of 25 and 59 days for CBU recipients compared to 19 and 27 days for BM recipients. In addition, 19% of CBT compared to 3% of BMT patients had primary graft failure. Although 100-day transplant related mortality (TRM) was greater in the patients receiving mismatched cord blood, rates of acute and chronic GVHD, and 5-year leukemia-free survival were similar when compared to patients receiving allele-matched BM. Fully matched CB recipients had improved survival compared to unrelated BM recipients.

Another large study in children compared 99 myeloablative CBT with 262 unrelated BM transplants for patients, ages 2–12, with acute leukemia [Rocha, 2001]. Overall results were comparable to the above study. Only 8% of the CBUs were 6/6 matches with the recipient (43% were 5/6 and 41% were 4/6 matches) compared with 80% of the BM products and the time from complete remission (CR) to transplant was faster in the patients receiving CBUs (84 days vs. 113 days). Neutrophil engraftment was also delayed in the CBU recipients (32 days vs. 18 days) as was platelet recovery (>20,000: 81 days vs. 29 days). Only 4% of BMT compared to 20% of CBT patients had not recovered their

TABLE I.1. Comparative Studies of *Unrelated* Myeloablative Single CB versus BM Transplant in Children with Leukemia[a]

			Engraftment			GVHD (%)	
Study	N (CB/BM)	% DFS (follow-up in months)	ANC (day)	Plts (day)	Primary Failure (%)	Acute II–IV	Chronic
Rocha, 2001	99/262	31/43 (19/30)	32/18	81/29	20/4	35/58	12/43
Eapen, 2007	503/282	33/38[b] (44/60)	27/19	59/25	19/3[b]	41/46[b]	15/32[b]

[a]Abbreviations: DFS = disease-free survival; ANC = absolute neutrophil count >500; Plts = platelets >20,000; GVHD = graft versus host disease.

[b]Two HLA antigen-mismatch CB/HLA allele-matched BM.

TABLE I.2. Comparative Studies of Unrelated Myeloablative Single CB versus 6/6 Matched BM Transplant in Adults[a]

			Engraftment			GVHD (%)	
Study	N (CB/BM)	% DFS (follow-up in months)	ANC (day)	Plts (day)	Primary Failure (%)	Acute II–IV	Chronic
Laughlin, 2004	150/367	23/33 (40/48)	27/18	60/29	30/10[b]	41/48	51/35[c]
Rocha, 2004	98/584	33/38 (27/24)	26/19	N/S	20/7	26/39	30/46
Takahashi, 2004[d]	68/45	74/44 (26/59)	22/18	40/25	8/0	50/67	78/74[c]
Takahashi related, 2007[d]	100/71	70/60 (22/32)	22/17	40/22	5/0	55/60	89/90[c]

[a]Abbreviations (also see Table I.1): N/S = not stated.
[b]Approximate.
[c]Extensive chronic GVHD greater in patients who received BM.
[d]Takahashi, 2004: results include 6 patients with 5/6 matched adult donors. Takahashi, 2007: results include 11 patients with 5/6, 6 patients with 4/6 matched. Adult donors and 16 related peripheral blood stem cell transplants.

neutrophil counts by day 60. In contrast to the above study, perhaps due to the lack of allele-level typing of BM, CBU recipients also had significantly less acute and chronic GVHD. When adjusted for prognostic factors, CBU recipients had an increased risk of 100-day TRM, but overall post-100-day survival was equivalent.

Although the majority of CBTs are performed for hematologic diseases, there also has been success treating metabolic disease in children. For example, Staba et al. treated 20 children with Hurler's disease with single cord blood transplant [Staba, 2004]. Hurler's disease is a mucopolysaccharidosis that leads to neurologic disease and death. Overall survival and rates of GVHD appeared better than historical reports of BMT for this condition. Improvements in neurological development were also demonstrated. Other nonmalignant diseases including thalassemia and X-linked adrenoleukodystrophy also have been treated with CBT [Beam, 2007; Jaing, 2005].

BONE MARROW VERSUS CORD BLOOD: ADULTS

Given the success in pediatric patients, ~50% of CBTs are now performed in adults [Netcord, 2007]. Similar to the pediatric literature, there are also observational studies in adults treated with ablative regimens comparing BM with CB (Table I.2). Rocha et al. compared 584 adult patients receiving unrelated donor BM to 98 patients receiving unrelated CBUs for acute leukemia [Rocha, 2004]. Only 6% of CBUs were a 6/6 match (cf. 100% of BM), 39% were a 4/6 match and 4% were a 3/6 match. Median time to neutrophil engraftment was delayed (19 days for BM and 26 days for CB) and graft failure occurred in 7% of patients in the BM group and 20% in the CB group ($p < 0.001$). The rate of acute GVHD (aGVHD) was significantly lower in the CB group with a similar trend for chronic GVHD (cGVHD). There were no significant differences in TRM, relapse, or disease free survival.

Laughlin et al. compared 150 single unrelated cord blood transplants to 450 unrelated bone marrow transplants for acute leukemia [Laughlin, 2004]. While 82% of BMs were 6/6 matches (the rest were 5/6 matches), 77% of CBUs were 4/6 matches with the rest being 5/6 matches. Both neutrophil and platelet recovery were delayed. The time to an ANC >500 occurred at a median of 27 days for CBT versus 18–20 days for BMT patients and the time to platelets >20,000 occurred at median of 60 versus 29 days. There was also an increase in graft failure with CB when compared to BM, but this only reached statistical significance when measured against 6/6 matched marrow ($p < 0.01$).

The rate of acute GVHD in the CB recipients was similar to that of patients who received 6/6 matched bone marrow and better than that of patients who received 5/6 BM. While there was significantly more chronic GVHD in CB compared to matched BM recipients, the rate of extensive chronic GVHD was lower in CB recipients when compared to all BM recipients. In regard to TRM and overall mortality, CB was essentially equivalent to 5/6 matched marrow, but worse than 6/6 matched marrow (OR = 1.53, 95% CI = 1.21–1.94).

A third retrospective study from Japan included adults (>16 years old) with leukemia, myelodysplastic syndrome, or non-Hodgkin's lymphoma [Takahashi, 2004]. All patients received a myeloablative conditioning regimen followed by infusion of unrelated CB ($n = 68$) or unrelated BM ($n = 45$). None of the CB units were a 6/6 match with the recipient and 22% were a 3/6 match. In contrast, 87% of the BM was a 6/6 match with the remainder being a 5/6 match with the recipient. The duration of the donor search was significantly shorter for CBU recipients (3 months vs. 11 months) with a similar trend for time from diagnosis to transplant (17 months vs. 20 months).

Neutrophil engraftment occurred at day 22 for CB and day 18 for BM recipients and short- and long-term platelet engraftment was also significantly delayed (>20,000: 40 vs. 25 days; >50,000: 48 vs. 28 days). The graft failure rate was also higher with CBT (8% vs. 0%). In regard to GVHD, the rate of grade III/IV acute GVHD was significantly higher with BM transplant with a similar trend for extensive chronic GVHD. The 1-year TRM (9% vs. 29%) and 2-year disease-free survival (74% vs. 44%) was significantly better with CB transplant.

A follow-up study from Japan included patients aged 16 years or older with a variety of hematologic malignancies who received a myeloablative unrelated CB transplant ($n = 100$), a related BM transplant ($n = 55$) or related PBSC transplant ($n = 16$) [Takahashi, 2007]. The majority of patients had acute or chronic leukemia and none of the CB units were a 6/6 match compared to 76% of the BM. Additionally, 28% of the CBUs were a 3/6 match. Likely due to better availability of related donors, the difference in time from diagnosis to transplant was shorter for BMT, but this did not reach statistical significance (17.5 months for CB vs. 15 months for BM, $p = 0.3$). Similar to the above studies, both neutrophil (22 vs. 17 days) and short-term platelet engraftment (>20,000; 40 days vs. 22.5 days) was delayed with CB compared to BM. There was also a significant delay in long-term platelet engraftment (>50,000; 46 vs. 27 days) and a slightly higher rate of graft failure rate (5% vs. 0%) in the CB group. The above engraftment issues may have led to the increase in hospital length of stay in recipients receiving CB (121 vs. 89 days, $p = 0.1$).

The rate of grade III/IV acute GVHD and extensive chronic GVHD was significantly lower with CB transplant. As a result, more BM recipients required steroid treatment for acute GVHD and had a slower taper of their immunosuppressive regimens. There were no significant differences in TRM, relapse or survival.

CORD BLOOD TRANSPLANT: ADVANTAGES AND DISADVANTAGES

Review of the above studies clearly demonstrates the advantages and disadvantages of single CBT. The CB allows faster time to transplant. In most studies in which it was measured, time from diagnosis or complete remission to transplant was months shorter than with an unrelated BM donor. As discussed above, this reflects the fact that all testing is done "up front" with CB and should allow more patients to obtain transplants.

Another advantage is that a less than perfect HLA match is adequate for CBT. The majority of units transplanted in the above studies were a 4/6 match with the recipient. Some were even 3/6 matches with all typing performed at the low-resolution level for class I. Even with this degree of mismatch, the rates of acute GVHD or extensive chronic GVHD were either the same or lower in the CBT patients. This increased flexibility with HLA matching will allow more individuals to be transplanted. Stevens et al. calculated, based on the ethnic distribution of donors and patients in New York City, that $<170{,}000$ units would be required to guarantee a minimum 5/6 match for 80% of adult transplant candidates [Stevens, 2005]. While some studies indicate better HLA matching may lead to improved engraftment and survival, it is clear that CBU 4/6 matches, unlike for BM and PBSC, are acceptable and can lead to good results with no increase in GVHD [Eapen, 2007; Kamani, 2008]. Further research is needed on the effect of HLA mismatch and high- versus low-resolution HLA matching on CBT outcomes [Delaney, 2007].

The above studies demonstrate that the major disadvantage of CB is the low-cell dose. In all studies, TNC dose was approximately 10-fold lower in the CBUs and this likely led to the delays in engraftment, increased rates of primary graft failure, as well as increases in TRM. Several groups have directly correlated infused cell dose to engraftment and TRM [Gluckman, 2000; Wagner, 2002].

Despite worse engraftment with CBT, disease-free survival generally showed no difference when compared to BM. Perhaps, any increase in early mortality is compensated by the decrease in late mortality due to GVHD. The study by Laughlin et al., in which CBT was equivalent to one-antigen mismatched marrow, but worse than fully matched BM, is an exception among the above studies [Laughlin, 2004]. In this study, however, transplants were performed in an earlier time period when only the sickest adult patients with no other options were transplanted with CB [Sanz, 2004].

Another exception is the unrelated transplant trial by Takahashi et al. [Takahashi, 2004]. The authors found that CB recipients did significantly better than unrelated BM recipients. This study involved only a single institution and perhaps there were differences in care or patient makeup that led to this result. For example, Japan has a more homogeneous population than Europe and North America and HLA disparity may have less impact on outcomes. Note that the same group did not find this difference in their study of related BM donors. These results might be explained by differences in GVHD related deaths in the two studies. With unrelated BM transplants, 10 of the 24 BM recipients died of GVHD (42%) as compared to none of the 16 expired CB recipients. In the study with related BM donors, 26 in the BM group and 25 in the CB group died. In both groups, GVHD was the cause of death for 12% of patients. The sharp drop in GVHD related deaths in the BM recipients may reflect better (allele-level) HLA matching when there is a related BM donor.

DOUBLE CORD BLOOD TRANSPLANTS: ABLATIVE REGIMENS

Prior studies indicate that cell dose is an important factor for engraftment and survival in the setting of CBT. As such, several groups have attempted infusion of two cord blood units to overcome this limitation (Table I.3).

In 2005, Barker et al. published the largest study of myeloablative DCBTs to date [Barker, 2005]. Twenty-three patients aged 13–53 with leukemia received myeloablative conditioning followed by infusion of two CBUs (median TNC dose of 3.5×10^7/kg).

TABLE I.3. Double Cord Blood Transplant (DCBT) Studies[a]

				Engraftment			GVHD (%)	
Study	N-DCBT	Diagnosis	% DFS (follow-up in months)	ANC (day)	Plts (day)	1° Failure (%)	Acute II–IV	Chronic
Barker, 2005	23	ALL, AML, CML	57 (10)	23	N/S	0	65	23
Brunstein, 2007, RIC[b]	110	High-risk heme malignancy	38 (19)	12	N/S	6	59	23
Ballen, 2007 RIC	21	AML, ALL, NHL, CLL, HD, MDS, AA	55 (18)	20	41	10	40	31
Cutler, 2007 RIC	27	AML, ALL, NHL, CLL, HD, MDS, CML	54 (15)	21	42	0	11	7
Majhail, 2006 RIC[a]	9	HL	25 (17)	10	N/S	0	33 (III/IV)	11

[a]Abbreviations (also see Table I.1): RIC = reduced intensity conditioning; AML = acute myeloid leukemia; ALL = acute lymphocytic leukemia; CLL = chronic lymphocytic leukemia; CML = chronic myelogenous leukemia, NHL = non-Hodgkin's lymphoma; HD = Hodgkin's disease; MDS = myelodysplastic syndrome, AA = aplastic anemia.

[b]Outcomes also reflect several single CB transplants (Majhail: 2, Brunstein: 17).

Most patients received 4 or 5/6 matched units with only two receiving a 6/6 matched CBU. All the units were a minimum 4/6 match with each other. The median time to neutrophil engraftment was 23 days (range: 15–41), which is slightly better than the times reported in most retrospective studies of myeloablative single cord blood transplants. While there was no graft failure, median time to platelets >50,000 was ~90 days. Thirteen percent of patients had severe grade III or IV acute GVHD and 23% of patients had chronic GVHD. The predicted 1-year survival was 57%.

DOUBLE CORD BLOOD TRANSPLANT: NON-MYELOABLATIVE REGIMENS

A number of studies also have investigated the use of RIC regimens followed by DCBT (Table I.3). Although DLI cannot be administered following CBT, these studies have shown promise in regard to transplant outcomes. Considering at least 40% of allogeneic transplants utilize RIC regimens, CBT is an important option for patients without a matched adult donor [CIBMTR, 2007].

Brunstein et al. have reported the largest group of patients undergoing CBT with a RIC regimen [Brunstein, 2007]. Patients with hematologic malignancies were conditioned with fludarabine, cytoxan, and total body irradiation (TBI). Of 110 patients, 85% received two CBUs. Patient were selected to receive DCBT if the total TNC dose from a single unit was $<3 \times 10^7$/kg. The TNC and CD34+ cell dose were statistically equivalent in recipients of one or two CBUs, but the CD3+ cell dose was two times higher in recipients of two units. The majority of units (64%) were a 4/6 HLA match with the recipient. In most DCBT patients, both units were a 4/6 match (43%) or one was a 4/6 match and one was a 5/6 match (33%). Rates of neutrophil recovery were similar for single and DCBT recipients

(94% vs. 91%). Neutrophil and platelet recovery >50,000 occurred at a median of 12 and 49 days, respectively, but data was not stratified for recipients of one or two CBUs. The overall rate of grades II–IV aGVHD was 59% and of cGVHD it was 23%. The DCBT patients had a statistically significant higher risk of aGVHD and a trend for better event-free survival. For all patients, 3-year survival was 45%.

In our initial study of DCBT, we enrolled 21 adult DCBT recipients conditioned with fludarabine, melphalan, and anti-thymocyte globulin [Ballen, 2007]. The median combined TNC dose was 4×10^7 and 78% of the CBUs were a 4/6 match with the recipient at the allele level for both class I and class II MHC. Median times of neutrophil and platelet (>20,000) recovery were 20 and 41 days, respectively. Two patients had primary graft failure and one had a successful second DCBT using a different conditioning regimen. Only one patient developed severe (grade III) aGVHD and 5 of 16 evaluable patients (31%) developed cGVHD with two having extensive disease. With a median follow up of 18 months, 2-year projected overall survival was 55%. There were no significant associations between cell dose and degree of HLA match with GVHD or survival.

In a second study, we utilized the same conditioning regimen, but altered GVHD prophylaxis from cyclosporine and mycophenolate mofetil to sirolimus and tacrolimus, based on studies with this GVHD regimen in the related and unrelated donor setting [Cutler, 2004]. Twenty-seven patients with hematologic malignancies have been enrolled. Median times to engraftment were 21 days for ANC >500 and 42 days for platelets >20,000. All patients initially engrafted, although three were not platelet transfusion independent by day 100 and three had late graft failure. The rate of aGVHD was lower than in our previous trial (11.1%) and only two patients developed cGVHD. Two-year disease-free survival was 54.4% [Cutler, 2007].

Only one small study utilizing a reduced intensity conditioning regimen has compared DCBT to BM transplants [Majhail, 2006b]. Patients either received a busulfan/fludarabine/TBI or a cytoxan/fludarabine/TBI regimen. Twelve patients with Hodgkin's disease who received related BM were compared to nine patients who received unrelated CB (seven of these patients received two CBUs). While there were no graft failures, neutrophil recovery was delayed for the CB patients (10 vs. 7 days). There were no significant differences in rates of acute or chronic GVHD or progression-free survival. Although the authors suggest that BM is comparable to CB, the numbers are too small to generalize this conclusion.

The above results are very promising and indicate there can still be significant graft versus tumor effect without high rates of GVHD in CBTs. Comparison of these trials is difficult due to variations in conditioning regimens [Barker, 2003].

CHIMERISM

In the setting of HSC transplant, chimerism refers to the percentage of donor versus recipient contribution to hematopoiesis. Typically, DNA fingerprinting techniques (i.e., amplification of short tandem repeat loci) are used to measure chimerism [Baron, 2005]. As opposed to standard transplants in which there is only one donor and one recipient, in the setting of DCBT there are two donors and the recipient. Initially, there was a concern that DCBT could lead to a "graft versus graft" effect and impair engraftment, but fortunately this does not appear to commonly occur. Although several early case reports demonstrated double chimerism post-transplant, subsequent studies indicate that generally only one CBU usually contributes to long-term engraftment [Barker, 2001; de Lima, 2002; Haspel, 2007].

The first DCBT study that analyzed chimerism was the myeloablative study of Barker et al. [Barker, 2005]. By day 21, 76% of the patients showed hematopoiesis from only a single cord and this increase to 100% by day 100. In the study by Brunstein et al., for 81 patients with sustained chimerism that received a RIC regimen and DCBT, 57% of patients had complete single donor chimerism at day 21, 91% at day 100 and 100% at 1 year. The median percentage contribution of the predominant unit was 83% (range: 8–100%) at day 21 and 100% (range: 34–100%) after day 100 [Brunstein, 2007].

In our study of 38 patients, we reported three chimerism patterns [Haspel, 2007]. Group 1 consisted of patients in whom a single CBU contributed to >= 95% of hematopoiesis. In group 2, both CB units contributed to hematopoiesis. The third group consisted of patients with contributions of one CBU and the recipient. At day 30, 29% of patients had single CBU hematopoiesis (group 1). As time progressed, patients transitioned to complete single unit chimerism with 50% in this group by day 60 and 66% in this group by day 100 (Fig. I.1).

Of the 16 patients followed over 1 year, 14 had chimerism from a single CBU by day 100 and all have retained complete single donor chimerism (five of these over 2 years and two over 3 years). The two other patients followed at least 1 year continue to have contributions of both CBU to hematopoiesis with 83 and 66% chimerism of the predominant CBU.

In our prior study of 21 patients, we also found a significantly different rate of chronic GVHD (0–100%) when comparing patients with complete single donor chimerism at 6 weeks to those with either contribution of two CBUs or one CBU and the recipient to hematopoiesis [Ballen, 2007]. Chimerism pattern, however, does not appear to affect engraftment. Small numbers may limit the above results and further studies are needed to determine the effect of chimerism patterns on outcomes.

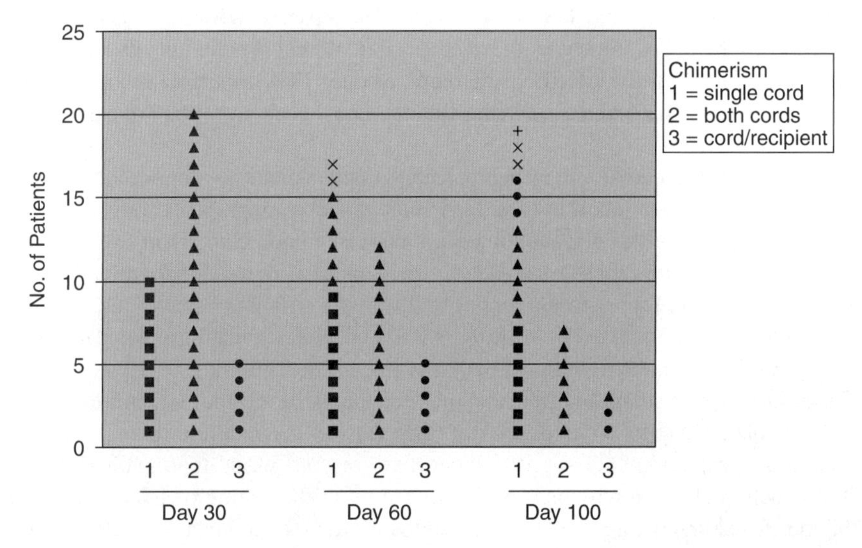

Figure I.1. Chimerism patterns. The numbers of patients with different chimerism patterns are shown at days 30, 60, and 100. Individual patients are marked by shape based on chimerism at day 30 (■ = group 1, ▲ = group 2, ● = group 3) or × = first chimerism result obtained at day 60, + = first chimerism result obtained at day 100. [First published in Haspel, 2007.]

PREDICTING THE WINNER

The DCBT offers a unique opportunity to study "competitive" transplants in humans. These studies may help us understand important and potentially modifiable preinfusion characteristics that may improve engraftment. The initial ablative Barker study found a link between a higher CD3+ cell dose and cord blood predominance [Barker, 2005]. The authors hypothesized an immunomodulatory effect, but this association disappeared as more patients were accrued [Majhail, 2006a]. In the Brunstein et al. study of 110 patients receiving a RIC regimen, neither TNC dose, CD34+ cell dose, CD3+ cell dose, ABO match, gender match or order of infusion predicted CBU predominance [Brunstein, 2007].

In our series of 38 patients, we also investigated preinfusion parameters (TNC dose, CD34+ cell dose, ABO match, gender match and age of the CBU) for an effect on CBU predominance [Haspel, 2007]. We found that both a higher TNC and CD34+ cell dose ($p = 0.06, p = 0.03$), as well as order of infusion ($p = 0.03$) were all independently associated with cord predominance. In 26/38 transplants (68%), the predominant cord blood unit was the first one infused.

What accounts for the differences in results? One possibility relates to recent findings on the nature of the hematopoietic stem cell (HSC) niche. First hypothesized by Schofield in 1978, the niche is a specific area in the BM that promotes the maintenance of the HSC [Schofield, 1978]. Early data, based on radiation studies and selective isolation of BM regions, suggested that the niche space might be near the endosteum [Gong, 1978; Schofield, 1978]. In 2001, Nilsson et al. formally demonstrated that injected HSCs, but not more mature progenitors home preferentially to near the endosteal surface [Nilsson, 2001].

The support cell in the niche is the osteoblast. Zhang et al. were able to image the HSC in contact with endosteal osteoblast [Zhang, 2003]. Through conditional inactivation of the BMP (bone morphogenetic protein) receptor, they also demonstrated that the subsequent increase in osteoblasts led to an increased number of HSCs. Similarly, Calvi et al. increased the number of osteoblasts and HSCs by injecting parathyroid hormone (PTH) into mice [Calvi, 2003]. The PTH treatment of mice prior to lethal irradiation also lowered the minimum transplant cell dose for hematopoietic rescue. This important proof-of-principle experiment suggests that stem cell niche quality can be an important factor in HSCT outcomes.

More recently, Czechowicz et al. demonstrated that the niche is saturable at a fixed quantity of HSCs in nonconditioned murine recipients [Czechowicz, 2007]. That is, once 250 HSCs are infused there is only a minimal increase in donor chimerism even at 70-fold higher doses. The authors then "cleared" the niche using a monoclonal antibody to c-kit, which is a receptor on the surface of mouse HSCs. Creating this "space" allowed greater engraftment in these animals. The authors hypothesized that creating greater niche space may allow HSC transplants in humans with less intense conditioning.

Taken together, this data suggests that niche size may be a limiting factor in the setting of HSC transplant. Due to differences in oxygen tension between the fetus and adult, the endosteal niche space may also be less supportive for cord blood than adult HSCs [Dao, 2007]. As such, niche size may be especially limiting in the setting of CBT leading to the unit that is infused first taking up the available HSC space and having an engraftment advantage. In addition, as only a certain percentage of stem cells home correctly to the BM, a higher cell dose would lead to a greater number of cells being able to reach and survive in the niche space.

The Minnesota group found no relationship between order of infusion or cell dose and cord predominance. This difference might reflect the fact that in those studies the CBU were infused <1 apart [Brunstein, 2007]. In our study, as the second unit was not thawed and processed until confirmation of successful infusion of the first unit, 94% of units were infused >4 h apart. The Nilsson study demonstrated that HSCs can home to the endosteal space in under 5 h [Nilsson, 2001]. The effect of timing of infusion on outcomes from DCBT needs to be further investigated.

There may be clinical consequences of predicting cord predominance. In our study of 38 patients, only the CD34 dose of the winning unit, not the combined dose nor dose of the losing unit, correlated with time to platelets >100,000 ($p = 0.05$) with similar trends for ANC >500 and platelets >20,000 [Haspel, 2007]. This result provides direct evidence that only the winning cord is able to successfully occupy the niche space and only characteristics of this cord can specifically impact long-term engraftment.

This finding also suggests that infusing the CBU with the higher CD34+ cell dose first may lead to that unit predominating and faster recovery of cell counts. In 20 transplants, we infused the CBU with the higher prethaw TNC dose first. In 14 of these transplants (70%), the unit with higher CD34+ cell dose, based on our postthaw determination, also was infused first. In 10 of these 14 transplants (71%), the CBU infused first (higher CD34+ cell dose) was predominant [Haspel, 2007]. Consistent with the known poor correlation between collection and transplant center CD34+ cell dose, using the prethaw CD34+ cell dose would have led to the unit with the higher postthaw CD34+ infused first only 65% of the time [Wagner, 2006]. Our results reflect the imperfect correlation between TNC and CD34+ cell counts and the variability in methods for determining cellular content of products. A method to accurately determine CD34+ cell dose prior to infusion might be beneficial to insure that the unit with the higher dose is infused first. In this way, that unit would be more likely to predominate with subsequent faster engraftment.

ARE TWO CORDS BETTER THAN ONE?

There is limited data to suggest DCBT is better than single CBT in regard to engraftment. In the largest study, Brunstein et al. compared patients treated with a RIC regimen followed by single or DCBT [Brunstein, 2007]. Although there was no difference in graft failure, the authors do not report the separate median engraftment times for the single versus the DCBTs. While there was a trend for better overall survival with DCBT, there was also a higher rate of acute GVHD, which may be a reflection of greater infused CD3+ cell dose. In agreement with the authors, an accompanying editorial concluded that DCBT "seems likely to emerge as a standard" for CBT in adults [Shpall, 2007]. This statement should be taken cautiously as this trial included only 15 single CBT patients and was not randomized. It is possible that if DCBT recipients were simply infused with the unit with the higher cell dose, similar results would have been obtained.

One might expect stem cell dose would still be an issue with DCBT as bone marrow still has approximately five times the TNC dose of two CBUs. In addition, only a single CBU is often detected by chimerism studies days to weeks prior to ANC and platelet engraftment. Consistent with these facts, we found that only the CD34+ cell dose of the predominant unit actually predicts engraftment.

If a second cord is promoting engraftment or enhancing survival, this benefit is likely not due to HSC support, but perhaps through some other accessory cell mechanism. For example, two groups have reported that infusion of mesenchymal cells with CB in mouse models of CBT leads to improved engraftment [Kim, 2004; Noort, 2002]. Improved survival may also be a reflection of increased numbers of infused T-cells in DCBT leading to a greater graft versus leukemia effect. A randomized trial is needed to determine whether DCBT is truly better than single CBT and is underway in the pediatric population.

OTHER EXPERIMENTAL STRATEGIES

Other approaches to improve engraftment are being investigated. A Spanish group has attempted coinfusion of adult HSCs with a single CBU in 27 patients receiving an ablative regimen [Magro, 2006]. The adult HSCs were CD34+ cell selected and infused at a median dose of 2.3×10^6 cells/kg. The majority (85%) of the adult HSCs were from haploidentical donors while the remainder were from unrelated donors. The median TNC dose of the CBUs was 2.37×10^7/kg. Neutrophil engraftment occurred rapidly at a median of 10 days. Engraftment of platelets >20,000 and 50,000 occurred at a median of 33 and 57 days, respectively. At the time of engraftment, chimerism studies showed a predominance of hematopoiesis from the adult HSCs. Over time, there was a gradual increase in contribution to hematopoiesis by the CBU such that 93% of the patients eventually achieved full donor CBU chimerism at a median of 55 days.

Other groups have attempted to expand the HSCs in CBUs *ex vivo* [Hofmeister, 2007]. Systems generally involve addition of stromal cells and/or cytokines. Patients in human trials, however, have not had faster engraftment possibly due to loss of HSCs during culture. The HSC expansion is further discussed in Chapter 2.

In contrast to increasing the number of infused HSCs, a unique approach to improving outcomes in CBT involves enhancing the ability of cord blood stem cells to engraft. As noted above, PTH has been shown to enhance the HSC niche space and minimize the number of stem cells required to rescue mice following lethal irradiation [Calvi, 2003]. A study is currently underway to determine whether PTH may enhance engraftment in the setting of DCBT. In another approach, North et al. have used prostaglandin E2 to expand HSCs in mouse and zebrafish models. This treatment may also prove useful in cord blood transplants [North, 2007].

SUMMARY

Cord blood transplant is a viable option for patients with hematologic malignancies. Many clinical and scientific questions remain, however, regarding strategies to improve engraftment and other outcomes. As the only procedure involving perinatal stem cells in widespread clinical use, cord blood transplant serves as a paradigm for new approaches in this area of regenerative medicine.

REFERENCES

Appelbaum FR. 2007. Hematopoietic-cell transplantation at 50. N Engl J Med. 357:1472–1475.

Ballen KK, Spitzer TR, Yeap BY, McAfee S, Dey BR, Attar E, Haspel R, Kao G, Liney D, Alyea E, Lee S, Cutler C, Ho V, Soiffer R, Antin JH. 2007. Double unrelated reduced-intensity umbilical cord blood transplantation in adults. Biol Blood Marrow Transplant. 13:82–89.

Barker JN, Weisdorf DJ, Wagner JE. 2001. Creation of a double chimera after the transplantation of umbilical-cord blood from two partially matched unrelated donors. N Engl J Med. 344:1870–1871.

Barker JN, Weisdorf DJ, DeFor TE, Blazar BR, Miller JS, Wagner JE. 2003. Rapid and complete donor chimerism in adult recipients of unrelated donor umbilical cord blood transplantation after reduced-intensity conditioning. Blood. 102:1915–1919.

Barker JN, Weisdorf DJ, DeFor TE, Blazar BR, McGlave PB, Miller JS, Verfaillie CM, Wagner JE. 2005. Transplantation of 2 partially HLA-matched umbilical cord blood units to enhance engraftment in adults with hematologic malignancy. Blood. 105:1343–1347.

Baron F, Little MT, Storb R. 2005. Kinetics of engraftment following allogeneic hematopoietic cell transplantation with reduced-intensity or nonmyeloablative conditioning. Blood Rev. 19:153–164.

Beam D, Poe MD, Provenzale JM, Szabolcs P, Martin PL, Prasad V, Parikh S, Driscoll T, Mukundan S, Kurtzberg J, Escolar ML. 2007. Outcomes of unrelated umbilical cord blood transplantation for X-linked adrenoleukodystrophy. Biol Blood Marrow Transplant. 13:665–674.

Bensinger WI, Martin PJ, Storer B, Clift R, Forman SJ, Negrin R, Kashyap A, Flowers ME, Lilleby K, Chauncey TR, Storb R, Appelbaum FR. 2001. Transplantation of bone marrow as compared with peripheral-blood cells from HLA-identical relatives in patients with hematologic cancers. N Engl J Med. 344:175–181.

Blaise D, Kuentz M, Fortanier C, Bourhis JH, Milpied N, Sutton L, Jouet JP, Attal M, Bordigoni P, Cahn JY, Boiron JM, Schuller MP, Moatti JP, Michallet M. 2000. Randomized trial of bone marrow versus lenograstim-primed blood cell allogeneic transplantation in patients with early-stage leukemia: A report from the Societe Francaise de Greffe de Moelle. J Clin Oncol. 18:537–546.

Bredeson C, Leger C, Couban S, Simpson D, Huebsch L, Walker I, Shore T, Howson-Jan K, Panzarella T, Messner H, Barnett M, Lipton J. 2004. An evaluation of the donor experience in the Canadian multicenter randomized trial of bone marrow versus peripheral blood allografting. Biol Blood Marrow Transplant. 10:405–414.

Brunstein CG, Barker JN, Weisdorf DJ, DeFor TE, Miller JS, Blazar BR, McGlave PB, Wagner JE. 2007. Umbilical cord blood transplantation after nonmyeloablative conditioning: Impact on transplant outcomes in 110 adults with hematological disease. Blood. 110:3064–3070.

Calvi LM, Adams GB, Weibrecht KW, Weber JM, Olson DP, Knight MC, Martin RP, Schipani E, Divieti P, Bringhurst FR, Milner LA, Kronenberg HM, Scadden DT. 2003. Osteoblastic cells regulate the haematopoietic stem cell niche. Nature (London). 425:841–846.

Center for International Blood Marrow Transplant Research (CIBMTR). 2007. Allogeneic stem cell sources by recipient age 1997–2006. Available at https://campus.mcw.edu/AngelUploads/Content/CS_IBMTR2/_assoc/ECCBED0AF0A4492BB667FB6227DC7C06/summary05_Pt1_files/frame.htm. Accessed 1/24/08.

Chen Y-B, Spitzer TR. 2007. Current status of reduced-intensity allogeneic stem cell transplantation using alternative donors. Leukemia. 22:31–41.

Chow R, Nademanee A, Rosenthal J, Karanes C, Jaing TH, Graham ML, Tsukahara E, Wang B, Gjertson D, Tan P, Forman S, Petz LD. 2007. Analysis of hematopoietic cell transplants using plasma-depleted cord blood products that are not red blood cell reduced. Biol Blood Marrow Transplant. 13:1346–1357.

Couban S, Simpson DR, Barnett MJ, Bredeson C, Hubesch L, Howson-Jan K, Shore TB, Walker IR, Browett P, Messner HA, Panzarella T, Lipton JH. 2002. A randomized multicenter comparison of bone marrow and peripheral blood in recipients of matched sibling allogeneic transplants for myeloid malignancies. Blood. 100:1525–1531.

Cutler C, Antin JH. 2004. Sirolimus for GVHD prophylaxis in allogeneic stem cell transplantation. Bone Marrow Transplant. 34:471–476.

Cutler C, Mitrovitch R, Kao G, Ho V, Alyea E, Koreth J, Armand P, Dey B, Spitzer T, Soiffer R, Antin J, Ballen K. 2007. Double umbilical cord blood transplantation with reduced intensity conditioning and sirolimus-based GVHD prophylaxis. Blood. 118:600a.

Czechowicz A, Kraft D, Weissman IL, Bhattacharya D. 2007. Efficient transplantation via antibody-based clearance of hematopoietic stem cell niches. Science. 318:1296–1299.

Dalle JH, Duval M, Moghrabi A, Wagner E, Vachon MF, Barrette S, Bernstein M, Champagne J, David M, Demers J, Rousseau P, Winikoff R, Champagne MA. 2004. Results of an unrelated transplant search strategy using partially HLA-mismatched cord blood as an immediate alternative to HLA-matched bone marrow. Bone Marrow Transplant. 33:605–611.

Daly A, McAfee S, Dey B, Colby C, Schulte L, Yeap B, Sackstein R, Tarbell NJ, Sachs D, Sykes M, Spitzer TR. 2003. Nonmyeloablative bone marrow transplantation: Infectious complications in 65 recipients of HLA-identical and mismatched transplants. Biol Blood Marrow Transplant. 9:373–382.

Dao MA, Creer MH, Nolta JA, Verfaillie CM. 2007. Biology of umbilical cord blood progenitors in bone marrow niches. Blood. 110:74–81.

Delaney M, Yeap B, Haspel RL, Spitzer TR, McAfee SL, Dey B, Attar E, Kao G, Alyea E, Lee S, Cutler C, Ho V, Soiffer RJ, Antin JH, Ballen K. 2007. HLA locus-specific outcomes in double umbilical cord blood reduced intensity transplantation (DCBT) in adults. Blood. 118:605a.

de Lima M, St JL, Wieder ED, Lee MS, McMannis J, Karandish S, Giralt S, Beran M, Couriel D, Korbling M, Bibawi S, Champlin R, Komanduri KV. 2002. Double-chimaerism after transplantation of two human leucocyte antigen mismatched, unrelated cord blood units. Br J Haematol. 119:773–776.

Dey BR, McAfee S, Colby C, Sackstein R, Saidman S, Tarbell N, Sachs DH, Sykes M, Spitzer TR. 2003. Impact of prophylactic donor leukocyte infusions on mixed chimerism, graft-versus-host disease, and antitumor response in patients with advanced hematologic malignancies treated with nonmyeloablative conditioning and allogeneic bone marrow transplantation. Biol Blood Marrow Transplant. 9:320–329.

Eapen M, Rubinstein P, Zhang MJ, Stevens C, Kurtzberg J, Scaradavou A, Loberiza FR, Champlin RE, Klein JP, Horowitz MM, Wagner JE. 2007. Outcomes of transplantation of unrelated donor umbilical cord blood and bone marrow in children with acute leukaemia: A comparison study. Lancet. 369:1947–1954.

Favre G, Beksac M, Bacigalupo A, Ruutu T, Nagler A, Gluckman E, Russell N, Apperley J, Szer J, Bradstock K, Buzyn A, Matcham J, Gratwohl A, Schmitz N. 2003. Differences between graft product and donor side effects following bone marrow or stem cell donation. Bone Marrow Transplant. 32:873–880.

Food and Drug Administration (FDA). 2006. Draft Guidance for Industry: Minimally Manipulated, Unrelated, Allogeneic Placental/Umbilical Cord Blood Intended for Hematopoietic Reconstitution in Patients with Hematological Malignancies. Available at http://www.fda.gov/cber/gdlns/cordbld.pdf. Accessed 1/24/08.

Gluckman E, Broxmeyer HA, Auerbach AD, Friedman HS, Douglas GW, Devergie A, Esperou H, Thierry D, Socie G, Lehn P. 1989. Hematopoietic reconstitution in a patient with Fanconi's anemia by means of umbilical-cord blood from an HLA-identical sibling. N Engl J Med. 321:1174–1178.

Gluckman E. 2000. Current status of cord blood hematopoietic stem cell transplantation. Exp Hematol. 28:1197–1205.

Gong JK. 1978. Endosteal marrow: A rich source of hematopoietic stem cells. Science. 199:1443–1445.

Government Accounting Office (GAO). 2002. Bone Marrow Transplants. Available at http://www.gao.gov/new.items/d03182.pdf. Accessed 1/24/08.

Hahn T, Bunworasate U, George MC, Bir AS, Chinratanalab W, Alam AR, Bambach B, Baer MR, Slack JL, Wetzler M, Becker JL, McCarthy PL, Jr. 2003. Use of nonvolume-reduced (unmanipulated after thawing) umbilical cord blood stem cells for allogeneic transplantation results in safe engraftment. Bone Marrow Transplant. 32:145–150.

Haspel R, Kao G, Yeap BY, Cutler C, Soiffer RJ, Alyea EP, Ho VT, Koreth J, Dey BR, McAfee SL, Attar EC, Spitzer T, Antin JH, Ballen KK. 2007. Preinfusion variables predict the predominant unit in the setting of reduced-intensity double cord blood transplantation. Bone Marrow Transplant. Advance online publication November 26.

Heldal D, Brinch L, Tjonnfjord G, Solheim BG, Egeland T, Gadeholt G, Albrechtsen D, Aamodt G, Evensen SA. 2002. Donation of stem cells from blood or bone marrow: Results of a randomised study of safety and complaints. Bone Marrow Transplant. 29:479–486.

Hofmeister CC, Zhang J, Knight KL, Le P, Stiff PJ. 2007. Ex vivo expansion of umbilical cord blood stem cells for transplantation: Growing knowledge from the hematopoietic niche. Bone Marrow Transplant. 39:11–23.

Horowitz MM, Confer DL. 2005. Evaluation of hematopoietic stem cell donors. Am Soc Hematol Educ Program. 469–475.

Jaing TH, Hung IJ, Yang CP, Chen SH, Sun CF, Chow R. 2005. Rapid and complete donor chimerism after unrelated mismatched cord blood transplantation in 5 children with beta-thalassemia major. Biol Blood Marrow Transplant. 11:349–353.

Kamani N, Spellman S, Hurley CK, Barker JN, Smith FO, Oudshoorn M, Bray R, Smith A, Williams TM, Logan B, Eapen M, Anasetti C, Setterholm M, Confer DL. 2008. State of the art review: HLA matching and outcome of unrelated donor umbilical cord blood transplants. Biol Blood Marrow Transplant. 14:1–6.

Kim DW, Chung YJ, Kim TG, Kim YL, Oh IH. 2004. Cotransplantation of third-party mesenchymal stromal cells can alleviate single-donor predominance and increase engraftment from double cord transplantation. Blood. 103:1941–1948.

Kogler G, Enczmann J, Rocha V, Gluckman E, Wernet P. 2005. High-resolution HLA typing by sequencing for HLA-A, -B, -C, -DR, -DQ in 122 unrelated cord blood/patient pair transplants hardly improves long-term clinical outcome. Bone Marrow Transplant. 36:1033–1041.

Kolb HJ, Mittermuller J, Clemm C, Holler E, Ledderose G, Brehm G, Heim M, Wilmanns W. 1990. Donor leukocyte transfusions for treatment of recurrent chronic myelogenous leukemia in marrow transplant patients. Blood. 76:2462–2465.

Laroche V, McKenna DH, Moroff G, Schierman T, Kadidlo D, McCullough J. 2005. Cell loss and recovery in umbilical cord blood processing: A comparison of postthaw and postwash samples. Transfusion. 45:1909–1916.

Laughlin MJ, Eapen M, Rubinstein P, Wagner JE, Zhang MJ, Champlin RE, Stevens C, Barker JN, Gale RP, Lazarus HM, Marks DI, van Rood JJ, Scaradavou A, Horowitz MM. 2004. Outcomes after transplantation of cord blood or bone marrow from unrelated donors in adults with leukemia. N Engl J Med. 351:2265–2275.

Magro E, Regidor C, Cabrera R, Sanjuan I, Fores R, Garcia-Marco JA, Ruiz E, Gil S, Bautista G, Millan I, Madrigal A, Fernandez MN. 2006. Early hematopoietic recovery after single unit unrelated cord blood transplantation in adults supported by co-infusion of mobilized stem cells from a third party donor. Haematologica. 91:640–648.

Majhail NS, Brunstein CG, Wagner JE. 2006a. Double umbilical cord blood transplantation. Curr Opin Immunol. 18:571–575.

Majhail NS, Weisdorf DJ, Wagner JE, DeFor TE, Brunstein CG, Burns LJ. 2006b. Comparable results of umbilical cord blood and HLA-matched sibling donor hematopoietic stem cell transplantation after reduced-intensity preparative regimen for advanced Hodgkin lymphoma. Blood. 107:3804–3807.

Nagamura-Inoue T, Shioya M, Sugo M, Cui Y, Takahashi A, Tomita S, Zheng Y, Takada K, Kodo H, Asano S, Takahashi TA. 2003. Wash-out of DMSO does not improve the speed of engraftment of cord blood transplantation: Follow-up of 46 adult patients with units shipped from a single cord blood bank. Transfusion. 43:1285–1295.

Netcord. 2007. Inventory and Use November. Available at https://www.netcord.org/inventory.html. Accessed 1/24/08.

Nilsson SK, Johnston HM, Coverdale JA. 2001. Spatial localization of transplanted hemopoietic stem cells: Inferences for the localization of stem cell niches. Blood. 97:2293–2299.

Noort WA, Kruisselbrink AB, in't Anker PS, Kruger M, van Bezooijen RL, de Paus RA, Heemskerk MH, Lowik CW, Falkenburg JH, Willemze R, Fibbe WE. 2002. Mesenchymal stem cells promote engraftment of human umbilical cord blood-derived CD34(+) cells in NOD/SCID mice. Exp Hematol. 30:870–878.

North TE, Goessling W, Walkley CR, Lengerke C, Kopani KR, Lord AM, Weber GJ, Bowman TV, Jang IH, Grosser T, FitzGerald GA, Daley GQ, Orkin SH, Zon LI. 2007. Prostaglandin E2 regulates vertebrate haematopoietic stem cell homeostasis. Nature (London). 447:1007–1011.

Rocha V, Cornish J, Sievers EL, Filipovich A, Locatelli F, Peters C, Remberger M, Michel G, Arcese W, Dallorso S, Tiedemann K, Busca A, Chan KW, Kato S, Ortega J, Vowels M, Zander A, Souillet G, Oakill A, Woolfrey A, Pay AL, Green A, Garnier F, Ionescu I, Wernet P, Sirchia G, Rubinstein P, Chevret S, Gluckman E. 2001. Comparison of outcomes of unrelated bone marrow and umbilical cord blood transplants in children with acute leukemia. Blood. 97:2962–2971.

Rocha V, Labopin M, Sanz G, Arcese W, Schwerdtfeger R, Bosi A, Jacobsen N, Ruutu T, de LM, Finke J, Frassoni F, Gluckman E. 2004. Transplants of umbilical-cord blood or bone marrow from unrelated donors in adults with acute leukemia. N Engl J Med. 351:2276–2285.

Rowley SD, Donaldson G, Lilleby K, Bensinger WI, Appelbaum FR. 2001. Experiences of donors enrolled in a randomized study of allogeneic bone marrow or peripheral blood stem cell transplantation. Blood. 97:2541–2548.

Sanz MA. 2004. Cord-blood transplantation in patients with leukemia—a real alternative for adults. N Engl J Med. 351:2328–2330.

Schmitz N, Beksac M, Hasenclever D, Bacigalupo A, Ruutu T, Nagler A, Gluckman E, Russell N, Apperley JF, Gorin NC, Szer J, Bradstock K, Buzyn A, Clark P, Borkett K, Gratwohl A. 2002. Transplantation of mobilized peripheral blood cells to HLA-identical siblings with standard-risk leukemia. Blood. 100:761–767.

Schofield R. 1978. The relationship between the spleen colony-forming cell and the haemopoietic stem cell. Blood Cells. 4:7–25.

Shpall EJ, de Lima M, Jones R, Champlin R. 2007. Are 2 cords better than 1? Blood. 110:2789–2790.

Staba SL, Escolar ML, Poe M, Kim Y, Martin PL, Szabolcs P, lison-Thacker J, Wood S, Wenger DA, Rubinstein P, Hopwood JJ, Krivit W, Kurtzberg J. 2004. Cord-blood transplants from unrelated donors in patients with Hurler's syndrome. N Engl J Med. 350:1960–1969.

Stevens CE, Scaradavou A, Carrier C, Carpenter C, Rubinstein P. 2005. An Empirical Analysis of the Probability of Finding a Well Matched Cord Blood Unit: Implications for a National Cord Blood Inventory. Blood. 106:579a.

Takahashi S, Iseki T, Ooi J, Tomonari A, Takasugi K, Shimohakamada Y, Yamada T, Uchimaru K, Tojo A, Shirafuji N, Kodo H, Tani K, Takahashi T, Yamaguchi T, Asano S. 2004. Single-institute comparative analysis of unrelated bone marrow transplantation and cord blood transplantation for adult patients with hematologic malignancies. Blood. 104:3813–3820.

Takahashi S, Ooi J, Tomonari A, Konuma T, Tsukada N, Oiwa-Monna M, Fukuno K, Uchiyama M, Takasugi K, Iseki T, Tojo A, Yamaguchi T, Asano S. 2007. Comparative single-institute analysis of cord blood transplantation from unrelated donors with bone marrow or peripheral blood stem-cell transplants from related donors in adult patients with hematologic malignancies after myeloablative conditioning regimen. Blood. 109:1322–1330.

Wagner E, Duval M, Dalle JH, Morin H, Bizier S, Champagne J, Champagne MA. 2006. Assessment of cord blood unit characteristics on the day of transplant: Comparison with data issued by cord blood banks. Transfusion. 46:1190–1198.

Wagner JE, Barker JN, DeFor TE, Barker S, Blazar BR, Eide C, Goldman A, Kersey J, Krivit W, MacMillan ML, Orchard PJ, Peters C, Weisdor DJ, Ramsay NKC, Davies SM. 2002. Transplantation of unrelated donor umbilical cord blood in 102 patients with malignant and non-malignant diseases: Influence of CD34 cell dose and HLA disparity on treatment-related mortality and survival. Blood. 100:1611–1618.

Zhang J, Niu C, Ye L, Huang H, He X, Tong WG, Ross J, Haug J, Johnson T, Feng JQ, Harris S, Wiedemann LM, Mishina Y, Li L. 2003. Identification of the haematopoietic stem cell niche and control of the niche size. Nature (London). 425:836–841.

2

EXPANDING THE THERAPEUTIC POTENTIAL OF UMBILICAL CORD BLOOD HEMATOPOIETIC STEM CELLS

Rouzbeh R. Taghizadeh and James L. Sherley

Programs in Regenerative Biology and Cancer, Boston Biomedical Research Institute, Watertown, MA 02472

INTRODUCTION: HEMATOPOIETIC STEM CELLS, THE THERAPEUTIC DISTRIBUTED STEM CELLS IN UMBILICAL CORD BLOOD

"Distributed stem cells" (DSCs) is a new terminology recently coined to provide a more exact name for "adult stem cells" (Fig. II.1) [Sherley, 2008]. The descriptor "adult" was originally chosen to contrast embryonic stem cells (ESCs) [Watt, 1991]. However, subsequently, "adult" is often inaccurately applied to stem cells found in fetal tissues, newborns, and children. Other terms, for example, "somatic stem cells," also lack the desirable feature of being comprehensive for all nonembryonic stem cell types. The DSCs encompass all nonembryonic stem cell types independent of the stage of development of their tissue of residence. Moreover, the term "distributed" connotes the concept that the pluripotency attributed to ESCs is *distributed* during development among later unipotent and multipotent stem cell types.

The DSCs have the unique ability to maintain themselves through continual asymmetric self-renewal divisions that simultaneously replenish short-lived, tissue-specific mature differentiated cells (Fig. II.1). These cells are derived during development from currently uncertain embryonic precursors. However, it is possible that they derive from novel metakaryote cells recently identified in diverse embryonic tissues of the mouse and human [Gostjeva, 2006]. They are present as a small, but essential, fraction of the cells in all evaluated normal tissues, including hematopoietic [Becker, 1963; Boggs, 1983; Cheshier, 1999; Till, 1961; Urist, 1965; Wu, 1968], liver [Lee, 2003; Sell, 1994], intestines [Potten, 1988; Potten, 1990], muscle [Potten, 1990; Qu-Petersen, 2002], brain [Potten, 1990], and epithelia

Perinatal Stem Cells. Edited by C. L. Cetrulo, K. J. Cetrulo, and C. L. Cetrulo, Jr.

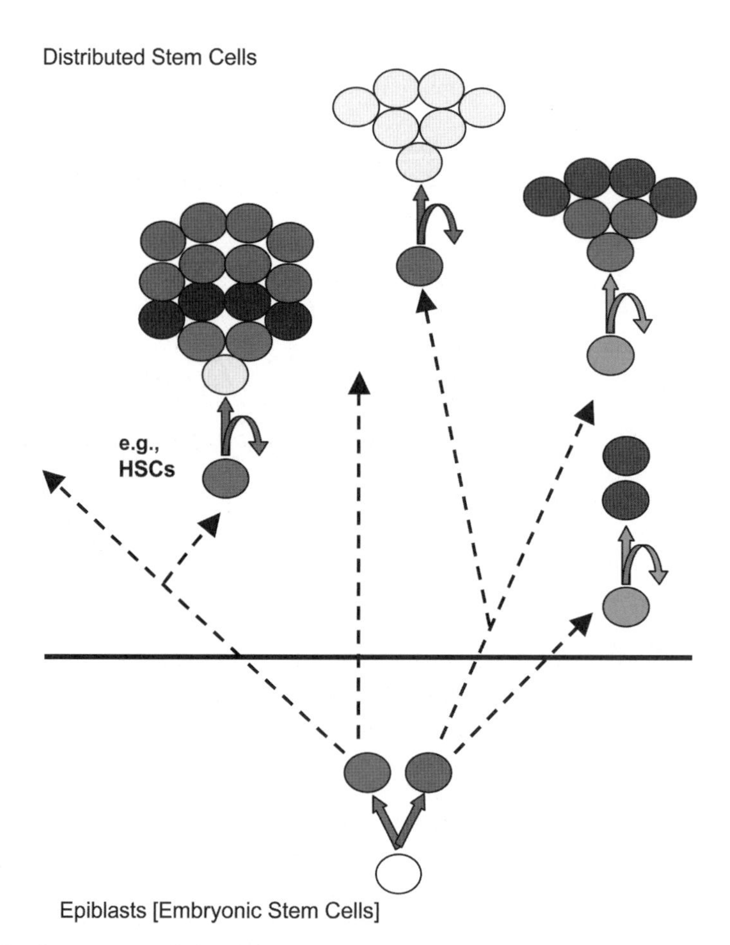

Figure II.1. The concept of distributed stem cells. Epiblasts in the early embryo divide asymmetrically to establish the complex cell differentiation plan of embryonic development. Their phenotypically asymmetric cell divisions do not preserve pluripotency during normal development. By preventing these divisions *in vitro*, embryonic stem cells are derived with the ability to preserve pluripotency. *In vivo* during embryonic development, the initiating embryonic pluripotency is "distributed" as multipotency and unipotency among postembryonic stem cells that become long lived in postnatal tissues [e.g., hematopietic stem cells (HSCs) found in umbilical cord blood (UCB)]. Unlike the phenotypically asymmetric divisions of epiblasts and other embryonic precursors, DSCs undergo asymmetric self-renewal divisions that preserve their specific potency even as they continuously renew tissue-specific differentiating cells. (See color insert.)

[Karam, 1993; Potten, 1988]. Although universal prospective markers for DSC identity continue to evade discovery, by definition, all DSCs share the exclusively defining property of asymmetric self-renewal [Hawkins, 1998; Lovell-Badge, 2001; Merok, 2001; Reya, 2001; Sherley, 2002].

Hematopoietic stem cells are a highly recognized type of DSCs because of their well-established successful clinical therapies and their potential for even more far-reaching biomedical advances in the future. The HSCs are the engines that drive the production and underpin the maintenance of the wide panoply of functional blood cells required for healthy human life. The human body consumes an astounding 400 billion mature blood cells every day. Even greater numbers are required when the body is under conditions of stress or trauma, for example, infection, low oxygen levels, or severe blood loss [Koller, 1999]. In order to meet this demand, a complex system of multilineage proliferation and differentiation of hematopoietic cells has evolved. The highly orchestrated production of ~12 different types of mature blood cells is sustained by primitive HSCs that reside primarily in the bone marrow (BM), but are also present in smaller numbers in the blood, in particular UCB.

During normal replenishment of mature blood cells, HSCs self-renew asymmetrically [Huang, 1999; Punzel, 2003] and thereby simultaneously maintain the HSC pool while producing two different classes of multipotent progenitor cells (MPCs). The MPCs are thought to divide with symmetric cell kinetics to give rise to either of two progenitor cell types, common myeloid progenitors (CMPs) or common lymphoid progenitors (CLPs). The CMPs continue to divide to give rise to erythrocytes, monocytes (i.e., macrophages, osteoclasts, and dendritic cells), granulocytes (i.e., neutrophils, eosinophils, basophils, and mast cells), and platelets that are derived from noncirculating megakaryocytes [Koller, 1999]. Thymus-derived T-lymphocytes, bone marrow-derived B-lymphocytes, and natural killer (NK) cells constitute the lymphoid lineage derived by CLPs. Although some lymphocytes are thought to survive for many years (i.e., memory T and B cells), other mature cells like erythrocytes and neutrophils have a limited lifespan (120 days and 8 h, respectively) [Cronkite, 1988]. As a result, hematopoiesis is a highly prolific process [Jansen, 2005; Storb, 2003], occurring throughout human life to fulfill this great demand while ensuring hematopoietic homeostasis. Because of their role in the continuous production of new mature blood cells to replace defunct and expired ones, HSCs are an absolute necessity for life.

The tremendous ability of HSCs for lifelong regulated production of mature blood cells drives great interest in adapting HSCs for cellular therapy applications, for example, advanced bone marrow transplantations (BMT), gene therapy, and tissue engineering. Currently, transplantation of HSC preparations is part of the standard of care for the treatment of both non-hematopoietic (e.g., breast cancer) and hematopoietic disorders [Jansen, 2005; Ueno, 2006]. The three sources of HSCs used to reconstitute hematopoiesis after myeloablation are bone marrow, UCB, and mobilized peripheral blood (MPB) [Rebel, 1996; Storb, 2003; Szilvassy, 2001]. Unfortunately, hematopoietic cell transplants are often not effective. In fact, for those patients for which a histocompatible allogeneic transplant source is available, as many as 35% do not survive [Tyndall, 2000]. One major cause of death for transplanted patients is prolonged time to engraftment. Due to insufficient numbers of short-term engrafting cells that provide immediate immunity, patients succumb to the effects of infection. In addition to poor engraftment, there are two other cellular factors that compromise HSC transplantation efficacy. These are HSC dose and the risk of graft versus host disease (GvHD) [Storb, 2003]. Currently, although there is no means for exact determination of HSC dose, it is evident from past studies that increasing the amount of HSCs or the HSC fraction of transplanted cell populations increases efficacy [Locatelli, 1999; Migliaccio, 2000; Schoemans, 2006; Wagner, 2002]. The GvHD is a particularly significant barrier to the realization of the full potential of HSC therapy, because it limits the most common type of HSC transplantation, which is allogeneic [Schoemans, 2006].

Of the three sources of human HSCs for transplantation therapy, UCB has the best prospects for advancing HSC therapy beyond its current limitations. About 200 common HLA haplotypes occur within the U.S. Caucasian population, which translates into a 1 in 1000 to 1 in several million chance of finding a donor-recipient match for BMT [Revazova, 2008]. An aggressive healthcare campaign to establish geographically diverse blood banks that cryopreserve and share strategically collected UCB donations that incorporate representative demographic groups could one day provide coverage for all but the most rare HLA haplotypes. Further advances in publicly funded banking of UCB for all births would provide comprehensive HSC transplant coverage.

A number of both private companies and public agencies now bank UCB routinely. However, from these enterprises, it is now abundantly clear that just the act of banking even multiple doses of typed UCB is not sufficient to resolve the current shortage of effective HSCs for transplantation therapy. Although UCB specimens have shown limited efficacy in children, they are for the most part inadequate for treatment of adults, who constitute the majority of patients. The key to unlocking the great therapeutic potential of UCB is *ex vivo* expansion of the HSCs that it contains. The main thrust of this section is to consider the biological, technical, clinical, and experimental issues that surround this longstanding biomedical research challenge.

BIOLOGICAL BARRIERS TO HEMATOPOIETIC STEM CELL THERAPY

Scarcity, identification, and expansion, these are the biological barriers that stand in the way of the translation of DSC therapeutic potentialities into clinical realities. The HSCs in UCB are in no way exempt from crossing these hurdles, too. However, they do start with two potential advantages over BM HSCs, a larger pool of potential donors and a higher HSC fraction. The current birth rate in the United States of $\sim$14/1000 (est. 2008) [Central Intelligence Agency, 2008] corresponds to >4 million potential UCB specimens yearly. Also, whereas estimates of the HSC fraction of human BM can be as low as 10^{-6}, UCB fractions are estimated to be as high as 10^{-4} [Mayani, 1998; Piacibello, 1999; Schoemans, 2006; Verfaillie, 2002; Wang, 1997]. The scarcity of adult donors and the low cell fraction of HSCs in adult BM combine to pose a major limitation to the use of adult-derived HSCs for cell therapy research and clinical treatments. The frequency of HSCs in adult BM is estimated to be 10^{-5}–10^{-6} [Bhatia, 1997; van der Loo, 1998].

Despite such low numbers, collectively, BM HSCs have sufficient proliferative capacity to last several lifetimes [Boggs, 1983; Spangrude, 1988]. It is remarkable that such a rare population of cells has the power and fidelity to fuel the production of $\sim$1 billion red blood cells and 100 million white blood cells per hour in the adult human [Stiff, 2000] and a total of 6 billion cells per kilogram of body weight per day [Beutler, 2001]. Though no hinder to superior endogenous hematopoiesis, the low fraction of HSCs has greatly impeded both research to define the biological mechanisms responsible for their essential functions and clinical development for transplantation therapies.

The scarcity of HSCs in tissues is also a major factor that has limited their purification and, thus far, precluded their specific prospective identification. Both of these crucial goals are consistently compromised by a high degree of contamination with coisolated differentiated cells and lineage-committed hematopoietic progenitor cells [hematopoietic progenitor cell (HPC); e.g., CMPs and CLPs]. Landmark methods have been developed to enrich for HSCs using markers expressed on the HSC plasma membrane. The cell surface markers, CD34 [Spangrude, 1989; Sutherland, 1989] and CD133 [de Wynter, 1998; Gallacher, 2000] identify and enrich for cell fractions that include HSCs. The current clinical biomarker CD34 is used to enrich for HSCs toward improving repopulation efficiency in patients receiving bone marrow transplantations [Verfaillie, 2002]. However, neither of these markers is expressed solely on HSCs. In addition to HSCs, CD34, and CD133 are expressed on both hematopoietic [de Wynter, 1998; Gallacher, 2000; Spangrude, 1988; Sutherland, 1989] and non-hematopoietic lineage-committed progenitor cells [Asahara, 1997; Peichev,

2000]. These markers are also selectively expressed on rare cells in other stem cell compartments like, for example, CD34 on cells in the hair follicle bulge region [Morris, 2004; Trempus, 2003]. However, neither marker shows expression exclusive for HSCs or any other DSCs [Pare, 2006].

The lack of exclusive marker specificity is particularly problematic for identification and purification of HSCs because of their exceedingly low cell fraction. In fact, with current enrichment methods, the estimated fraction of human HSCs in therapeutic transplant cell preparations is $\sim$0.1–1.0% [Bhatia, 1997; Taghizadeh, in preparation; Verfaillie, 2002]. The remaining cells are HPCs and other differentiated cells. This situation greatly limits research to discover and test candidates for exclusive HSC molecular markers, because a high degree of cell-type purity is essential for success in these studies.

The inexact quality of current so-called "HSC markers" is an important caveat for previous studies that considered CD34- or CD133-selected cells as pure HSC populations during molecular characterization of HSC enriched populations [Ivanova, 2002; Ramalho-Santos, 2002]. Without markers that identify HSCs exclusively, the ideal of direct prospective quantification of HSCs in any setting is wishful thinking. Thus, both clinical and basic HSC research must resort to evaluation of the ability of human HSCs toproduce long-term human myeloid and lymphoid cells in immunocompromised animal models [Bhatia, 1997; van der Loo, 1998] as the only available confident measure for HSCs. However, this detection assay, though certainly a breakthrough development in HSC research, provides only retrospective semiquantitative data, and it is much too cumbersome and slow for effective use in the treatment setting.

In combination with the ready availability of UCB, the ability to expand HSCs *ex vivo* could subdue the problem of HSC scarcity and possibly the problem of exclusive identification as well. In the limit, an effective method of selective expansion of HSCs could eventually yield pure populations of expanded HSCs, and certainly more highly enriched ones than are available currently. Unfortunately, such a strategy, though obvious in concept, has been formidable in practice. Whereas exclusive identification of HSCs has been challenging but hopeful, expanding them *ex vivo* has been deceptively unyielding. As will be considered in greater detail later, the field of HSC research is overrun with reports of presumed successful *ex vivo* expansion of HSCs. However, the problem with these reports is their definition of "HSCs," which is often inexact, nonquantitative, and ambiguous. As noted earlier, many investigators use nonexclusive molecular markers as identifiers and do not account for the presence and proliferation of non-HSC contaminants, which are invariably the predominant cell type.

As a general rule, all DSCs are difficult to expand *ex vivo*. But HSCs have proved particularly recalcitrant. The hypotheses advanced in the past to explain this vexing character are as numerous as the strategies attempted to overcome it. These include deficits in special growth factors, cytokines, survival factors, chromatin modifiers, feeder cells, extracellular matrices, synthetic materials, three-dimensionality (3D), byproduct removal, and removal of differentiation factors. Finally in this chapter, we will consider in greater detail a newer hypothesis based on the property of asymmetric self-renewal, which recently inspired a new method that has proven effective for expansion of other types of DSCs tested to date. The method, called suppression of asymmetric cell kinetics (SACK), is based on the fundamental precept that, as a default division mode, DSCs adopt asymmetric self-renewal (Fig. II.2). This program of producing a replacement stem cell and a non-stem cell from each DSC division is also referred to as asymmetric cell kinetics. The latter terminology highlights the distinction in cell kinetics between the two cell lineages produced by each kinetically asymmetrical DSC division. The DSC lineage retains division capacity for

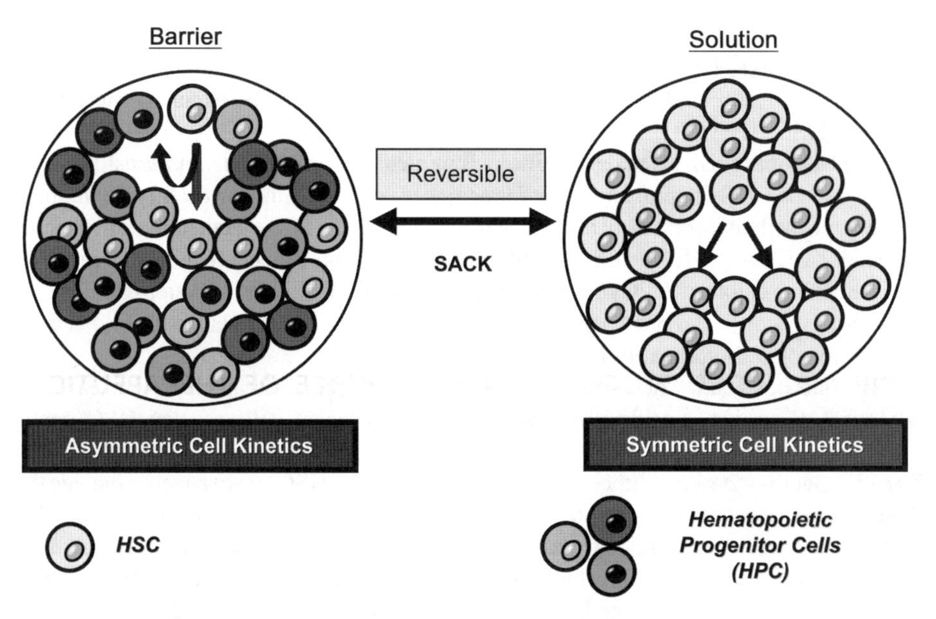

Figure II.2. Suppression of asymmetric cell kinetics: a solution to overcoming the cell kinetics barrier to hematopoietic stem cell expansion. Asymmetrically self-renewing HSCs divide with asymmetric cell kinetics to give rise to two daughter cells, a new HSC, and a hematopoietic progenitor cell (HPC). *Ex vivo*, the HPC divides to give rise to differentiated, maturing blood cells. As the culture continues, the HSC is predicted to become diluted and lost among the accumulating progeny from active HPCs and differentiated cell proliferation. The solution to overcome this cell kinetics barrier is SACK, which promotes symmetric self-renewal of HSCs. In this state, not only does the HSC number increase exponentially, but consequently the production of HPCs and differentiated cells is also reduced. In order for this solution to be ideally useful, it must be reversible; so that upon removal of SACK agents (see Fig. II.3), expanded HSCs can revert back to their asymmetrically self-renewing state to give rise to differentiated, functional blood cells, *in vitro* and *in vivo*. (See color insert.)

the entire lifespan, whereas the non-stem lineage has a limited division capacity. In some tissues, the non-stem cell may not divide further. In others, its descendants may undergo an extensive, but finite, number of divisions before maturing into a permanent postmitotic division arrest.

Asymmetric cell kinetics by DSCs was recognized as posing a major biological barrier to DSC expansion *ex vivo* [Pare, 2006; Rambhatla, 2001; Sherley, 2002]. If explanted DSCs maintained their *in vivo* asymmetric cell kinetics, there own numbers were expected to be constant as they continuously produced non-stem progeny cells that would accumulate in culture. With increasing culture time and progressive culture passages, DSCs were predicted to be obscured gradually by the accumulating larger numbers of their differentiating progeny. Moreover, they would be diluted as a consequence of passaging; and, as a result, serial cultures would always eventually progress to a point when they were devoid of any DSCs. In the absence of mitigating gene mutations, thereafter, cultures now containing only non-stem cell progeny would head inexorably to complete cessation of cell division (i.e., cellular senescence by asymmetric DSC kinetics).

Thus, the SACK concept also represented a novel alternative explanation for the well-known phenomenon of cellular senescence in culture [Hayflick, 1965; Rambhatla, 2001].

More importantly, it introduced a previously overlooked critical barrier to the *ex vivo* production of DSCs, like HSCs, regardless of their source. Thus, we now advance that any path leading to effective use of UCB HSCs for adult hematopoietic therapy must address the need for SACK. Specifically, SACK is the reversible shift of DSCs from asymmetric self-renewal to symmetric self-renewal. Symmetric self-renewal divisions yield two DSCs, and therefore simultaneously reduce the production of non-stem differentiating progeny cells. This pattern of DSC self-renewal is necessary for developmental increases in body size; and it is also predicted to occur during tissue repair [Sherley, 2002]. In culture, it promotes the exponential expansion of DSCs (Fig. II.2) [Pare, 2006].

UMBILICAL CORD BLOOD: AN IDEAL SOURCE OF THERAPEUTIC HEMATOPOIETIC STEM CELLS

The UCB cells represent the best source of therapeutic HSCs for several reasons. First, neonatal UCB contains a higher percentage of HSCs than adult blood or marrow primarily as a consequence of the site of hematopoiesis transitioning from the fetal liver to the bone marrow at the time of birth. This relocation of hematopoiesis could account for the greater proliferative and expansion properties of UCB cells, compared to adult HSC sources [Lansdorp, 1995a, b; Vaziri, 1994]. This difference may reflect a higher fraction of symmetric self-renewal divisions by UCB HSCs to support expansion of the hematopoiesis system during fetal and neonatal development. In the pediatric transplant setting, UCB cells allow up to a 1000-fold lower number of infused cells compared to BMTs with similar, if not better, outcomes [Schoemans, 2006].

Second, HSCs from UCB engraft with a lower frequency and severity of GvHD even with incomplete recipient—donor HLA-match [Meyer, 2005]. One possible reason for the reduced level of GvHD with UCB is that neonatal immune cells are relatively naïve for antigen encounter compared to adult cells [Bradley, 2005]. As a result, UCB cells are less able to activate adverse reactions against recipient antigens [Sirchia, 1999]. This allows for greater degrees of donor–patient HLA disparity (cf. BM and MPB sources for HSCs) [Gluckman, 2004; Gluckman, 2005; Rubinstein, 1998], which results in increased transplantation efficacy. In fact, transplantation survival is primarily sensitive to the degree of HLA matching in UCB grafts at the level of HLA-A and HLA-B, and high-resolution allelic typing for HLA-DRB1 [Kögler, 2005]. Thus, further high-resolution HLA characterization does not yield better survival. This finding is in sharp contrast to unrelated allogeneic BMTs, where any single serological mismatch is considered a risk factor [Kögler, 2005]. However, even with these advantages, as in the case for BM or MPB transplantations, immunosuppressive therapy is an important adjunct to UCB transplantation to reduce both GvHD and graft rejection.

Third, the diversity of cryo-preserved HSCs in UCB banks augments possibilities for greater patient–donor matches, since they are more representative of the general population, as compared to bone marrow donor programs. In fact, it was recently shown that patients have a 99% chance of finding a 4/6 HLA-match UCB cell source and a 70% chance of finding a 5/6 or 6/6 HLA matched UCB unit [Stevens, 2005]. Furthermore, since UCB cells are collected and stored cryogenically, UCB does not have the same donor attrition issues as adult sources of HSCs. In fact, the time from unit identification to transplantation is significantly reduced to days-to-weeks for UCB sources, rather than weeks-to-months for MPB or BM sources. Even before storage, UCB units are available for transplantation, as they are fully HLA typed without risking donor attrition or morbidity. The significance of this is evident in the median time to donor identification, 13.5 days for UCB units compared to 49 days for BM [Barker, 2002].

Although UCB is the ideal source for therapeutic HSCs, it has several limitations. One therapeutic drawback for UCB is that it can only be collected and cryo-preserved at birth. This limitation is more significant for patients who, because of rare HLA-haplotypes, require autologous transplantations. How long cryo-preserved HSCs can be stored without losing their engraftment capacity is also unclear. It is evident that cryo-preserved cells maintain their clonogenic capacity for up to 15 years of storage [Broxmeyer, 2003; Spurr, 2002], but the extent of their frozen lifetime for engraftment must await experience derived from routine clinical practice.

Another disadvantage is that time-to-neutrophil recovery is slower in patients transplanted with UCB compared to BM. In fact, recent clinical trials comparing UCB to BM transplantation exhibited time-to-neutrophil recovery ranging from 22 to 27 days, as compared to 18 days for an unrelated BM transplant cohort [Laughlin, 2004; Takahashi, 2004]. In a related study, by post-transplantation day 60, 75% of patients transplanted with UCB grafts exhibited engraftment to normal neutrophil levels, compared to 89% of patients transplanted with BM derived grafts [Rocha, 2004]. Clinical trials comparing UCB and BM transplantations in patients with acute and chronic leukemia indicated slower overall engraftment in UCB cohorts. The median overall survival percentage of patients transplanted after 2 years with UCB or BM grafts was 36 and 42%, respectively [Rocha, 2004]. However, after 3 years, the median overall percentage survival was observed to be 26 and 35% in UCB and BM transplanted patients, respectively [Laughlin, 2004]. Despite these disappointing clinical results, UCB cells are still the focus of intense research, primarily due to their availability, reduced stringency for HLA matching, and higher HSC fraction.

The cause of engraftment delay is possibly due to a lower effective infused cell dose, as engraftment and the rate of neutrophil and platelet recovery directly correlate with infused cell dose [Locatelli, 1999; Migliaccio, 2000; Wagner, 2002]. The current empirical threshold for transplanted UCB is 1.7×10^5 infused CD34+ cells/kg or, on average, 2.5×10^7 cryo-preserved nucleated cells [Barker, 2005a, b]. This dose is at least a log lower than that in BM transplantations. Since the volume of each UCB preparation is intrinsically limited to $\sim$100 mL, the number of HSCs obtained from UCB collection is restricted. Although sufficient to engraft recipients weighing less than 50 kg (i.e., small children), single UCB preparations are not efficacious for transplanting adults, or even larger children.

Several approaches have been evaluated for circumventing the slow engraftment kinetics and high transplant related morbidity (TRM) of conventional postmyeloablation UCB transplantations, including coinfusion of pooled UCB units and attempted *ex vivo* expansion of UCB HSCs. These strategies aim to increase the total number of infused HSCs with the expectation of accelerated engraftment kinetics and reduced TRM. Trials conducted using combined units of UCB have shown a wide range of neutrophil engraftment rates (dependent on the pretransplantation conditioning used) low TRM rates (14–48%), slightly higher rates of acute GvHD, previously observed levels of chronic GvHD, and low frequencies of graft failure (0–22%) with 1-year overall survival ranging from 31 to 79% [Ballen, 2005; Barker, 2005a, b; Brunstein, 2005a, b].

There are significant trade-offs for the modest gains in engraftment rates achieved in patients receiving combined unit grafts [Verneris, 2005]. The slower rates of immune reconstitution from pooled UCB grafts result in higher rates of infection [Meyer, 2005], and the risk of acute GvHD is elevated [Verneris, 2005]. The increased GvHD risk may reflect an increased overall diversity in the immune cell population as a result of combining UCB units. Interestingly, patients infused with combined UCB units exhibited engraftment by HSCs from only one of the combined units by day 100. This finding may indicate that a small number of HSCs (i.e., potentially 1) is sufficient for human engraftment (as has been shown for mouse HSCs [Cao, 2004; Moore, 1997; Takano, 2004], or that some units have factors that allow for preferential homing to the bone marrow [Schoemans, 2006]. However, the underlying mechanism postulated to account for the modest increases in success of combined UCB units is the increased total number of infused primitive HSCs in the combined transplant sample.

One method used to circumvent these obstacles has been to pool UCB samples from several HLA matched donors. A strategy that would obviate the need for multiple UCB units and enable routine UCB transplantations is *ex vivo* expansion of HSCs from UCB units, as previously discussed. Such a cell expansion technology would enhance hematopoietic engraftment, reduce the risk of infection, and increase the number of potentially available allogeneic donors. Additionally, successful *ex vivo* expansion of HSCs would enable a variety of clinical applications including permitting efficacious UCB transplantations for adult patients, facilitating greater efficiency in repairing genetic errors in proposed gene therapy applications, and mass producing large quantities of functional, differentiated blood cells. To date, such potential breakthroughs have been soundly thwarted by the lack of a successful method for the *ex vivo* expansion of HSCs, despite many years of effort documented in thousands of published research articles from the past 30 years on the subject.

Despite the generally universal lack of success, several clinical studies [Jaroscak, 2003; Shpall, 2002] investigated transplantation of putative *ex vivo* expanded UCB units with the primary aim of evaluating safety and feasibility. In addition, reduction of the commonly encountered post-transplant pancytopenia to achieve decreased TRM and time to engraftment were evaluated as a secondary aim. If obtained, this outcome would have indicated expansion of committed HPCs, since expansion of primitive, long-term HSCs was not demonstrated. However, these clinical trials failed to show better recovery kinetics over historical controls, and the incidence of both acute and chronic GvHD was high. Moreover, there was high TRM that resulted in high relapse rates and poor survival [Devine, 2003]. Thus, these studies place in greater relief the persistent need for a substantial advance in understanding the exact nature of the UCB HSC expansion problem.

SHORTCOMINGS OF *EX VIVO* HSC EXPANSION STRATEGIES

There are several fundamental biological features of past and current HSC expansion studies that are frequently overlooked by investigators, which commonly lead to overstatements and erroneous conclusions. One such feature is the inherent problem of simply maintaining HSCs in culture. Invariably, *ex vivo* culture of HSC containing populations, including UCB cells, under a variety of conditions, leads to an inexplicable loss of HSCs, resulting in reductions in overall hematopoietic reconstitution ability [Devine, 2003; Eaves, 1992; Taghizadeh, in preparation]. Stimulation of proliferation by combinations of various growth factors in attempts to achieve *ex vivo* expansion of HSCs results in extensive overall cell production due to the proliferative response of HPCs. However, no reported growth factor cocktail combination or coculture conditions with bone marrow stromal cells has achieved reproducible net expansion of HSCs *ex vivo*. In fact, these conditions result in an unexplained reduction in HSC activity, as measured by *in vivo* engraftment studies [Devine, 2003; Eaves, 1992; Harrison, 1987; Taghizadeh, in preparation].

Possible reasons for HSC depletion in culture are (1) negative effects of the growth factor cocktails on HSC viability and/or function; (2) similar negative effects from non-HSC cells whose numbers are expanded by growth factors; and/or (3) the need for absent HSC survival factors. Before any *ex vivo* HSC expansion technology succeeds in the clinic, a fundamental understanding of the basis for the loss of HSC function in culture is essential.

Another critical feature of HSC expansion studies that many investigators overlook, misstate, or ignore is absolute bookkeeping of HSC yield. When comparative studies are conducted to investigate *ex vivo* HSC expansion assayed by *in vivo* animal transplantation models, results from cultured cell populations must be compared to results from uncultured control cells. Note that culturing with growth factors results in large increases in total cell number. Thus, it is imperative that investigators develop a cell number-independent basis for evaluation of compared cell samples. Ideally, good HSC bookkeeping requires a means for tracking the yield of HSC activity units (AU) during the course of an experiment. This point is crucial for expansion studies, because, without this consideration, small variances in the excessively abundant HPC compartment can create the erroneous appearance of significant differences in HSC number or frequency.

Despite this important caveat, cell number-independent analyses were rarely employed in previously reported HSC expansion studies. The failure to apply such analyses is further aggravated by the Poisson nature of HSC repopulation statistics. Developing a cell number-independent analysis is accomplished straightforwardly by keeping track of cell culture units and relating all transplantation doses to the culture units that contained all of the investigated cell population at the start of expansion studies. In our own work, we call this metric the "cell well-equivalent (CWE)." All expansion trials begin with tested cell populations in single-culture wells. Thereafter, all cell derivatives of the input cells at the start of a culture are related based on their projected fraction of the initial culture well. For example, consider that 100,000 viable cells were put into a well at time zero, but after culture only 10,000 viable cells remained. If the entire final culture (est. 10,000 cells) were transplanted, it would be designated as 1 CWE. This designation would be the same for the uncultured starting input population of 100,000 viable cells. By using this cell number-independent denominator (i.e., 1 CWE for both compared samples) for comparison of transplantation activity, the effect of culture on the starting total HSC function can be determined directly without incorporating the variance of cell number determinations.

Similarly, if 1/10 of the final culture were transplanted, it would be designated as 0.1 CWE. If three replicate wells were combined and transplanted, their union would be designated as 3 CWE. If 1/5 of the culture were removed, diluted 10-fold, and 1/3 of the dilution transplanted, it would be designated as 0.06 CWE. Although inaccurate cell number determinations can be tracked in parallel, the CWE approach does not presume any proportionality or dependence between total cell number and HSC number or function. In fact, by comparing confident cell number data to CWE normalized data, the existence of such biological dependencies can be inferred objectively. Thus, a cell number-independent approach allows one to monitor the yield of HSC activity specifically, which is the critical value for evaluation of success in HSC expansion trials.

Because BM HSCs reside in a complex environment consisting of an assortment of cell types producing numerous cytokines and stimulatory factors, a "niche hypothesis" has been proposed and investigated [Schofield, 1978]. This hypothesis postulates that HSCs reside in stable microenvironments in association with various supporting cell types. One of these supporting cell types, stromal cells, has been assigned credit for forming niches that engender and preserve the self-renewal capabilities of HSCs, while carefully controlling differentiation and maturation of derivative HPCs. Experimental evidence supports the notion that stromal cells provide a complex microenvironment of stimulatory cytokines, growth factors, extracellular matrix proteins, and adhesion molecules that regulate and balance HSC self-renewal, survival, and differentiation [Verfaillie, 2002].

Many investigators have pursued the HSC niche as the missing element needed for effective HSC expansion *ex vivo* [Calvi, 2003, 2006; Hofmeister, 2007; Martin, 2005; Wilson, 2006]. However, despite many attempts, no definitive evidence has emerged to show that the presumptive niche provides factors necessary for expansion of HSCs in culture. However, there is one well-described report in this regard involving cultured BM stromal cells that is often misstated or misrepresented. Clonal mouse BM stromal cell lines were first developed by immortalization with a temperature-sensitive simian virus-40 T-antigen [Wineman, 1996]. Among these lines, AFT024 was selected for additional study, because it was reported to show the best cobblestone area-forming assay result for maintaining HSCs. In addition to maintaining HSCs *in vitro*, it was further concluded that AFT024 cells stimulated active proliferation of HSC/HPC cells [Moore, 1997]. Like human HSCs, mouse HSCs show a largely unexplained loss after culture periods on the order of days. The report of Moore and colleagues [Moore, 1997] is often cited as the basis for the belief that stromal cells like AFT024 can preserve HSC activity in culture. However, the report contained an anomalous result that went unexplained experimentally. Although the authors observed nearly quantitative maintenance of competitive repopulating units (CRU) as a measure of HSC activity in 5-week cultures with AFT024 cells, the same stromal cell line show no ability to preserve CRUs after only 5 days of culture. The authors suggested that this situation might be explained by initial differentiation of the majority of HSCs followed by the expansion of the remaining undifferentiated HSCs.

Whatever the cause of the apparent Moore paradox, the needed time scale for immediate clinical expansion of HSCs is days. Lewis and co-workers reported the evaluation of AFT024 cells for support of UCB HSC expansion trials on this time scale [Lewis, 2001]. Although they concluded that the cells were effective for maintaining UCB HSCs in culture for 7 days, the data presented were loosely interpreted. Because the authors did not use a cell number-independent basis for analysis, dilution effects could account for the modest effects reported; and none of the reported effects were statistically significant. A later report on coculture of AFT024 and human UCB cells, from Nolta and others [Nolta, 2002] is similarly disappointing, as these authors did not evaluate human hematopoietic cell engraftment in mouse models and did not include stromal cell-free controls in their study.

Thus, the question of a possible advantage from stromal cell components in human HSC expansion research is still an open one. If a robust effect of a stromal cell type is demonstrable, it will be important to discover isolable cellular factors that are responsible. In theory, the discovery of such factors would enable cell-free, defined conditions for *ex vivo* maintenance and production of human HSCs that reside in UCB. The earliest stromal cell line studies suggest that, if possible, the development of cell-free components will be challenging, as the effects described to date required direct cell-to-cell access [Wineman, 1996]. This experience with BM stromal cell lines brings into highlight two important points: (1) AFT024 cells and other transformed stromal cell lines are likely to provide only a small slice of the complexity of whole BM stromal cells; and (2) while they may perhaps preserve HSC number and/or function to some extent, they have not demonstrably enabled HSC expansion. Another overlooked consideration is that any gains made in stabilizing human UCB HSCs using stromal cell lines may necessarily compromise their subsequent expansion. If stromal cells that form the HSC niche function to maintain homeostatic HSC asymmetric self-renewal, as envisioned, then their addition to HSC cultures is predicted to limit HSC number, precluding the desired expansion.

A NOVEL STRATEGY WITH POTENTIAL TO ENABLE *EX VIVO* EXPANSION OF UMBILICAL CORD BLOOD HEMATOPOIETIC STEM CELLS

Previous research with the goal of enabling *ex vivo* expansion of HSCs, and DSCs in general, has focused on features and factors that might be *lost* by the act of *ex vivo* explant of tissue cells. However, more recently, a few studies have been based on the idea that the major barrier to *ex vivo* DSC expansion is an intrinsic property of DSCs, which is, in fact, *retained* in culture (Fig. II.2). This property is asymmetric self-renewal. Sherley and co-workers [Merok, 2001; Rambhatla, 2001; Sherley, 2002] initially proposed that asymmetric self-renewal, the defining property of DSCs, was also the major impediment to their *ex vivo* expansion. In later studies, they demonstrated that pharmacological suppression of asymmetric self-renewal by DSCs in adult rat liver could be used to efficiently expand several clonal adult hepatic DSC strains [Lee, 2003]. Their method has recently shown effectiveness for expansion of DSCs for adult human liver as well [Panchalingam, in preparation].

In vivo, asymmetric self-renewal maintains a constant HSC pool, and it may also limit their accumulation of mutation [Merok, 2002; Sherley, 2006]. In contrast, *ex vivo*, the continuation of asymmetric self-renewal by HSCs will prevent an increase in their number, while the numbers of HPCs and their more differentiated progenitors progressively increase dramatically (Fig. II.2). This problematic relationship is only worsened by growth factor cocktails that stimulate the proliferation of HPCs and differentiated cells without causing any change in the basic asymmetric cell kinetics of HSCs. The only solution to this cell kinetics barrier is a shift of HSCs from asymmetric cell kinetics to symmetric cell kinetics. The HSC divisions that produce two HSCs characterize symmetric cell kinetics. Not only do symmetric HSC kinetics promote exponential expansion of HSCs, they also preclude the production of committed HPCs, which arise from asymmetric HSC divisions (Fig. II.2).

Although asymmetric cell kinetics are the default state for DSCs in adult tissues, symmetric cell kinetics must occur to support fetal-to-adult body growth (e.g., crypt fission in the maturing intestinal tract). Symmetric cell kinetics may also occur during repair of injured tissues, as is suggested from the apparent expansion of murine HSC number in BM transplantation studies using irradiated mice [Lemischka, 1992]. In some tissues, DSCs may undergo intrinsic symmetric self-renewal divisions that are balanced by symmetric differentiation after some divisions, the latter which completely extinguishes DSC function [Sherley, 2002]. However, whether due to the autonomous asymmetric self-renewal of individual HSCs or due to the balanced cell kinetics of populations of HSCs, the intrinsic asymmetric cell kinetics of HSCs must be converted to symmetric cell kinetics to achieve their *ex vivo* expansion. This induced conversion of the self-renewal pattern of DSCs, called "SACK" for suppression of asymmetric cell kinetics (as introduced earlier), was recently described for expansion of two different types of adult rat hepatic DSCs, hepatocytic and cholangiocytic [Lee, 2003; Pare, 2006].

The SACK approach is based on the discovery of a p53-dependent biochemical pathway that controls the choice of asymmetric self-renewal versus symmetric self-renewal by DSCs in culture (reviewed in Pare, 2006). In normal cells, wild-type p53 protein regulates the expression of inosine monophosphate dehydrogenase (IMPDH; EC 1.2.1.14) to a level that limits production of cellular guanine ribonucleotides (rGNPs). The IMPDH is an essential cellular enzyme that catalyzes the rate-limiting step for rGNP biosynthesis (Fig. II.3a). This pathway, which includes p53, IMPDH, and yet to be specified critical rGNPs, functions

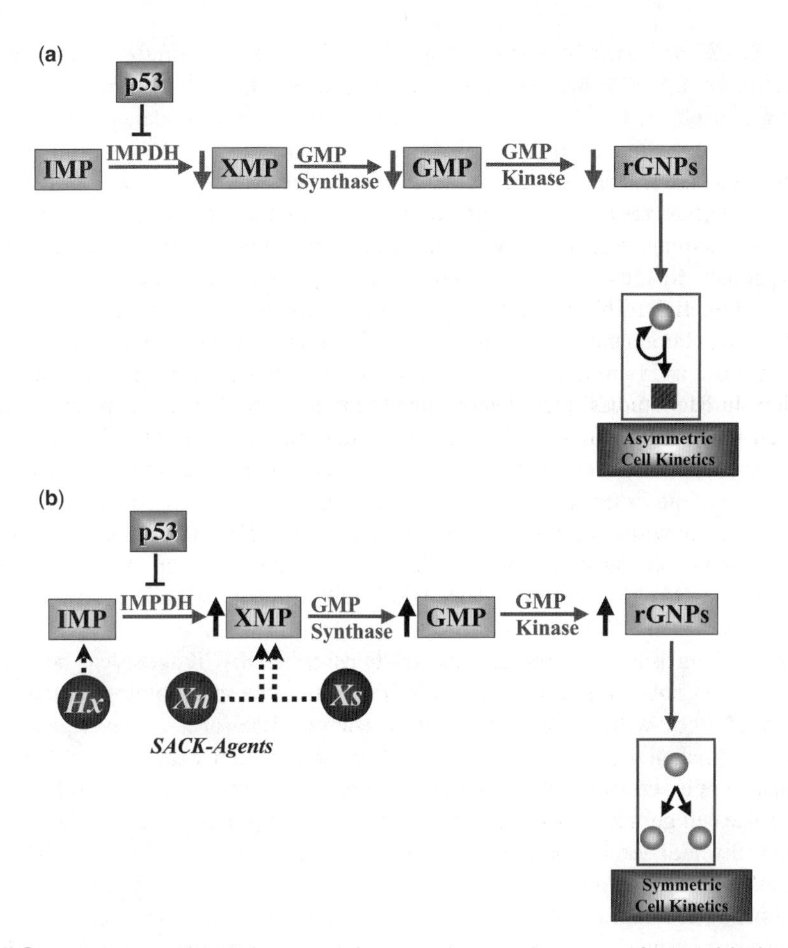

Figure II.3. A cellular pathway that controls the self-renewal pattern of DSCs. (a) The cellular pathway in its basal state in DSCs. The IMPDH is the rate-determining enzyme for the production of cellular guanine ribonucleotides (rGNPs). In normal DSCs, the p53 tumor suppressor protein downregulates IMPDH expression. This regulation keeps the rate of rGNP synthesis and the level of rGNP pools sufficiently low so that DSCs retain a state of asymmetric self-renewal. Inosine monophosphate (IMP); XMP, xanthosine monophosphate; GMP, guanosine monophosphate. (b) The cellular pathway's response to addition of SACK agents. Hypoxanthine (Hx), xanthine (Xn), or xanthosine (Xs) overcome or bypass the regulation of p53 by promoting increased synthesis and levels of rGNPs. Higher rGNP levels cause a shift of DSCs from asymmetric self-renewal to symmetric self-renewal. Mutations that lead to loss of p53 regulation or upregulation of cellular IMPDH activity are predicted to lock DSCs into a chronic state of symmetric exponential proliferation via this mechanism. (See color insert.)

to maintain DSCs in asymmetric self-renewal. Loss of p53 function or up-regulation of IMPDH activity are predicted to result in inappropriate expansive symmetric self-renewal by affected DSCs. Such a change in DSC kinetics is predicted to be carcinogenic [Sherley, 1996, 2000, 2006]. Consistent with this prediction, liver cancers are well known for up-regulation of IMPDH activity [Weber, 1996] and loss of wild-type p53 function is the most commonly observed defect in diverse human tumors [Levine, 1994].

The SACK approach is based on putting DSCs in a state similar to loss of p53 or upregulation of IMPDH, but reversibly. By supplementing DSCs in culture with purine nucleosides or purine bases that can be utilized by purine nucleotide salvage pathways to expand rGNP pools independent of IMPDH, it is possible to shift DSCs from asymmetric self-renewal to symmetric self-renewal (Fig. II.3b). The effects of these SACK agents are reversible, because their action requires only changes in cellular nucleotide pools whose normal levels are rapidly restored when the agents are removed. Nucleotide salvage precursors that do not expand rGNP pools do not confer SACK effects [Sherley, 1991].

Although time-lapse microscopy showed that individual cells in human HSC enriched populations that cycled with asymmetric self-renewal kinetics were the only cells to give rise to both myeloid and lymphoid cell types *in vitro* [Huang, 1999], no investigations of HSC expansion based on SACK have been reported to date. Thus far, the expansion rationales have been based on general concepts in hematopoiesis and the BM microenvironment, including hematopoietic cytokines and microenvironmental factors (e.g., cell-to-cell or cell-to-matrices interactions). However, culture manipulations based on these concepts have been unsuccessful in promoting *ex vivo* expansion of human HSCs from all sources, including UCB. So, an obvious next direction for the field of HSC therapeutics is investigation of SACK agents for their effectiveness in expanding HSCs from UCB and other sources.

Though there is no guarantee that previously described SACK agents can be as effective for HSCs as they have been for liver-derived DSCs, there are sound biological rationales for optimism that they will be effective. Asymmetric self-renewal is a universal property of DSCs in mammalian tissues; IMPDH is an essential mammalian enzyme; rGNPs are universal regulators of essential cellular processes; and p53 is the quintessential regulator of mammalian tissue cell growth. Thus, it is highly predicted that pathways targeted by SACK will function in diverse human DSC types, including UCB HSCs. Even if the exact SACK agents defined in earlier studies are inefficient for HSCs, it is certainly the case that SACK by some means must be accomplished before successful *ex vivo* expansion can ever be achieved.

There is, however, one aspect of the HSC expansion problem that may also limit the effectiveness of SACK. This is the earlier discussed unexplained loss of HSCs in short-term culture. Since there are no effective means to assay liver-derived DSCs in short-term primary cell culture, it is not known whether they show a similar loss. If they do, and SACK agents effectively overcome such losses, then SACK agents may prove particularly effective for HSC expansion. Yet, there is nothing in SACK theory that would anticipate such an effect on DSC retention independent of symmetric self-renewal. If loss in short-term culture is a HSC specific feature unrelated to self-renewal pattern, it will also compromise expansion effects of SACK agents. The answer to this question will require SACK trials for UCB HSCs. The investigation of these new ideas and methods with human UCB is an exciting future prospect. A convenient and safe method that provides a modest, but reliable expansion of the HSCs from UCB would have a tremendous impact on the health and quality of life of countless patients and their families.

ACKNOWLEDGMENTS

We thank Ms. K. Panchalingam and Dr. J.-F. Pare for reviewing the manuscript. JLS and RRT were supported by NIH Director's Pioneer Award #5DP10D000805 from the National Institute of General Medical Sciences; and JLS was supported by grant No.1 P50 HG003170 from the National Human Genome Research Institute.

REFERENCES

Asahara T, Murohara T, Sullivan A, Silver M, van der Zee R, Li T, Witzenbichler B, Schatteman G, Isner JM. 1997. Isolation of Putative Progenitor Endothelial Cells for Angiogenesis. Science. 275:964–966.

Ballen KK, Spitzer TR, Yeap B, Steve M, Dey BR, Attar E, Alyea E, Cutler C, Ho V, Lee S, Soiffer R, Antin JH. 2005. Excellent Disease-Free Survival after Double Cord Blood Transplantation Using a Reduced Intensity Chemotherapy Only Conditioning Regimen in a Diverse Adult Population. ASH Annual Meeting Abstracts. 106:2048.

Barker JN, Krepski TP, DeFor TE, Davies SM, Wagner JE, Weisdorf DJ. 2002. Searching for unrelated donor hematopoietic stem cells: Availability and speed of umbilical cord blood versus bone marrow. Biology of blood and marrow transplantation: Journal of the American Society for Blood and Marrow Transplantation. 8:257–260.

Barker JN, Scaradavou A, Stevens CE, Rubinstein P. 2005a. Analysis of 608 Umbilical Cord Blood (UCB) Transplants: HLA-Match Is a Critical Determinant of Transplant-Related Mortality (TRM) in the Post-Engraftment Period Even in the Absence of Acute Graft-vs-Host Disease (aGVHD). ASH Annual Meeting Abstracts. 106:303.

Barker JN, Weisdorf DJ, DeFor TE, Brunstein CG, Wagner JE. 2005b. Umbilical Cord Blood (UCB) Transplantation after Non-Myeloablative (NMA) Conditioning for Advanced Follicular Lymphoma, Mantle Cell Lymphoma and Chronic Lymphocytic Leukemia: Low Transplant-Related Mortality and High Progression-Free Survival. ASH Annual Meeting Abstracts. 106:2900.

Becker A, McCulloch E, Till, J. 1963. Cytological demonstration of the clonal nature of spleen colonies derived from transplanted mouse marrow cells. Nature (London). 197:452–454.

Beutler E, Lichtman MA, Coller BS, Kipps TJ, Seligsohn U. 2001. Williams Hematology. New York: McGraw Hill. 29–58 pp.

Bhatia M, Wang JCY, Kapp U, Bonnet D, Dick JE. 1997. Purification of primitive human hematopoietic cells capable of repopulating immune-deficient mice. PNAS. 94:5320–5325.

Boggs D, Boggs SS, Saxe DF, Gress LA, Canfield DR. 1983. Hematopoietic stem cells with high proliferative potential. Assay of their concentration in marrow by the frequency and duration of cure of W/Wv mice. J Clin Invest. 70:242–253.

Bradley MB, Cairoa MS. 2005. Cord Blood Immunology and Stem Cell Transplantation. Human Immunol. 66:431–446.

Broxmeyer HE, Srour EF, Hangoc G, Cooper S, Anderson SA, Bodine DM. 2003. High-efficiency recovery of functional hematopoietic progenitor and stem cells from human cord blood cryopreserved for 15 years. Proc Natl Acad Sci USA. 100:645–650.

Brunstein CG, Barker JN, DeFor TE, French K, Weisdorf DJ, Wagner JE. 2005a. Non-Myeloablative (NMA) Umbilical Cord Blood Transplantation (UCBT): Promising Disease-Free Survival in 95 Consecutive Patients. ASH Annual Meeting Abstracts. 106:559.

Brunstein CG, Larson S, Rodgers T, Eastlund T, Symons K, Weisdorf D. 2005b. Red Blood Cell (RBC) and Platelet (PLT) Transfusion after Allogeneic Transplant for Leukemia: Similar Utilization after Sibling (SIB) Peripheral Blood Stem Cell (PBSC) and Umbilical Cord Blood (UCB) Graft. ASH Annual Meeting Abstracts. 106:1755.

Calvi LM. 2006. Osteoblastic Activation in the Hematopoietic Stem Cell Niche. Ann NY Acad Sci. 1068:477–488.

Calvi LM, Adams GB, Weibrecht KW, Weber JM, Olson DP, Knight MC, Martin RP, Schipani E, Divieti P, Bringhurst FR, Milner LA, Kronenberg HM, Scadden DT. 2003. Osteoblastic cells regulate the haematopoietic stem cell niche. Nature. 425:841–846.

Cao Y, Wagers A, Beilhack A, Dusich J, Bachmann M, Negrin R, Weissman I, Contag C. 2004. Shifting foci of hematopoiesis during reconstitution from single stem cells. Proc Natl Acad Sci USA. 101:221–226.

Central Intelligence Agency C. 2008. The World Factbook—United States.

Cheshier S, Morrison SJ, Liao X, Weissman IL. 1999. In vivo proliferation and cell cycle kinetics of long-term self-renewing hematopoietic stem cells. Proc Natl Acad Sci USA. 96:3120–3125.

Cronkite E. 1988. Analytical review of structure and regulation of hemopoiesis. Blood Cells. 14:313–328.

de Wynter EA, Buck D, Hart C, Heywood R, Coutinho LH, Clayton A, Rafferty JA, Burt D, Guenechea G, Bueren JA, Gagen D, Fairbairn LJ, Lord BI, Testa NG. 1998. CD34 + AC133+ Cells Isolated from Cord Blood are Highly Enriched in Long-Term Culture-Initiating Cells, NOD/SCID-Repopulating Cells and Dendritic Cell Progenitors. Stem Cells. 16:387–396.

Devine SM, Lazarus HM, Emerson SG. 2003. Clinical application of hematopoietic progenitor cell expansion: Current status and future prospects. Bone Marrow Transplant. 31:241–252.

Eaves C, Fraser C, Udomsakdi C, Sutherland H, Barnett M, Szilvassy S, Hogge D, Lansdorp P, Eaves A. 1992. Manipulation of the hematopoietic stem cell in vitro. Leukemia 6 Suppl. 1:27–30.

Gallacher L, Murdoch B, Wu DM, Karanu FN, Keeney M, Bhatia M. 2000. Isolation and characterization of human CD34-Lin- and CD34 + Lin- hematopoietic stem cells using cell surface markers AC133 and CD7. Blood. 95:2813–2820.

Gluckman E. 2004. Ex vivo expansion of cord blood cells. Exp Hematol. 32:410–412.

Gluckman E, Koegler G, Rocha V. 2005. Human leukocyte antigen matching in cord blood transplantation. Semin Hematol. 42:85–90.

Gostjeva EV, Zukerberg L, Chung D, Thilly WG. 2006. Bell-shaped nuclei dividing by symmetrical and asymmetrical nuclear fission have qualities of stem cells in human colonic embryogenesis and carcinogenesis. Cancer Genet Cytogenet. 164:16–24.

Harrison DE, Lerner CP, Spooncer E. 1987. Erythropoietic repopulating ability of stem cells from long-term marrow culture. Blood. 69:1021–1025.

Hawkins N, Garriga G. 1998. Asymmetric cell division: From A to Z. Genes Dev. 12:3625–3638.

Hayflick L. 1965. The Limited in vitro Lifetime of Human Diploid Cell Strains. Exp Cell Res. 614–636.

Hofmeister CC, Zhang J, Knight KL, Le P, Stiff PJ. 2007. Ex vivo expansion of umbilical cord blood stem cells for transplantation: Growing knowledge from the hematopoietic niche. Bone Marrow Transplant. 39:11–23.

Huang S, Law P, Francis K, Palsson BO, Ho AD. 1999. Symmetry of initial cell divisions among primitive hematopoietic progenitors is independent of ontogenic age and regulatory molecules. Blood. 94:2595–2604.

Ivanova NB, Dimos JT, Schaniel C, Hackney JA, Moore KA, Lemischka IR. 2002. A Stem Cell Molecular Signature. Science. 298:601–604.

Jansen J, Hanks S, Thompson J, Dugan M, Akard L. 2005. Transplantation of hematopoietic stem cells from the peripheral blood. J Cell Mol Med. 9:37–50.

Jaroscak J, Goltry K, Smith A, Waters-Pick B, Martin PL, Driscoll TA, Howrey R, Chao N, Douville J, Burhop S, Fu P, Kurtzberg J. 2003. Augmentation of umbilical cord blood (UCB) transplantation with ex vivo-expanded UCB cells: Results of a phase 1 trial using the AastromReplicell System. Blood. 101:5061–5067.

Karam S, Leblond CP. 1993. Dynamics of epithelial cells in the corpus of the mouse stomach. I. Identification of proliferative cell types and pinpointing of the stem cell. Anat Rec. 236:259–279.

Kögler G, Enczmann J, Rocha V, Gluckman E, Wernet P. 2005. High-resolution HLA typing by sequencing for HLA-A, -B, -C, -DR, -DQ in 122 unrelated cord blood/patient pair transplants hardly improves long-term clinical outcome. Bone Marrow Transplantation. 36:1033–1041.

Koller M, Master JR, Palsson BO. 1999. Primary Hematopoietic Cells. New York: Kluwer Academic Publishers. 342 p.

Lansdorp P. 1995a. Developmental changes in the function of hematopoietic stem cells. Exp Hematol. 23:187–191.

Lansdorp PM. 1995b. Telomere length and proliferation potential of hematopoietic stem cells. J Cell Sci Suppl. 108:1–6.

Laughlin MJ, Eapen M, Rubinstein P, Wagner JE, Zhang M-J, Champlin RE, Stevens C, Barker JN, Gale RP, Lazarus HM, Marks DI, van Rood JJ, Scaradavou A, Horowitz MM. 2004. Outcomes after Transplantation of Cord Blood or Bone Marrow from Unrelated Donors in Adults with Leukemia. N Engl J Med. 351:2265–2275.

Lee H, Crane CG, Merok JR, Tunstead JR, Hatch NL, Panchalingam K, Powers MJ, Griffith LG, Sherley JL. 2003. Clonal expansion of adult rat hepatic stem cell lines by suppression of asymmetric cell kinetics (SACK). Biotechnol Bioeng. 83:760–771.

Lemischka I. 1992. What we have learned from retroviral marking of hematopoietic stem cells. Curr Top Microbiol Immunol. 177:59–71.

Levine A, Perry ME, Chang A, Silver A, Dittmer D, Wu M, Welsh D. 1994. The 1993 Walter Hubert Lecture: The role of the p53 tumour-suppressor gene in tumorigenesis. Br J Cancer. 69:409–416.

Lewis ID, Almeida-Porada G, Du J, Lemischka IR, Moore KA, Zanjani ED, Verfaillie CM. 2001. Umbilical cord blood cells capable of engrafting in primary, secondary, and tertiary xenogeneic hosts are preserved after ex vivo culture in a noncontact system. Blood. 97:3441–3449.

Locatelli F, Rocha V, Chastang C, Arcese W, Michel G, Abecasis M, Messina C, Ortega J, Badell-Serra I, Plouvier E, Souillet G, Jouet J-P, Pasquini R, Ferreira E, Garnier F, Gluckman E. 1999. Factors Associated With Outcome After Cord Blood Transplantation in Children With Acute Leukemia. Blood. 93:3662–3671.

Lovell-Badge R. 2001. The future for stem cell research. Nature (London). 414:88–91.

Martin MA, Bhatia M. 2005. Analysis of the Human Fetal Liver Hematopoietic Microenvironment. Stem Cells Devel. 14:493–504.

Mayani H, Lansdorp PM. 1998. Biology of Human Umbilical Cord Blood-Derived Hematopoietic Stem/Progenitor Cells. Stem Cells. 16:153–165.

Merok JR, Lansita JA, Tunstead JR, Sherley JL. 2002. Cosegregation of Chromosomes Containing Immortal DNA Strands in Cells That Cycle with Asymmetric Stem Cell Kinetics. Cancer Res. 62:6791–6795.

Merok JR, Sherley JL. 2001. Breaching the Kinetic Barrier to In Vitro Somatic Stem Cell Propagation. J Biomed Biotechnol. 1:25–27.

Meyer EA, Gebbie KM, Hanna KE. 2005. Cord Blood: Establishing a National Hematopoietic Stem Cell Bank Program.

Migliaccio AR, Adamson JW, Stevens CE, Dobrila NL, Carrier CM, Rubinstein P. 2000. Cell dose and speed of engraftment in placental/umbilical cord blood transplantation: Graft progenitor cell content is a better predictor than nucleated cell quantity. Blood. 96:2717–2722.

Moore K, Ema H, Lemischka IR. 1997. In vitro maintenance of highly purified, transplantable hematopoietic stem cells. Blood. 89:4337–4347.

Morris RJ, Liu Y, Marles L, Yang Z, Trempus C, Li S, Lin JS, Sawicki JA, Cotsarelis G. 2004. Capturing and profiling adult hair follicle stem cells. Nature Biotech. 22:411–417.

Nolta J, Thiemann F, Arakawa-Hoyt J, Dao M, Barsky L, Moore K, Lemischka I, Crooks G. 2002. The AFT024 stromal cell line supports long-term ex vivo maintenance of engrafting multipotent human hematopoietic progenitors. Leukemia. 16:352–361.

Panchalingam K et al., in preparation.

Pare J-F, Sherley JL. 2006. Biological Principles for Ex Vivo Adult Stem Cell Expansion. Curr Top Dev Biol. 73:141–171.

Peichev M, Naiyer AJ, Pereira D, Zhu Z, Lane WJ, Williams M, Oz MC, Hicklin DJ, Witte L, Moore MAS, Rafii S. 2000. Expression of VEGFR-2 and AC133 by circulating human CD34+ cells identifies a population of functional endothelial precursors. Blood. 95:952–958.

Piacibello W, Sanavio F, Severino A, Dane A, Gammaitoni L, Fagioli F, Perissinotto E, Cavalloni G, Kollet O, Lapidot T, Aglietta M. 1999. Engraftment in Nonobese Diabetic Severe Combined Immunodeficient Mice of Human CD34+ Cord Blood Cells After Ex Vivo Expansion: Evidence for the Amplification and Self-Renewal of Repopulating Stem Cells. Blood. 93:3736–3749.

Potten C, Loeffler M. 1990. Stem cells: Attributes, cycles, spirals, pitfalls and uncertainties. Lessons for and from the crypt. Development. 110:1001–1020.

Potten C, Morris RJ. 1988. Epithelial stem cells in vivo. J Cell Sci Suppl. 10:45–62.

Punzel M, Liu D, Zhang T, Eckstein V, Miesala K, Ho AD. 2003. The symmetry of initial divisions of human hematopoietic progenitors is altered only by the cellular microenvironment. Experl Hematol. 31:339–347.

Qu-Petersen Z, Deasy B, Jankowski R, Ikezawa M, Cummins J, Pruchnic R, Mytinger J, Cao B, Gates C, Wernig A, Huard J. 2002. Identification of a novel population of muscle stem cells in mice: Potential for muscle regeneration. J Cell Biol. 157:851–864.

Ramalho-Santos M, Yoon S, Matsuzaki Y, Mulligan RC, Melton DA. 2002. "Stemness": Transcriptional Profiling of Embryonic and Adult Stem Cells. Science. 298:597–600.

Rambhatla L, Bohn SA, Stadler PB, Boyd JT, Coss RA, Sherley JL. 2001. Cellular Senescence: Ex Vivo p53-Dependent Asymmetric Cell Kinetics. J Biomed Biotechnol. 1:28–37.

Rebel V, Miller C, Eaves C, Lansdorp P. 1996. The repopulation potential of fetal liver hematopoietic stem cells in mice exceeds that of their liver adult bone marrow counterparts. Blood. 87:3500–3507.

Revazova ES, Turovets NA, Kochetkova OD, Agapova LS, Sebastian JL, Pryzhkova MV, Smolnikova VI, Kuzmichev LN, Janus JD. 2008. HLA Homozygous Stem Cell Lines Derived from Human Parthenogenetic Blastocysts. Cloning Stem Cells. 10:11–24.

Reya T, Morrison SJ, Clarke MF, Weissman IL. 2001. Stem cells, cancer, and cancer stem cells. Nature (London). 414:105–111.

Rocha V, Labopin M, Sanz G, Arcese W, Schwerdtfeger R, Bosi A, Jacobsen N, Ruutu T, de Lima M, Finke J, Frassoni F, Gluckman E. 2004. Transplants of Umbilical-Cord Blood or Bone Marrow from Unrelated Donors in Adults with Acute Leukemia. N Engl J Med. 351:2276–2285.

Rubinstein P, Carrier C, Scaradavou A, Kurtzberg J, Adamson J, Migliaccio AR, Berkowitz RL, Cabbad M, Dobrila NL, Taylor PE, Rosenfield RE, Stevens CE. 1998. Outcomes among 562 Recipients of Placental-Blood Transplants from Unrelated Donors. N Engl J Med. 339:1565–1577.

Schoemans H, Theunissen K, Maertens J, Boogaerts M, Verfaillie C, Wagner J. 2006. Adult umbilical cord blood transplantation: A comprehensive review. Bone Marrow Transplantation. 38:83–93.

Schofield R. 1978. The relationship between the spleen colony-forming cell and the haemopoietic stem cell. Blood Cells. 4:7–25.

Sell S. 1994. Liver Stem Cells. Mod Pathol. 7:105–112.

Sherley J. 1991. Guanine nucleotide biosynthesis is regulated by the cellular p53 concentration. J Biol Chem. 266:24815–24828.

Sherley J. 2000. Tumor Suppressor Genes and Cell Kinetics in the Etiology of Malignant Mesothelioma. In: Peters GPB, editor. Sourcebook of Asbestos Diseases. Santa Monica: Peters & Peters. pp. 91–141.

Sherley JL. 1996. The p53 Tumor Suppressor Gene as Regulator of Somatic Stem Cell Renewal Division. Cope. 12:9–10.

Sherley JL. 2002. Asymmetric Cell Kinetics Genes: The Key to Expansion of Adult Stem Cells in Culture. Stem Cells. 20:561–572.

Sherley JL. 2006. Mechanisms of Genetic Fidelity in Mammalian Adult Stem Cells. Tissue Stem Cells eds. C. S. Potten, R. B. Clarke, J. Wilson, and A. G. Renehan, Taylor Francis, New York: 37–54.

Sherley JL. 2008. A New Mechanism for Aging: Chemical Age Spots in Immortal DNA Strands in Distributed Stem Cells. Breast Disease 29:37–46.

Shpall E, Quinones JR, Giller R, Zeng C, Baron AE, Jones RB, Bearman SI, Nieto Y, Freed B, Madinger N, Hogan CJ, Slat-Vasquez V, Russell P, Blunk B, Schissel D, Hild E, Malcolm J, Ward W, McNiece IK. 2002. Transplantation of ex vivo expanded cord blood. Biology of blood and marrow transplantation: Journal of the American Society for Blood and Marrow Transplantation. 8:368–376.

Sirchia G, Rebulla P. 1999. Placental/umbilical cord blood transplantation. Haematologica. 84:738–747.

Spangrude G, Heimfeld S, Weissman I. 1989. Purification and characterization of mouse hematopoietic stem cells. Science. 241:58–62.

Spangrude G, Heimfeld S, Weissman IL. 1988. Purification and characterization of mouse hematopoietic stem cells. Science. 241:58–62.

Spurr E, Wiggins N, Marsden K, Lowenthal R, Ragg S. 2002. Cryopreserved human haematopoietic stem cells retain engraftment potential after extended (5–14 years) cryostorage. Cryobiology. 44:210–217.

Stevens CE, Scaradavou A, Carrier C, Carpenter C, Rubinstein P. 2005. An Empirical Analysis of the Probability of Finding a Well Matched Cord Blood Unit: Implications for a National Cord Blood Inventory. Blood. 106:2047.

Stiff P, Chen B, Franklin W, Oldenberg D, Hsi E, Bayer R, Shpall E, Douville J, Mandalam R, Malhotra D, Muller T, Armstrong RD, Smith A. 2000. Autologous transplantation of ex vivo expanded bone marrow cells grown from small aliquots after high-dose chemotherapy for breast cancer. Blood. 95:2169–2174.

Storb R. 2003. Allogeneic hematopoietic stem cell transplantation-Yesterday, today, and tomorrow. Exp Hematol. 31:1–10.

Sutherland H, Eaves CJ, Eaves AC, Dragowska W, Lansdorp PM. 1989. Characterization and partial purification of human marrow cells capable of initiating long-term hematopoiesis in vitro. Blood. 74:1563–1570.

Szilvassy SJ, Meyerrose TE, Ragland PL, Grimes B. 2001. Differential homing and engraftment properties of hematopoietic progenitor cells from murine bone marrow, mobilized peripheral blood, and fetal liver. Blood. 98:2108–2115.

Taghizadeh RR, Sherley JL. In preparation.

Takahashi S, Iseki T, Ooi J, Tomonari A, Takasugi K, Shimohakamada Y, Yamada T, Uchimaru K, Tojo A, Shirafuji N, Kodo H, Tani K, Takahashi T, Yamaguchi T, Asano S. 2004. Single-institute comparative analysis of unrelated bone marrow transplantation and cord blood transplantation for adult patients with hematologic malignancies. Blood. 104:3813–3820.

Takano H, Ema H, Sudo K, H. N. 2004. Asymmetric Division and Lineage Commitment at the Level of Hematopoietic Stem Cells: Inference from Differentiation in Daughter Cell and Granddaughter Cell Pairs. J Exp Med. 199:295–302.

Till J, McCulloch, E. 1961. A direct measurement of the radiation sensitivity of normal mouse bone marrow cells. Radiat Res. 14:1419–1430.

Trempus CS, Morris RJ, Bortner CD, Cotsarelis G, Faircloth RS, Reece JM, Tennant RW. 2003. Enrichment for Living Murine Keratinocytes from the Hair Follicle Bulge with the Cell Surface Marker CD34. J Invest Derm Atol. 120:501–511.

Tyndall A, Gratwohl A. 2000. Immune ablation and stem-cell therapy in autoimmune disease: Clinical experience. Arthritis Res. 2:276–280.

Ueno N, Konoplev S, Buchholz T, Smith T, Rondon G, Anderlini P, Giralt S, Gajewski J, Donato M, Cristofanilli M, Champlin R. 2006. High-dose chemotherapy and autologous peripheral blood stem cell transplantation for primary breast cancer refractory to neoadjuvant chemotherapy. Bone Marrow Transplantation. 37:929–935.

Urist M. 1965. Bone: Formation by autoinduction. Science. 150:893–899.

van der Loo JCM, Hanenberg H, Cooper RJ, Luo F-Y, Lazaridis EN, Williams DA. 1998. Nonobese Diabetic/Severe Combined Immunodeficiency (NOD/SCID) Mouse as a Model System to Study the Engraftment and Mobilization of Human Peripheral Blood Stem Cells. Blood. 92:2556–2570.

Vaziri H, Dragowska W, Allsopp R, Thomas T, Harley C, Lansdorp P. 1994. Evidence for a Mitotic Clock in Human Hematopoietic Stem Cells: Loss of Telomeric DNA with Age. Proc Natl Acad Sci USA. 91:9857–9860.

Verfaillie CM. 2002. Hematopoietic stem cells for transplantation. Nature Immunol. 3:314–317.

Verneris MR, Brunstein C, DeFor TE, Barker J, Weisdorf DJ, Blazar BR, Miller JS, Wagner JE. 2005. Risk of Relapse (REL) after Umbilical Cord Blood Transplantation (UCBT) in Patients with Acute Leukemia: Marked Reduction in Recipients of Two Units. ASH Annual Meeting Abstracts. 106:305.

Wagner JE, Barker JN, DeFor TE, Baker KS, Blazar BR, Eide C, Goldman A, Kersey J, Krivit W, MacMillan ML, Orchard PJ, Peters C, Weisdorf DJ, Ramsay NKC, Davies SM. 2002. Transplantation of unrelated donor umbilical cord blood in 102 patients with malignant and non-malignant diseases: Influence of CD34 cell dose and HLA disparity on treatment-related mortality and survival. Blood. 100:1611–1618.

Wang JCY, Doedens M, Dick JE. 1997. Primitive Human Hematopoietic Cells Are Enriched in Cord Blood Compared With Adult Bone Marrow or Mobilized Peripheral Blood as Measured by the Quantitative In Vivo SCID-Repopulating Cell Assay. Blood. 89:3919–3924.

Watt F. 1991. Cell culture models of differentiation. FASEB J. 5:287–294.

Weber G, Shen F, Prajda N, Yeh Y, Yang H, Herenyiova M, Look K. 1996. Increased signal transduction activity and down-regulation in human cancer cells. Anticancer Res. 16:3271–3282.

Wilson A, Trumpp A. 2006. Bone-marrow haematopoietic-stem-cell niches. Nat Rev Immunol. 6:93–106.

Wineman J, Moore K, Lemischka I, Muller-Sieburg C. 1996. Functional heterogeneity of the hematopoietic microenvironment: Rare stromal elements maintain long-term repopulating stem cells. Blood. 87:4082–4090.

Wu A, Till JE, Siminovitch L, McCulloch EA. 1968. Cytological evidence for a relationship between normal hemotopoietic colony-forming cells and cells of the lymphoid system. J Exp Med. 127:455–464.

3

USE OF FETAL CELLS IN REGENERATIVE MEDICINE

Christian Breymann

Feto Maternal Haematology Unit, Obstetric Research, University Hospital Zurich, Switzerland

INTRODUCTION

Since decades, cells from cord blood, amniotic fluid, and the chorion were used in perinatal medicine mainly for invasive diagnostic purposes, such as detection of fetal infections, rare metabolic diseases, and mainly fetal karyotyping. With the introduction of the use of umbilical cord stem cells for the treatment of haematologic diseases many years ago, a new future for the use of perinatal cells has begun, namely, the use of perinatal (amniotic, placental, chorionic, umbilical) stem cells for therapeutic purposes, such as the treatment of hematologic and metabolic diseases. Lately, the use for regenerative medicine applications is also called tissue engineering.

Today, it is well known and evident that adult stem cells, also called progenitor cells, play an important role in regenerative medicine and even enable the mechanisms of tissue regeneration. After tissue damage or destruction in an *in vivo* environment, the stem cell can be recruited from several sources (e.g., bone marrow, blood, or as resident stem cell from the injured tissue itself in order to start the cascade for tissue repair). The use of these stem cells, which are predominantly "adult" or multipotent for tissue engineering purposes, also opens a new horizon and possibly justification for autologous banking of these cells that is still under wide rejection and criticism in the medical community. The reason is because of the low number of possible indications for the use of autologous perinatal cells. It is clear that the possible use of perinatal stem cells in regenerative medicine would change the opinion concerning the sense of private cell banking dramatically. For example concerning the use of perinatal stem cells in cardiac repair, the autologous lifetime chances for the use of these cells up to age 70 would lie around 0.15% (Netherlands) to 0.25% (USA)

Perinatal Stem Cells. Edited by C. L. Cetrulo, K. J. Cetrulo, and C. L. Cetrulo, Jr.

[Nietfeld, 2004]. Approximately 3000 articles on the use of stem cells for tissue engineering can be found at present by Medline research concerning many organs, such as skin, cardiovascular, lungs, bones, urinary tract, liver, eyes, and teeth. Furthermore, there is increasing interest of the public field and the mass media and it seems only to be a matter of time until tissue engineering reaches routine medical practice [Fauza, 2004; Kunisaki, 2007; Siegel, 2007].

PRINCIPLES OF TISSUE ENGINEERING

Tissue engineering, focusing on the *in vitro* fabrication of autologous, living tissues with the potential for regeneration is a promising scientific field, addressing the so far unmet medical need for growing replacements, particularly for the repair of congenital malformation. The main principles or systems used for tissue engineering purposes are cell suspensions (autogenic, allogenic, and xenogenic), *in vitro* cultures (static and dynamic), and scaffolds (natural, synthetic, and biodegradable). Characteristics and requirements of biodegradable scaffolds are listed in Table III.1. In the so-called open systems, the cells (e.g., fibroblasts) are attached to the scaffold that can be natural, polymers, or ceramics (e.g., these scaffolds enable the cells to grow and to organize in a specific matter). The diffusion of nutrients and waste must be possible and the material should be biodegradable, that is, after a certain period after degradation of a scaffold, cells are in complete contact with host's body and there is full incorporation and replacement of a specific organ. Possible problems to face are immunologic rejection, growth limitations, differentiation and function restraints, incorporation barriers, and cell or tissue delivery difficulties.

Currently used replacements in various fields are mainly nonliving and based on foreign materials, They have a lack of growth and remodeling and carry the risks for thromboembolic complications and infections. For example, in today's congenital heart surgery, there is a substantial need for appropriate, growing replacement materials for the repair of congenital cardiac defects. The surgical treatment is commonly based on nonautologous valves or conduits with disadvantages including obstructive tissue ingrowths and calcification of the replacement. These limitations and the lack of growth typically necessitate various reoperations of the pediatric patients with cardiovascular defects associated with increased morbidity and mortality each time [Horstrup, 2002; Mayer, 1995; Schoen, 1999].

Tissue engineered constructs based on perinatal derived cells could be an attractive alternative to the currently replacement materials for the repair of congenital malformations being living material and due to their progenitor characteristics having the capacity of self-remodeling. Here, perinatal derived cells would be an ideal setting for tissue engineered constructs as they can be harvested without harming intact donor structures and high risks for the child. The advantages of using perinatal derived cells are summarized in Table III.2.

TABLE III.1. Requirements of Scaffolds

Cells closely attached to *scaffolds* (natural, polymers, ceramics)
Enables cell organization and growth
Possible diffusion of nutrients and waste
Biodegradable material
Cells should be in contact with hosts body
Full incorporation and replacement of target organ possible

TABLE III.2. Possible Advantages of Using Fetal Derived Cells for Tissue Engineering

Autologous cells source
Cryopreservation possible
Easy obtainable at variuos stages (pre- and perinatal)
Low risk of bacterial/viral contamination
Low immunogenic properties
Cell source providing different types of cells including progenitor cells
Excellent cell growth capacities
Tumorogenic potential lower compared to omnipotent cells

CELL SOURCES FOR TISSUE ENGINEERING

In contrast to the highly standardized and industrially fabricated scaffolds, the quality of cells varies from patient to patient, depending on the individual tissue characteristics and comorbidities. In order to create a functional, living tissue replacement the choice of the cell source is critical. Besides cell growth and expansion capacity, an important issue is the possibility of developing a cell phenotype that matches the native counterparts. This finding is expected to have a major impact on the long-term functionality of the replacements. By using cells originating from the tissue to be replaced would be the safest approach. For example, in the case of heart valve tissue engineering, the usage of valvular interstitial cells obtained by biopsy has been shown to be feasible.

However, with respect to clinical applications, these cells are difficult to obtain and the approach bears substantial risks. Therefore, several alternative human cell sources have been investigated for their use in heart valve tissue engineering. Among the most promising are vascular-, bone-marrow, blood-, and umbilical cord-derived cells, particularly for pediatric application. Of particularly importance for the pediatric cardiovascular tissue engineering is cell-harvesting at an early stage that allows having the tissue-engineered replacement ready for implantation at the birth of the patient. Therefore, cells should be harvested prenataly at an early stage of pregnancy without harming any intact child donor structures [Schmidt, 2004, 2005a, 2006, 2007]. As used in routine prenatal diagnostics, both the amniotic fluid and the placenta could provide easy access to fetal cells. It has been recently shown that amniotic derived mesenchymal stem cells have the highest proliferation rates *in vitro* compared to fetal or adult stem cells from subcutaneous tissues [Fauza, 2004].

EXAMPLES OF POSSIBLE USE AND IMPLICATIONS OF PERINATAL STEM CELLS

Umbilical Cord Derived Stem Cells

Characteristics of umbilical cord derived stem cells (UCSC) have been extensively described in Chapter 1. Umbilical cord derived stem cells can be used for differentiation into all three layers, such as ekto-, endo-, and mesoderm. Main examples for their use are cardiac repair; that is, cell suspensions for transplantation in myocardial regeneration [Guetta, 2005; Ma, 2006] and scaffolds for vessel and valve replacement. Both mesenchymal stem cells differentiating in myofibroblasts and endothelial progenitor cells differentiating into endothelium can be isolated from umbilical cord blood and finally used to create a tissue engineered heart valve.

Use of umbilical cord stem cells for the use of cardiovascular tissue engineering has been described in detail recently [Horstrup, 2002; Kadner, 2002, 2004; Schmidt, 2004, 2005b; Korbling, 2005; Newman, 2004a].

Concerning the nervous system, UCSC have been used for induction of neurons and glial cells, and recently matrix stem cells as transplants in a rodent model of Parkinson's disease [Newman, 2004a]. Cells also have been used preclinically in animal models of brain and spinal cord injuries, in which functional recovery have been shown [Newman, 2004b]. Hunter and Hurler syndrome and Krabbe disease have been treated successfully using UCSC [Korbling, 2005].

In addition, cord blood derived mesenchymal stem cells are presently studied in animal models for bone and skin regeneration [Dai, 2006; Waese, 2007]. Lee et al. recently performed a human clinical trial using UCSC for periurethral injections in female incontinence in 32 women, showing improvement of life quality in 80% of patients over several months [Lee, 2007]. It also has been shown that adipocitic differentiation is possible using mesenchymal stem cells from UC [Parolini, 2007].

PLACENTAL DERIVED STEM CELLS (CHORIONIC, AMNIOTIC MEMBRANES)

Four regions of fetal placenta can be distinguished: amniotic epithelial, amniotic mesenchymal, chorionic mesenchymal, and chorionic trophoblastic. Concerning tissue engineering aims, chorionic mesenchymal stromal cells (hCMSC) are of major importance.

Particularly in the fetal part of the placenta, the development of secondary chorionic villi to vascularized mesenchymal tertiary chorionic villi is an important time-dependent reconstruction (starting at around the third week of pregnancy). These villi represent an attractive cell source of various different progenitor cells until they are again transformed into mature intermediate villi, rather than into immature ones by approximately the twenty-third week of pregnancy. Harvesting the cells at an earlier state of pregnancy would increase the quality of cells and consequently the tissue of the engineered cardiovascular replacements immensely. Chorionic villi can be sampled beginning from around week 12 of gestation by chorionic villous sampling techniques. Normally between 5 and 40 mg of cells are gained by one procedure. Examples for the use of mesenchymal progenitor cells derived from chorionic villi of human placenta recently have been published concerning cartilage tissue engineering [Zhang, 2005].

Concerning cardiovascular, Schmidt et al. showed that prenatal fetal progenitors obtained from routine chorionic villus sampling (CVS) were successfully used as an exclusive cell source for the engineering of living heart valve leaflets. Fresh and cryopreserved cells showed myofibroblast-like phenotypes and genotyping confirmed pure fetal origin. Mechanical profiles approximated those of native heart valves. The authors conclude that, combined with the use of cell banking technology, this approach also might be applied to postnatal applications [Schmidt, 2006].

AMNIOTIC DERIVED STEM CELLS (*hAMSC/PLACENTAL!*)

The amniotic membrane itself contains multipotent cells that are able to differentiate in the various layers. Studies have described their potential in neural and glial cells, cardiac repair, and also hepatocyte cells [Parolini, 2007].

AMNIOTIC FLUID-DERIVED STEM CELLS (*NONPLACENTAL!*)

In normal pregnancies, amniotic cells can be sampled by amniocentesis at any time point starting from week 14 until the end of pregnancy. Amniocentesis should not be performed during the first trimester since an increased number of fetal limb defects were shown after "early" amniocentesis.

In routine amniocentesis 16–20 mL of amniotic fluid are sampled. It is not known whether amniotic fluid cells are purely from fetal origin or are also derived from the placenta in part. Amniotic fluid-derived stem cells (AFS) have physical characteristics of both embryonic and adult stem cells. Based on their morphological and growth characteristics, amniotic fluid cells can be classified into three types: epitheloid, amniotic-fluid specific, and fibroblastoid. Descending from a single cell line, differentiation along six distinct lineages (adipogenic, osteogenic, myogenic, endothelial, neurogenic, and hepatic) can be induced [De Coppi, 2007; Kim, 2007]. Amniotic stem cells express Oct-4 like embryonic stem cells do. They replicate rapidly in culture, do not require support of so-called "feeder" cells that can cause contamination and they do not form tumors *in vivo*. Compared to embryonic stem cells they seem therefore to be more stable [De Coppi, 2007]. Isolation of mesenchymal stem cells (MSC) from amniotic fluid has been described by various authors. It seems that mesenchymal amniocytes might be immunologically privileged, since they show depression of immunogenic antigens after several passages [Fauza, 2004], which makes them especially interesting for regenerative implantations. In this context, it is interesting that it was shown that MSC derived from human placenta suppress allogenic UC blood lymphocyte proliferation [Li, 2005]. It was shown by Kunisaki et al. that human mesenchymal amniocytes retain their progenitor phenotype and can be dependably expanded *ex vivo* in the absence of animal serum that will be important for feasibility of regulatory guidelines [Kunisaki, 2007].

Besides giving rise to the three embryonic layers, AFS cells are able to perform specialized functions, such as secretion of neurotransmitters (brain), production of calcium (bone), and production of liver specific proteins when transformed to hepatocytes [De Coppi, 2007].

Apart from adult-derived MSC, the potential of nonautologous cell sources for therapeutic applications is also being explored. For example, cells of all three germ layers have been detected in human amniotic fluid, but it was not until recently that AFS cell lines were established. These AFS cells express both somatic and embryonic stem cells (ESC) markers, and established cell lines have been reported to differentiate *in vitro* along the adipogenic, osteogenic, myogenic, endothelial, neurogenic, and hepatic pathways, covering all three germ layers. Human AFS cells seeded onto scaffolds and cultured in osteogenic inductive medium supported the formation of highly mineralized tissue upon implantation in immunodeficient mice [Waese, 2007]. Despite the success demonstrated in immunodeficient mice, a recent study showed that human AFS cells were rejected in the myocardium of normal and ischemic rats, regardless of the competency of the rats' immune system [Chiavegato, 2007]. In addition, amniotic fluid derived cells spontaneously developed chondro-osteogenic masses in the rat heart, indicating differentiation in a nondesired cell type.

However, concerning the use of AFS in cardiac tissue engineering Schmidt et al. recently showed successful use of amniotic fluid derived progenitor cells as a single-cell source for fabricating human living heart valves. After 28 days, tissue engineered leaflets were intact and densely covered with amniotic fluid-derived cells. In addition, it was shown that genotype of the cells was confirming pure fetal origin of the cells as it was shown by the same group for chorionic cells [Schmidt, 2006, 2007].

De Coppi et al. use amniotic derived mesenchymal stem cells for the treatment of bladder injury in a rat model and compared it with bone marrow derived MSC. Both cell types had limited effects on smooth muscle cell regeneration, but showed a positive effect on preventing hypertrophy of the surviving tissue. The AFS were not superior to bone marrow cells in this model [De Coppi, 2007].

Recently Cipriani et al. showed that amniotic fluid mesenchymal stem cells (AF–MSC) also survive and migrate after transplantation in the rat brain, indicating the potential for therapeutic strategies in brain disease [Cipriani, 2007]. Perin et al. used AFS in a murine kidney model and showed renal differentiation of the cells with development of primordial kidney structures, such as renal vesicles [Perin, 2007].

CONCLUSION

In conclusion, fetal or perinatal stem cells are a promising cell source for tissue engineering since they have demonstrated excellent growth properties and phenotypes similar to the native counterparts of the tissue they should replace. Not only the human umbilical cord, but also placental or amniotic fluid cells are an excellent and unique source of various types of cells (differentiated and nondifferentiated) that can be used for tissue engineering and regenerative medicine. There are numerous advantages for these autologous cells, such as high immune tolerance, no major risk of transmitting bacterial of viral infections, capacity of growth and remodeling, less risk of thromboembolic complications, and no sacrifice of intact donor structures. Fetal cells also might be useful for allogenic therapies, if they possibly would match between human leukocyte antigen (HLA) genes of donor and recipients. Concerning tissue engineering purposes, it must be pointed out that at present, fetal cells, especially with origin from amniotic fluid, seem even more promising concerning quantity and "quality" of MSC cells and their proliferation capacity in comparison with cord blood cells. Surprisingly a quantity of 5-mL amniotic fluid reveals enough stem cells for the production of a tissue engineered heart valve.

Possible problems concerning the use of fetal cells are maternal contamination, quality of samples, gene stability and tolerance after *ex vivo* transplantation, inflammatory reactions, influence of fetal diseases.

The potential of neoplastic proliferation of "immature" cells in the host's microenvironment is not yet really understood. It is not known how gene expression profiles change after *ex vivo* expansion. Also, at present there is no international consensus regarding the use for regenerative medical purposes under GMP related guidelines.

However, international networks (e.g., KIDESTEM = stem cells in kidney disease), which focus on special issues and applications of different stem cells in the various medical fields are upcoming, and will probably define protocols for sampling, processing, and application of the cells in animal or even human studies under defined conditions [Siegel, 2007].

Influence of cryoconservation on cell stability and quality has to be determined in the future. Further challenges are the clinical and animal application trials, investigation of future cell sources and their viability, optimization of scaffolds, scaled up bioreactors, prevention of tissue rejection, and product preservation.

A close cooperation between obstetric services (cell harvesting and acquisition) and the biotechnology laboratory (cell processing and tissue engineering) is needed to establish standards in quantity and quality control, sterility purposes, and sampling techniques. In addition, various factors, such as influence of type of birth (vaginal vs. caesarean),

TABLE III.3. Pre- and Perinatal Factors Influencing Stem Cell Quality and Quantity

Technics and skills of cell sampling (especially prenatal sampling)
Maternal contamination at sampling
Cell quantity at sampling
Bacterial and viral contamination
Genetic homogeneity of fetal cells
Cryopreservation
Obstetric factors
Type of birth, maternal diseases, fetal diseases, age, weight, gender, sex, etc.

vaginal contamination (meconium, bacteria, etc.), influence of maternal infections and certainly fetal factors (age, weight, infection, gender, sex, etc.) have to be closely evaluated (Table III.3) [Sanatloya, 2005; Solves, 2005]. If most of these questions are answered, there will be a great opportunity and perspective to use fetal cells for generating tissue engineered organs, as well as for the repair of congenital malformations. In a future scenario, after

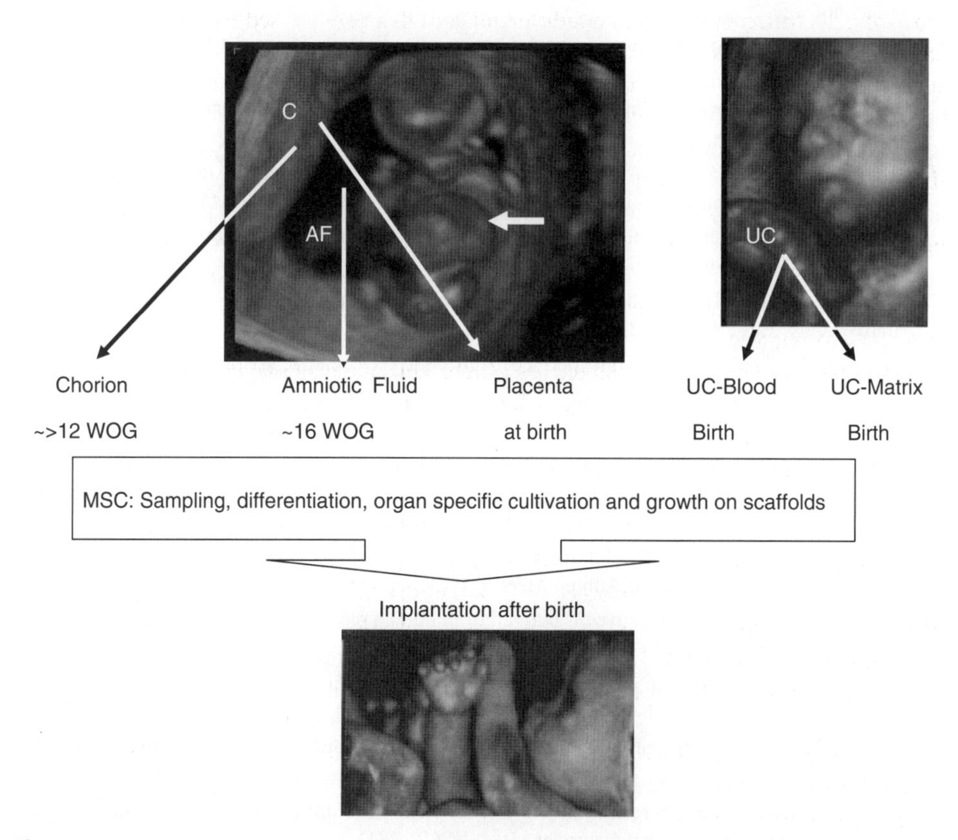

Figure III.1. Scenario of the use of fetal cells in regenerative medicine. This fetus shows an anterior body-wall defect (gastroschisis) that is, early detected (arrow = defect). Various cells sources can possibly be used for aquisition of progenitor cells. According to the needs for organ replacement (e.g., skin for abdominal wall closure after reposition of gastroschisis after birth) cells are then differentiated and used for regenerative organ growth. (See color insert.)

prenatal diagnosis, fetal cells would be sampled and processed immediately, and the tissue engineered organ would be available after birth at the time of programmed surgery some weeks later (Fig. III.1).

REFERENCES

Chiavegato A. et al. 2007. Human amniotic fluid-derived stem cells are rejected after transplantation in the myocardium of normal, ischemic, immuno-suppressed or immuno-deficient rat. J Mol Cell Card. 42:746–759.

Cipriani S. et al. 2007. Mesenchymal cells from human amniotic fluid survive and migrate after transplantation into adult rat brain. Cell Biology Int. 31:845–850.

Dai Y. et al. 2006. Skin epithelial cells in mice from umbilical cord blood mesenchymal stem cells. Burns. 33:418–428.

De Coppi PD. et al. 2007. Amniotic fluid and bone marrow derived mesenchymal stem cells can be coverted to smooth muscle cells in the cryo-injured rat bladder and prevent compensatory hypertrophy of surviving smooth muscle cells. J Urology. 177:369–376.

De Coppi PD. et al. 2007. Isolation of amniotic stem cell lines with potential for therapy. Nature Biotechnol. 25.

Fauza D. 2004. Amniotic fluid and placental stem cells. Best Practice in Clin Obstet Gynaecol. 18:877–891.

Guetta L. et al. 2005. Human umbilical cord blood derived CD 133+ cells enhance function and repair of the infarcted myocardium. Stem Cells.

Horstrup S, Kadner A, Breymann C. Living. 2002. Autologous Pulmonary Artery Conduits Tissue Engineered from Human Umbilical Cord Cells. Ann Thor Surg. 74(1):46–52.

Kadner A. et al. 2002. Human umbilical cord cells: A new cell source for cardiovascular tissue engineering. Ann Thorac Surg. 74:S1422–28.

Kadner A. et al. 2004. Human umbilical cord cells for cardiovascular tissue engineering: A comparative study. Eur J Cardio-Thoracic Surg. 25:635–641.

Kim J. et al. 2007. Human amniotic derived stem cells have characteristics of multipotent stem cells. Cell Prolif. 40:75–90.

Korbling M, Robinson S, Estrov Z. 2005. Umbilical cord blood derived cells for tissue repair. Cytotherapy. 7(3):258–261.

Kunisaki S, Armant M, Kao G, Stevenson K, Kim H, Fauza D. 2007. Tissue engineering from human mesenchymal amniocytes: A prelude to clinical trials. J Ped Surg. 42:974–980.

Lee J. 2007. Early phase I/II clinical trial results for human cord blood stem cell injection therapy for stress urinary incontinence. In: Annual Meeting American Urology Association; Anaheim, CA.

Li C. et al. 2005. Mesenchymal stem cells derived from human placenta suppress aloogeneic umbilical cord blood lymphocyte proliferation. Cell Res. 15:539–547.

Ma N. et al. 2006. Umbilical Cord Blood Cell Transplantation for Myocardial Regeneration. Transplantation Proc. 38:771–773.

Mayer J. 1995. Uses of homograft conduits for right ventricle to pulmonary artery connections in the neonatal period. Seminars in Thoracic and Cardiovascular Surg. 7:130–132.

Newman M, Emerich D, Sanberg C, Sanberg P. 2004a. Use of human umbilical cord blood cells to repair the damaged brain. Curr Neurovasc Res. 1:269–281.

Newman M, Emerich D, Borlongan C, Sanberg C, Sanberg P. 2004b. Use of human umbilical cord blood cells to repair damaged brain. Curr Neurovasc Res. 13:269–281.

Nietfeld V. 2004. Autologous transplant lifetime chances up to age 70. In: Meeting of the International Cord Blood Society. Tuft's University, Boston, MA.

Parolini O. 2007. First International Work Shop on Placental derived Stem Cells: Isolation and Characterization of cells from Human Term Placenta. Stem Cells. (DOI: 10.1634/stemcells. 2007-0594).

Perin L. et al. 2007. Renal differentiation of amniotic fluid stem cells. Cell Prolif. 40:936–948.

Sanatloya J, Leon JD. 2005. Is the number of haematopoietic stem cells recovered from umbilical cord blood affected by perinatal clinical and laboratory factors? Journal of the Society for Gynecological Investigation. 12(2):237A, Suppl. S.

Schmidt D. et al. 2004. Umbilical cord blood derived endothelial progenitor cells for tissue engineering of vascular grafts. Ann Thorac Surg. 78:2094–2098.

Schmidt D. et al. 2005a. Prenatal human progenitor cells used for tissue engineering of living heart valves. Circulation. 112(7):U504, S2147.

Schmidt D. et al. 2005b. Living patches engineered from human umbilical cord derived fibrioblasts and endothelial progenitor cells. Eur J Cardio-Thorac Surg. 27:795–800.

Schmidt D. et al. 2006. Living autologous heart valves engineered from human prenatally harvested progenitors. Circulation. 114; S I:I-125–131.

Schmidt D. et al. 2007. Prenatally fabricatde autologous human living heart valves based on amniotic fluid-derived progenitor cells as single cell source. Circulation. 116; Suppl. I:I 64–I 70.

Schoen FJ. 1999. Tissue heart valves: Current challanges and future research perspectives. J Biomed Mater Res. 47:439–465.

Siegel N, Rosner M, Hanneder M, Freilinger A, Hengstschläger M. 2007. Human amniotic fluid stem cells: A new perspective. Amino Acids.

Solves P, Perales A, Moraga R. 2005. Maternal, neonatal and collection factors influencing the haematopoietic content of cord blood units. Acta Haematol. 113(4):241–246.

Waese E, Kandel R, Stanford W. 2007. Application of stem cells in bone repair. Skeletal Radiol.

Zhang X. et al. 2005. Mesenchymal progenitor cells derived from chorionic villi of human placenta for cartilage tissue engineering. Biochemical and Biophysical Res Commun. 340:944–952.

4

PERINATAL STEM CELL THERAPY

Daniel Surbek, Anna Wagner, and Andreina Schoeberlein

Department of Obstetrics and Gynecology, and Department of Research, Inselspital, Bern University Hospital and University of Bern, Switzerland

THE FETUS AS RECIPIENT FOR PRE- AND PERINATAL STEM CELL TRANSPLANTATION

The fetus is amenable to pre- and perinatal stem cell therapy. For this, early prenatal diagnosis of the fetal disorder is a prerequisite. Today, fetal-placental material can be obtained for genetic analysis in the first trimester by chorionic villus sampling (CVS) [Surbek, 2001]. Diagnosis of chromosomal defects or single gene disorders is based on molecular biology techniques, such as polymerase chain reaction (PCR) using either fetal material from an invasive test (e.g., CVS), or could use fetal cells or fetal cell-free DNA from maternal blood [Holzgreve, 1997]. To date, noninvasive diagnostic techniques using maternal blood have been shown to be feasible in prenatal diagnosis of single gene disorders, such as hemoglobinopathies [Cheung, 1996; Li, 2005], although not in detection of abnormal fetal karyotype.

By using these techniques, many genetic diseases of the fetus including immunohematopoietic, connective tissue, and skeletal disease, can now be diagnosed as early as the first trimester of pregnancy. Examples include thalassemias, sickle cell anemia, immunodeficiencies like SCID (severe combined immunodeficiency syndrome), osteogenesis imperfecta, and storage diseases like Hurler's syndrome or Globoidcell-leucodystrophy (Krabbe's disease). With the advent of new molecular methods including DNA-chip technology, it is conceivable that noninvasive screening of populations at risk might rapidly increase the number of such diseases diagnosed prenatally.

The treatment success of these diseases in children is usually limited. In some of the diseases, stem cell transplantation from bone marrow, mobilized peripheral blood, or

Perinatal Stem Cells. Edited by C. L. Cetrulo, K. J. Cetrulo, and C. L. Cetrulo, Jr.

umbilical cord blood has the potential for definitive cure. Nevertheless, promising results have been achieved with allogeneic stem cell transplantation in some of these diseases [Buckley, 1999; Escolar, 2005; Issaragrisil, 1995; Krivit, 1998; Lucarelli, 1999; Shapiro, 2000], significant obstacles with postnatal stem cell transplantation remain. Only for a minority (up to one-third) of the patients a human leukocyte antigen (HLA) compatible donor can be found. Substantial treatment-associated morbidity results from immunosuppression and marrow ablation in the recipient (except in SCID patients), which are usually necessary. In particular in nonrelated donor stem cell transplantation, graft versus host disease (GvHD) or graft failure are common. Furthermore, many of these diseases lead to irreversible damage to the fetus before birth (e.g., storage diseases, osteogenesis imperfecta, or severe α-thalassemia). Because of these limitations, many parents choose the option to terminate pregnancy when facing the diagnosis of a severe fetal genetic defect early in pregnancy. However, termination of pregnancy when the fetus is affected by a disease cannot be the ultimate goal of prenatal diagnosis. Instead, the aim must be to improve treatment success in order to ameliorate long-term health for the future child.

PRENATAL CELL TRANSPLANTATION: CURRENT EXPERIENCE

Prenatal transplantation of stem cell might offer a therapeutic alternative to abortion for families affected with severe congenital disorders diagnosed in the fetus. The ontogenetic properties in the development of the hematopoietic system and the immune system—characterized by the transition of primary sites of hematopoiesis from yolk sac and the aorta-gonadal mesonephron region to the fetal liver, and later to the bone marrow [Tavian, 1996] and by the immunological incompetence of the fetus until the end of the first trimester of pregnancy—provides a unique window of opportunity. This therapeutic window is believed to exist until the beginning of the second trimester of pregnancy. During this developmental stage, stem cells could be transplanted across HLA barriers and without prior marrow ablation or immunosuppression, leads to stable engraftment and long-term hematopoietic chimerism [Flake, 1997]. Treatment early in gestation could prevent organ damage prenatally and therafter development of full clinical disease after birth [Blau, 1996]. There is a whole range of disorders; Table IV.1 summarizes the range of disorders theoretically amenable to prenatal stem cell transplantation.

Over 30 experimental prenatal stem cell transplantations in human fetuses affected by a variety of disorders have been reported (Table IV.1) [Flake, 1999]. In many cases, no clinical effect on disease course has been achieved. The first successful case of prenatal transplantation of allogeneic hematopoietic stem cells in humans have been reported [Flake, 1996]. Further successful cases of *in utero* transplantation for immunodeficiency disorders have been published. Nevertheless, success is limited to diseases with severely compromised fetal immune system [Pirovano, 2004]. Prenatal stem cell transplantation in hemoglobinopathies has not yielded clinically significant engraftment [Sanna, 1999; Westgren, 1996], treatment of fetuses with different storage diseases showed similar lack of clinical effect [Bambach, 1997].

The failure of significant engraftment in fetuses without deficient immune system is thought to be due to competitive fetal marrow population by host stem cells and/or poorly defined immunological barriers [Surbek, 1999]. It is conceivable that the window of opportunity, where the fetal immune system is not yet able to recognize and react against foreign cells, is earlier than previously believed, probably even before the twelfth week of gestation. The involvement of cytotoxic NK cells has been postulated, although it is not supported by recent experimental evidence [Milner, 1998a]. Recent experimental

TABLE IV.1. Genetic Disorders That Might Benefit from Prenatal Stem Cell Transplantation

Immunodeficiency disorders	Bare lymphocyte syndrome[a]
	Cartilage-hair hypoplasia
	Chediak–Higashi syndrome[a]
	Chronic granulomatous disease[a]
	Kostman's syndrome
	Leukocyte adhesion defiency
	Omenn syndrome[a]
	Severe combined immunodeficiency syndrome[a]
	Wiskott–Aldrich syndrome
	X-Linked immunodeficiency with hyperimmunoglobin M
	X-Linked Bruton agammaglobulinemia
Hemoglobinopathies and Rh disease	Congenital erythropoietic porphyria
	α-Thalassemia[a]
	β-Thalassemia[a]
	Sickle cell disease[a]
	Rhesus isoimmunization[a]
Enzyme storage diseases	α-Mannosidosis
	Adrenoleukodystrophy
	Gaucher disease
	Globoid cell leukodystrophy[a]
	Metachromatic leukodystrophy[a]
	Mucopolysaccharidoses
	Niemann–Pick disease[a]
	Wolmans disease
Others	Dyskeratosis congenital
	Familial hemaphagocytic lymphohistiocytosis
	Hemophilia A[a]
	Infantile osteopetrosis
	Osteogenesis imperfecta[a]
	Shwachman–Diamond syndrome

[a]Attempted prenatal transplantation in human fetuses.

data, however, again suggest the existence of an immunologic barrier very early in pregnancy [Peranteau, 2007]. An alternative explanation could be that endogenous nondefective hematopoietic precursors of the host successfully compete with donor cells for niches within the bone marrow microenvironment during development, possibly supported by host stem cell specific homing–mediating adhesion molecules [Flake, 1999].

HOW DO WE IMPROVE THE SUCCESS OF PRENATAL STEM CELL TRANSPLANTATION?

Due to limited success, several strategies to improve the selective advantage for donor cells in competition with host HSCs are being considered and explored in animal models of prenatal stem cell transplantation. These strategies address questions concerning the improvement of the level of long-term engraftment after *in utero* stem cell transplantation [Touraine, 2004; Troeger, 2007]. Animal models used include sheep, [Young, 2003; Zanjani, 1994] nonhuman primates [Shields, 1995] and mice [Archer, 1997]. Long-term chimerism has been achieved recently in a human-to-mouse model of *in utero* transplantation using NOD/LtSz *scid* mice (NOD/SCID mice) [Turner, 1998].

In the following list, different strategies for improvement of long-term donor stem cell engraftment are mentioned, some of which have provided partial success:

Modification of the source and dose of stem cells transplanted [Bambach, 1997].

Use of an alternative route of stem cell administration, including the novel extracoelomic route, which is possible even earlier than the usually used intraperitoneal transplantation [Noia, 2004; Santolaya-Forgas, 2007].

Improvement of competitive advantage by cotransplantation of stroma or mesenchymal stem cells [Almeida-Porada, 1999, 2000; Portmann-Lanz, 2006, 2007; Schoeberlein, 2005].

Induction of microchimerism *in utero* for later (postnatal) transplantation without rejection of donor cells [Carrier, 1995; Hayward, 1998; Hayashi, 2002; Peranteau, 2002].

Graft modulation by T-cell dosage [Petersen, 2007], T-cell modulation [Bhattacharyya, 2002; Peranteau, 2006; Shields, 2003, 2004], cytokine graft stimulation [Shaaban, 2006] or antibody-mediated immunosuppression [Peranteau, 2006].

Suppression of hematopoiesis in the fetal host using a parvovirus B19 capsid protein [Norbeck, 2004].

Several of these modifications lead to high-level engraftment in nonimmunodeficient recipients in animal models. If they prove to be effective in nonhuman primates, they should be translated into clinical use in human fetuses.

MESENCHYMAL STEM CELLS FOR PRENATAL TRANSPLANTATION

Mesenchymal stem cells have been shown to reconstitute a range of different tissues after *in utero* transplantation [Liechty, 2000]. These cells hold promise to be a source to treat bone genetic disorders, such as osteogenesis imperfecta, which can be diagnosed in pregnancy by ultrasound and molecular analysis. Furthermore, mesenchymal stem cells in fact lead to improved engraftment if cotransplanted with hematopoietic stem cells. Even after HSC transplantation, mesenchymal stem cells have been shown to be effective as treatment if graft-versus-host disease develops. Mesenchymal stem cells are less immunogeneic and can therefore also proably be transplanted at a later stage in pregnancy, and with better engraftment in nonimmunodeficient recipients. Recently, the first case report about a fetus affected by severe osteogenesis imperfecta who has been treated in midgestation by *in utero* transplantation of mesenchymal stem cells [Le Blanc, 2005]. Postnatally, up to 5% donor cell chimerism has been reported. If this novel treatment is able to significantly improve clinical outcome as compared to conventional postnatal treatment remains to be seen. Initial studies of postnatal mesenchymal stem cell transplantation suggest in fact that this might be possible in the future [Horwitz, 2001]. This case clearly shows the potential of mesenchymal stem cell *in utero* transplantation for genetic disease of bones, connective tissues, skeletal muscle, such as Duchenne dystrophy, as well as disorders leading to neurodegeneration [Portmann-Lanz, 2006].

FETAL GENE THERAPY AS A NEW STRATEGY

One of the promising strategies to circumvent allogeneic HLA barriers is to use genetically corrected autologous HSCs in the fetus, either *in vivo* or *in vitro* [Surbek, 2001, 2007]. Until now, success of gene therapy of genetic diseases using stem cells has been hampered mainly by inefficient transduction of stem cells [Anderson, 1998] and inability to achieve long-term expression of the therapeutic gene. The recent development of methods to overcome failure of efficient transduction of quiescent hematopoietic cells using new vectors and transduction protocols opens new perspectives for gene therapy in general, as well as for prenatal gene transfer in particular. The fetus is particularily susceptible for successful gene therapy approaches because of the expanding hematopoietic system during gestation and the immunologic naiveté early in gestation, precluding immune reaction toward the transgene. Long-term tolerance to the transgene is induced prenatally. The development of lentiviral human immunodeficiency virus (HIV)-based vectors, able to transduce nondividing human stem cells, and progress in fetal gene therapy approaches, offering a possibility for introduction of a therapeutic gene before fetal immunocompetence, may enhance the efficiency and duration of gene expression [Zanjani, 1999]. We will reviewed these issues in more detail in the following section.

CURRENT ISSUES OF POSTNATAL GENE THERAPY

Many gene therapy approaches target a range of non-hematopoietic cells, such as gene therapy for hemophilic disorders (e.g., factor IX deficiency) or for cystic fibrosis. In some of these disorders, initial clinical experience has already been obtained, with variable success [Manno, 2003].

Hematopoietic stem cells are attractive targets for somatic cell-based gene therapy, because they have the potential to produce progeny cells containing a therapeutic gene lifelong. Current clinical protocols of postnatal gene therapy in pediatric patients with genetic diseases are based on *ex vivo* retroviral transduction of lymphocytes [Blaese, 1995] or hematopoietic stem–progenitor cells from cord blood or bone marrow, followed by autologous transplantation of the engineered cells back into the patient. Initial trials showed the feasibility and safety of gene therapy using cord blood cells in patients with adenosine deaminase (ADA)-deficiency, although only very limited clinical efficiency has been achieved as reported in a follow-up study [Kohn, 1995, 1998]. Recently, however, clinical success has been reported 10 months after gene therapy in X-SCID disease using autologous retrovirally transduced bone marrow $CD34^+$ cells [Cavazzana-Calvo, 2000; Buckley, 2000]. Nevertheless, several obstacles must be overcome before gene therapy can gain broad clinical application [Anderson, 1998; Blau, 1997]. In particular, the problem of insertional mutagenesis that occurred in the recent series [Hacein-Bey, 2003a, b] must be addressed as a primary safety issue.

The main obstacles concern transduction efficiency, random integration of vector gene construct into host genome, duration of expression of the therapeutic gene (gene silencing), host immune response against vector, gene or gene product, and reproducible production of safe replication-free high-titer vectors [Verma, 1997]. Gene expression can be severely impaired by spontaneous cessation of regulatory sequence activity that control gene expression, by inactivation of promoters (e.g., by methylation) in the transduced host cell, by specific host defense mechanisms or by elimination of the transduced cells by the host immune system recognizing the foreign gene product [Bestor, 2000]. Inflammatory cytokines, such as TNF-alpha or IFN-γ have been shown to be involved in the host immune response toward the foreign gene or gene product by direct inhibition of the expression of transgenes [Qin, 1997]. Although some success has been achieved in mouse models, [Bunting, 1998] recent results in large animal models reveal that gene expression *in vivo* can still be severely impaired despite successful engraftment of genetically modified autologous hematopoietic progenitor–stem cells [Lutzko, 1999]. A further difficulty is the identification of the target for gene transfer; that is, hematopoietic stem cells, because definitive markers for undifferentiated, quiescent hematopoietic stem cells are still lacking. Recent reports suggest that there is no single hematopoietic stem cell marker, leading to the assumption that the hematopietic stem cell compartment is heterogeneous [Bhatia, 1998; Goodell, 1997; Ziegler, 1999].

Gene delivery systems to hematopoietic stem cells include retroviral and adenoviral vectors, adeno-associated vectors, lentiviral (HIV-based) vectors, and nonviral (liposome) vectors (reviewed in Verma, 1997). New generation adeno-associated viral vectors just recently have been reported as leading to successful long-term gene expression and correction of hemophilia B in canine and mouse models after intrahepatic [Snyder, 1999] or intramuscular [Herzog, 1999] injection. Because retroviruses (as opposed to adenoviruses) have the property to stably integrate into the host genome, they are preferentially used to deliver genes into hematopoietic stem cells in experimental animal models and clinical settings. However, besides the lack of control of long-term transgene regulation and expression in transduced cells, the major problem is the inefficiency of retroviral gene transfer into nondividing cells like hematopoietic stem cells [Brenner, 1996]. Recent developments in gene transfer technique address this problem partly successfully [Kalberer, 2000]. Efficient transduction with murine retroviral vectors requires host DNA replication, which can be achieved only after stem cells are released from quiescence by prestimulation with growth factors in culture, which in turn leads to cell differentiation and loss of self-renewal and multilineage differentiation capacity [Anderson, 1998]. New strategies for efficient and stable retroviral transduction of hematopoietic stem–progenitor cells by new generation of retroviral vectors and optimized transduction conditions, including prestimulation with novel growth factors, such as thrombopoietin and flt3-ligand, centrifugation (spinoculation) or transduction on fibronectin fragments, are evolving and yield promising results [Abonour, 2000; Conneally, 1998]. Nevertheless, because noncycling cells cannot be transduced efficiently, release from quiescence *in vitro* is still necessary. The development in the field of nonviral vectors, such as human artificial microchromosomes, as vectors for gene delivery into mammalian cells [Harrington, 1997] is rapidly evolving, though these vectors are still far away from application.

New viral vector generations include replication-deficient lentiviral HIV-based vector systems, which can integrate stably into the host genome in dividing and nondividing cells [Naldini, 1996]. Different replication-defective lentiviral vectors based on HIV have been constructed, including some in which one or more of these accessory genes are deleted because of their association with intracellular pathological effects [Zufferey, 1997]. Requirements for HIV regulatory and accessory genes depends on the nature of target cells and can have a major impact on transduction efficiency, as has been shown by *in vivo–in situ* gene-transfer experiments into rat liver and muscle cells, neurons and retina [Kafri, 1997]. Because cellular receptors for HIV-1 and HIV-2 have been shown to be absent on primitive hematopoietic precursors, pseudotyping lentiviral vectors with vesicular stomatitis virus G (VSV-G) protein is necessary, usually provided by a VSV-G plasmid. Safety aspects of the replication-defective HIV based vectors concern primarily the risk of recombination potentially leading to replication-competent viruses, which can be minimized if three-plasmid systems are used in the packaging.

Lentiviral vectors have recently been used for gene transfer into hematopoietic cells [Sutton, 1998]. Optimization of transduction conditions can be achieved by using Fibronectin fragment CH 296, protamine sulfate and spinoculation, with or without prestimulation with different cytokines. Besides that, several groups have shown efficient transduction of nondividing hematopoietic stem cells by HIV derived vectors in comparison to conventional murine retroviruses [Case, 1999; Uchida, 1998], it has now been proven in a NOD/SCID mouse model that long-term repopulating stem cells (SCID repopulating cells, SRCs) can be transduced without growth factor prestimulation [Miyoshi, 1999]. In a landmark paper, long-term expression of a therapeutic human β-globin gene in a lentiviral vector construct introduced in normal and β-thalassemic mice via transduced bone marrow cells has been achieved [May, 2000].

In conclusion, lentiviral HIV based vector systems are clearly superior to conventional murine oncoretroviruses concerning transduction efficacy of nondividing hematopoietic cells. They hold great promise to achieve the goal of stable long-term transduction of hematopoietic stem cells and their progeny.

PRENATAL GENE THERAPY

Several *in utero* gene transfer models targeted to different cells and organs have been developed in the last 10 years, including pulmonary epithelial cells, [Larsen, 1997; Sekhon, 1995], hepatocytes, [Wang, 1998], skin, [Hayashi, 1996], intestine, [Wu, 1999], heart, [Woo, 1997] and ductus arteriosus [Mason, 1999]. In these models, adenoviral or conventional murine retroviral vectors were used.

While postnatal approach by *in vivo* lentiviral vector-based gene transfer has been shown to be feasible in rats, [Kafri, 1997] direct *in utero* administration of lentiviral vectors has also been reported in mice and monkeys [Jimenez, 2005; Lee, 2005; Seppen, 2003].

The models mentioned above used intraamniotic, intratracheal, intraperitoneal, intrahepatic, intravascular, intraplacental, or *in situ* (Ductus arteriosus region) administration of the vector-gene construct. Recently, a minimal-invasive prenatal gene transfer into the fetal trachea has been described in the fetal sheep model [David, 2003; Peebles, 2004]. Besides these administration routes, transplacental transfer of the vector-gene construct after injection into maternal circulation has also been shown to be possible [Tsukamoto, 1995]. This route, however, is unlikely to be feasible for use in humans because of the high risk of maternal transduction.

Presently, the first proofs of principle for therapeutic *in utero* gene application and for the induction of tolerance to life-long expression of a transgenic human protein have been achieved in mouse and/or rat models of human genetic disease. Several recent studies have shown this independently [Dejneka, 2004; Rucker, 2004; Seppen, 2003; Waddington, 2004]. Furthermore, clear experimental evidence of induction of tolerance toward a transgene product after *in utero* gene transfer has been provided, [Waddington, 2003] supporting the application of gene therapy to the preimmune fetus.

A recent report suggests an alternative, completely different approach where placental tissue instead of hematopoietic cells are targeted: autologous trophoblast cells of the rodent placenta were removed, genetically altered *in vitro* and retransplanted into the placenta *in utero*. They could show that the transplanted cells survive and that the gene product is expressed and can be detected in the fetal circulation [Senut, 1998]. This approach has recently become supported by data providing evidence that the human placenta in the first trimester contains multipotent cells able to differentiate into multiple lineages including neural cells [Portmann-Lanz, 2006]. Placental cells can be harvested using a standard minimal invasive procedure (chorionic villus sampling), and the cells can be expanded *in vitro* before or after gene transfer.

ANIMAL MODELS OF PRENATAL GENE THERAPY

Most experiments in fetal gene therapy have been performed in mice and sheep, some in dogs and nonhuman primates [Coutelle, 2005]. Mice offer the great advantage of ease of maintenance and breeding and the availability of a range of mouse models for human genetic disease. However, their small size and differences in physiology limit their usefulness with respect to procedures applicable in the human fetus. We [Schoeberlein, 2004] and others have addressed possible fetal therapy approaches. Fetal sheep are known to be immunoincompetent up to around gestational day 60–70 (term, 145 days) [Miyasaka, 1988]. Direct vector-gene administration into the peritoneal cavity or the liver (or the bone marrow at a later stage of pregnancy) of the fetus can be performed in fetal mice and sheep. It is feasible to determine if long-term gene expression can be improved by exposing the preimmune fetus to the foreign antigen (the transgene), early (<60 day) versus late (>120 day) versus postnatal gene transfer (*ex vivo* and *in vivo*).

FETAL GENE THERAPY TARGETING HEMATOPOIETIC STEM CELLS

As mentioned previously, fetal gene therapy is a further strategy to circumvent the limitations of prenatal allogeneic stem cell transplantation described above [Coutelle, 1995]. Similar to *in utero* stem cell transplantation, prenatal gene therapy has the potential to prevent irreversible damage to the fetus before birth [Douar, 1996]. Several characteristics of the developing fetal hematopoietic system may prove beneficial and render it highly susceptible to gene transfer, including the high proliferative status and expansion of the stem cell pool, the relatively small amount of necessary gene-engineered stem cells, and the presumed lack of immune response towards the vector that could lead to cessation of gene expression and/or tissue damage [Zanjani, 1999]. Additionally, fetal umbilical cord stem cells have been shown to have a high expansion potential [Luther-Wyrsch, 2001; Piacibello, 1997; Wyrsch, 1999] and a superior retroviral transduction potential compared to adult cells [Ektherae, 1990]. Furthermore, recent data from experiments using stem cell transplantation into mouse blastocysts show that gene expression of donor cells largely depends on the developmental stage of the host microenvironment [Geiger, 1998], which could therefore lead to an enhanced gene expression after prenatal gene transfer compared to postnatal gene therapy.

Two different strategies with the aim to introduce a foreign therapeutic gene into fetal hematopoietic stem cells are under investigation:

Ex Vivo Gene Therapy. This strategy implies autologous transplantation of *in vitro* transduced stem cells, which has already been done with partial success in human neonates with ADA deficiency using cord blood stem cells [Kohn, 1998] or later with full success using autologous bone marrow [Cavazzana-Calvo, 2000]. Animal studies have demonstrated the feasibility of *ex vivo* gene therapy with circulating hematopoietic stem cells that were from the fetal circulation by cordocentesis, subjected to *ex vivo* expansion and transduction, and transplanted back into the fetus [Kantoff, 1989; Lutzko, 1999; Omori, 1999; Winkler, 1999]. The major advantage of this method is that the risk for germ-line transduction is substantially smaller as compared to *in vivo* gene delivery where the whole fetus is directly exposed to a high-titer gene-vector construct. On the other hand, major technical difficulties in obtaining enough stem cells from the fetus during pregnancy remain a challenge and must be overcome [Surbek, 2000]. One of the critical issue will be to achieve competitive advantage of transduced autologous cells, which, after *ex vivo* gene transfer, will represent only a fraction of HSCs of the recipient. However, recently reported preliminary results of *in utero* HSC transplantation in a thalassemic mouse model suggest that nonaffected donor HSC transplanted *in utero* into thalassemic mice provide a selective competitive advantage over deficient host cells and can lead to successful long-term engraftment [Archer, 1998; Milner, 1998]. Similarily, genetically corrected autologous HSCs might gain selective advantage compared to noncorrected cells. Alternatively, *in vivo* HSC selection after transfer of a drug-resistance gene might be possible [Allay, 1998]. If the selection drug is used in the newborn. Expansion could provide advantage in the prenatal setting as opposed to postnatal transplantation, where expansion prior to transplantation has been shown to delay engraftment, [Güenechea, 1999] and could additionally provide a competitive advantage of autologous donor cells.

Recent evidence shows that fetal liver cell aspiration during the early second trimester gestation in ongoing pregnancy is feasible and yields enough stem cells for *ex vivo* gene transfer and autologous transplantation [Schoeberlein, 2004; Surbek, 2002]. A further option might be (as mentioned above) to use placental cells obtained by chorionic villus sampling [Portmann-Lanz, 2006].

In Vivo Gene Therapy. In this approach the gene-containing vector is transferred directly to the fetus, leading to *in vivo* transduction of fetal hematopoietic stem cells. Direct transfer to the fetus is considered as a feasible alternative approach [Clapp, 1995]. Recently, successful long-term gene expression in a sheep model after intraperitoneal gene transfer to hematopoietic cells has been reported, [Porada, 1998; Tran, 2000] and *in utero* vector-gene administration has proven feasible and effective in other animal models [Baumgartner, 1999; Lipshutz, 1999; Schachtner, 1999; Themis, 1999]. While this approach is technically much easier, a higher risk regarding germ line transduction must be expected. To date, no comparisons between the two approaches using similar delivery systems have been performed.

Hemoglobinopathies pose a specific challenge to fetal gene therapy [Persons, 2001]. A big step toward efficiency using lentiviral vectors has been taken recently [May, 2000]. The same research group now shows evidence for efficiency of fetal gene therapy for α-thalassemia in a mouse model [Han, 2007]. This finding represents a significant step toward the future *in utero* correction of hemoglobinopathies.

SAFETY ASPECTS OF PRENATAL GENE THERAPY

Important safety concerns exist whenever gene transfer is used to treat a fetus [Billings, 1999; King, 1999]. One major concern is the risk of transduction of gonadal cells that might result in genetic germ line transduction. There is also a potential risk for the mother because transduction of her somatic or germinal cells through transplacental migration of the vector-gene construct could theoretically occur.

Another important safety aspect is the possibility of insertional mutagenesis in fetal cells resulting in a functional gene defect leading to a genetic disease or to the formation of a malignant tumor. Malignant uncontrolled proliferation of T-cells (Leukemia-like syndrome) has occurred in two children in whom gene therapy was used to treat severe-combined immunodeficiency syndrome [Hacein-Bey, 2003a,b]. In mice, induction of liver tumors eventually induced by insertional mutagenesis after *in utero* gene transfer has been observed [Themis, 2005]. This finding is of importance, as the fetal system may be particularly sensitive to such events, and integrating vectors prefer to insert their genomes into chromatin in unpredictable configuration and loci.

Ex vivo gene therapy might be safer as compared to *in vivo* vector-gene administration *in utero* regarding these risks. Although it does not exclude insertional mutagenesis in cells retrovirally transduced *in vitro*, mutageneic gene insertion might be better detectable and controllable before autologous transplantation of the cells. However, it will never be possible to fully exclude these possible complications if fetal gene therapy is performed. Studies are now being done to determine the risk for germline or maternal cell transduction; some of them suggest this risk to be very low [Tran, 2000; Ye, 1998].

ETHICAL CONSIDERATIONS

Based on the recently published successful gene transfer *in utero*, the question of a trial in human fetuses is being raised. However, general concerns of fetal somatic gene therapy are shared not only by a number of scientist exponents, but also by many ethical committees and commissions, and lay persons, because the possibility of germ line transduction raises a number of important ethical, moral, societal, and legal questions that concern the society as a whole. Some argue that as long as germ line transduction cannot be excluded absolutely (which obviously will never be possible, at least with the *in vivo* approach), this treatment should be prohibited. On the other hand, advocates of this approach argue that there is a risk-benefit ratio not only for the individual patient, but also on a broader level of public health, and if this ratio is acceptable, fetal gene therapy should be envisaged [Fletcher, 1996]. It certainly is very important to carry out these discussions among scientists, as well as with the public and on an individual level during the preclinical development, before these new therapeutic modalities enter the field of clinical application. Major ethical questions must be addressed regarding immediate and long-term consequences. There is therefore a need for more scientific data regarding advantages, safety, and risks of fetal gene therapy, which can only be acquired *in vivo* in animal studies [Caplan, 2000; Schneider, 1999].

REFERENCES

Abonour R, Williams DA, Einhorn L, Hall KM, Chen J, Coffman J, Traycoff CM, Bank A, Kato I, Ward M, Williams SD, Hromas R, Robertson MJ, Smith FO, Woo D, Mills B, Srour EF, Cornetta K. 2000. Efficient retrovirus-mediated transfer of the multidrug resistance 1 gene into autologous human long-term repopulating hematopoietic stem cells. Nat Med. 6:652–658.

Allay JA, Persons DA, Galipeau J, Riberdy JM, Ashmun RA, Blakley RL, Sorrentino BP. 1998. In vivo selection of retrovirally transduiced hematopoietic stem cells. Nat Med. 4:1136–1143.

Almeida-Porada G, Flake AW, Glimp HA, Zanjani ED. 1999. Cotransplantation of stroma results in enhancement and early expression of donor hematopoietic stem cells in utero. Exp Hematol. 27:1569–1575.

Almeida-Porada G, Porada CD, Tran N, Zanjani ED. 2000. Cotransplantation of human stromal cell progenitors into preimmune fetal sheep results in early appearance of human donor cells in circulation and boosts cell levels in bone marrow at later time points after transplantation. Blood. 95:3620–3634.

Anderson WF. 1998. Human gene therapy. Nature (London). 392(Supp):25–30.

Archer DR, Turner CW, Yeager AM, Fleming WH. 1997. Sustained multilineage engraftment of allogeneic hematopoietic stem cells in NOD/SCID mice after in utero transplantation. Blood. 90:3222–3229.

Archer DR, Hester LE, Gu Y, Fleming WH, Yeager AM, Hsu LL. 1998. Successful in utero engraftment of allogeneic hematopoietic cells in murine β-thalassemia. Blood. 92:267a.

Bambach BJ et al. 1997. Engraftment following in utero bone marrow transplantation for globoid cell leukodystrophy. Bone Marrow Transplant. 19:399–402.

Baumgartner TL, Baumgartner BJ, Hudon L, Moise KJ. 1999. Ultrasonographically guided direct gene transfer in utero: Successful induction of beta-galactosidase in a rabbit model. Am J Obstet Gynecol. 181:848–852.

Bestor TH. 2000. Gene silencing as a threat to the success of gene therapy. J Clin Invest. 105:409–411.

Bhatia M, Bonnet D, Murdoch B, Gan OI, Dick JE. 1998. A newly discovered class of human hematopoietic cells with SCID-repopulating activity. Nat Med. 4:1038–1045.

Bhattacharyya S et al. 2002. Multilineage engraftment with minimal graft-versus-host disease following in utero transplantation of S-59 psoralen/ultraviolet A light treated, sensitized T cells and adult T cell-depleted bone marrow in fetal mice. J Immunol. 169:6133–6140.

Billings PR. 1999. In utero gene therapy. The case against. Nat Med. 5:255–257.

Blaese RM et al. 1995. T lymphocyte-directed gene therapy for ADA-SCID: Initial results after 4 years. Science. 270:475–480.

Blau CA, Stamatoyannopoulos G. 1996. Preemptive therapy for genetic disease. Nat Med. 2:161–162.

Blau H, Khavari P. 1997. Gene therapy: Progress, problems, prospects. Nat Med. 3:612–613.

Brenner M. 1996. Gene transfer to hematopoietic cells. N Engl J Med. 335:337–339.

Buckley RH, Schiff SE, Schiff RI, Markert ML, Williams LW, Roberts JL, Myers LA, Ward FE. 1999. Hematopoietic stem cell transplantation for the treatment of severe combined immunodeficiency syndrome. N Engl J Med. 340:508–516.

Buckley RH. 2000. Gene therapy for human SCID: Dreams become reality. Nat Med. 6:623–624.

Bunting KD, Sangster MY, Ihle JN, Sorrentino BP. 1998. Restoration of lymphocyte function in Janus kinase 3-deficient mice by retroviral-mediated gene transfer. Nat Med. 4:58–64.

Caplan AL, Wilson JM. 2000. The ethical challanges of in utero gene therapy. Nat Genet. 24:107.

Carrier E, Hae T, Busch MP, Cowan MJ. 1995. Induction of tolerance in nondefective mice after in utero transplantation of major histocompatibility complex-mismatched fetal hematopoietic stem cells. Blood. 86:4681–4690.

Case SS, Price MA, Jordan CT, Yu XJ, Wang L, Bauer G, Haas DL, Xu D, Stripecke R, Naldini L, Kohn DB, Crooks GM. 1999. Stable transduction of quiescent CD34+ CD38− human hematopoietic cells by HIV-1-based lentiviral vectors. Proc Natl cad Sci USA. 96:2988–2993.

Cavazzana-Calvo M, Hacein-Bey S, de Saint Basile G, Gross F, Yvon E, Nusbaum P, Selz F, Hue C, Certain S, Casanova JL, Bousso P, Deist FL, Fisher A. 2000. Gene therapy of human severe combined immunodeficiency (SCID)-X1 disease. Science. 288:669–672.

Cheung MC, Goldberg JD, Kan YW. 1996. Prenatal diagnosis of sickle cell anaemia and thalassaemia ba analysis of fetal cells in maternal blood. Nat Genet. 14:264–268.

Clapp DW, Freie B, Lee W-H, Zhang Y-Y. 1995. Molecular evidence that in-situ transduced fetal liver hematopoietic stem/progenitor cells give rise to medullary hematopoiesis in adult rats. Blood. 86:2113–2122.

Conneally E, Eaves CJ, Humphries RK. 1998. Efficient retroviral-mediated gene transfer to human cord blood stem cells with in vivo repopulating potential. Blood. 91:3487–3493.

Coutelle C, Themis M, Waddington SM, Buckley SMK, Gregory LG, Nivsarkar MS, David AL, Peebles D, Weisz B, Rodeck C. 2005. Gene Therapy Progress and Prospects: Fetal gene therapy—first proofs of concept—some adverse effects. Gene Ther. 12:1601–1607.

Coutelle C, Douar A-M, Colledge WH, Froster U. 1995. The challenge of fetal gene therapy. Nat Med. 1:864–866.

David AL, Peebles D, Gregory L, Themis M, Cook T, Miah M, Coutelle C, Kiserud T, Rodeck CH. 2003. Ultrasound-guided injection of the trachea in fetal sheep: A novel percutaneous technique to target the fetal airways in utero. Fet Diagn Ther. 18:385–390.

Dejneka NS et al. 2004. In utero gene therapy rescues vision in a murine model of congenital blindness. Mol Ther. 9:182–188.

Douar A-M, Themis M, Coutelle C. 1996. Fetal somatic gene therapy. Mol Hum Reprod. 2:633–641.

Ektherae D, Crombleholme T, Karson E, Harrison MR, Anderson WF, Zanjani ED. 1990. Retroviral vector-mediated transfer of the bacterial neomycin resistance gene into fetal and adult sheep and human hematopoietic progenitors in vitro. Blood. 75:365–369.

Escolar ML et al. 2005. Transplantation of umbilical-cord blood in babies with infantile Krabbe's disease. N Engl J Med. 352:2069–2081.

Flake AW, Roncarolo MG, Puck JM, Almeida-Porada G, Evans MI, Johnson MP, Abella EM, Harrison DD, Zanjani ED. 1996. Treatment of X-linked severe combined immunodeficiency by in utero transplantation of paternal bone marrow. N Engl J Med. 335:1806–1810.

Flake AW, Zanjani ED. 1997. In utero hematopoietic stem cell transplantation. A status report. JAMA. 278:932–937.

Flake AW, Zanjani ED. 1999. In utero hematopoietic stem cell transplantation: Ontogenetic opportunities and biologic barriers. Blood. 94:2179–2191.

Fletcher JC, Richter G. 1996. Human fetal gene therapy: Moral and ethical questions. Hum Gene Ther. 7:1605–1614.

Geiger H, Sick S, Bonifer C, Müller AM. 1998. Globin gene expression is reprogrammed in chimeras generated by injecting adult hematopoietic stem cells into mouse blastocysts. Cell. 93:1055–1065.

Goodell MA, Rosenzweig M, Kim H, Marks DF, DeMaria M, Paradis G, Grupp SA, Sieff CA, Mulligan RC, Johnson RP. 1997. Dye efflux studies suggest that hematopoietic stem cells expressing low or undetectable levels of CD34 antigen exist in multiple species. Nat Med. 1337–1345.

Güenechea G, Segovia JC, Albella B, Lamana M, Ramirez M, Regidor C, Fernández MN, Bueren JA. 1999. Delayed engraftment of nonobese diabetic/severe combined immunodeficient mice transpanted with ex vivo-expanded human CD34+ cord blood cells. Blood. 93:1097–1105.

Hacein-Bey S et al. 2003a. N Engl J Med. 348:255–256.

Hacein-Bey S et al. 2003b. Science. 302:415–419.

Han XD, Lin C, Chang J, Sadelain M, Kan YW. 2007. Fetal gene therapy of alpha-thalassemia in a mouse model. Proc Natl Acad Sci USA. 104:9007–9011.

Harrington JJ, Van Bokkelen G, Mays RW, Gustashaw K, Willard HF. 1997. Formation of de novo centromeres and construction of first-generation human artificial microchromosomes. Nat Genet. 15:345–355.

Hayashi S, Peranteau WH, Shaaban AF, Flake AW. 2002 Aug. 1. Complete allogeneic hematopoietic chimerism achieved by a combined strategy of in utero hematopoietic stem cell transplantation and postnatal donor lymphocyte infusion. Blood. 100(3):804–812.

Hayashi SI, Morishita R, Aoki M, Moriguchi A, Kida I, Nakajima M, Kaneda Y, Higaki J, Ogihara T. 1996. In vivo gene transfer of gene and oligo deoxynucleotides into skin of fetal rats by incubation in amniotic fluid. Gene Ther. 3:878–885.

Hayward A et al. 1998. Microchimerism and tolerance following intrauterine transplantation and transfusion for α-thalassemia-1. Fetal Diagn Ther. 13:8–14.

Herzog RW, Yang EY, Couto LB, Hagstrom JN, Elwell D, Fields PA, Burton M, Bellinger DA, Read MS, Brinkhous KM, Podsakoff GM, Nichols TC, Kurtzman GJ, High KA. 1999. Long-term correction of canine hemophilia B by gene transfer of blood coagulation factor IX mediated by adeno-associated viral vector. Nat Med. 5:56–63.

Holzgreve W. 1997. Will ultrasound-screening and ultrasound-guided procedures be replaced by non-invasive techniques for the diagnosis of fetal chromosomal anomalies? Ultrasound Obstet Gynecol. 9:217–219.

Horwitz EM et al. 2001. Clinical responses to bone marrow transplantation in children with severe osteogenesis imperfecta. Blood. 97:1227–1231.

Issaragrisil S, Visuthisakchai S, Suvatte V, Tanphaichitr VS, Chandanayingyong D, Schreiner T, Kanokpongsakdi S, Siritanaratkul N, Piankijagum A. 1995. Brief report: Transplantation of cord-blood stem cells into a patient with severe thalassemia. N Engl J Med. 332:367–369.

Jimenez DF, Lee CI, O'shea CE, Kohn DB, Tarantal AF. 2005. HIV-1-derived lentiviral vectors and fetal route of administration on transgene biodistribution and expression in rhesus monkeys. Gene Ther. 12:821–830.

Kafri T, Blomer U, Peterson DA, Gage FH, Verma IM. 1997. Sustained expression of genes delivered directly into liver and muscle by lentiviral vectors. Nat Genet. 17:314–317.

Kalberer CP, Pawliuk R, Imren S, Bachelot T, Takekoshi KJ, Fabry M, Eaves CJ, London IM, Humphries RK, Leboulch P. 2000. Preselection of retrovirally transduced bone marrow avoids subsequent stem cell gene silencing and age-dependent extinction of expression of human beta-globin in engrafted mice. Proc Natl Acad Sci USA. 97:5411–5415.

Kantoff PW, Flake AW, Eglitis MA, Scharf S, Bond S, Gilboa E, Erlich H, Harrison MR, Zanjani ED, Anderson WF. 1989. In utero gene transfer and expression: A sheep transplantation model. Blood. 73:1066–1073.

King D, Shakespeare T, Nicholson R, Clarke A, McLean S. 1999. Risks inherent in fetal gene therapy. Nature (London). 397:383.

Kohn DB et al. 1995. Engraftment of gene-modified umbilical cord blood cells in neonates with amino-deaminase deficiency. Nat Med. 1:1017–1023.

Kohn DB, Hershfield MS, Carbonaro D, Shigeoka A, Brooks J, Smogorzewska EM, Barsky LW, Chan R, Burotto F, Annett G, Nolta JA, Crooks G, Kapoor N, Elder M, Wara D, Bowen T, Madsen E, Snyer FF, Bastian J, Muul L, Blaese RM, Weinberg K, Parkman R. 1998. T lymhpocytes with a normal ADA gene accumulate after transplantation of transduced autologous umbilical cord blood CD34+ cells in ADA-deficient SCID neonates. Nat Med. 4:775–780.

Krivit W, Shapiro EG, Peters C, Wagner JE, Cornu G, Kurtzberg J, Wenger D, Kolodny EH, Vanier MT, Loes DJ, Dusenbery K, Lockman LA. 1998. Hematopoietic stem-cell transplantation in globoid-cell leukodystrophy. N Engl J Med. 338:1119–1126.

Larsen JE, Morrow SL, Happel L, Sharp JF, Cohen JC. 1997. Reversal of cystic fibrosis phenotype in mice by gene therapy in utero. Lancet. 349:619–620.

Le Blanc K, Gotherstrom C, Ringden O, Hassan M, McMahon R, Horwitz E, Anneren G, Axelsson O, Nunn J, Ewald U, Norden-Lindeberg S, Jansson M, Dalton A, Astrom E, Westgren M. 2005 June. Fetal mesenchymal stem-cell engraftment in bone after in utero transplantation in a patient with severe osteogenesis imperfecta. Transplantation. 15;79(11):1607–1614.

Lee CC et al. 2005. Fetal gene transfer using lentiviral vectors and the potential for germ cell transduction in rhesus monkeys (Macaca mulatta). Hum Gene Ther. 16:417–425.

Li Y et al. 2005. Detection of paternally inherited fetal point mutations for betal-thalassemai using size-fractionated cell-free DNA in maternal plasma. JAMA. 293:843–849.

Liechty KW et al. 2000. Human mesenchymal stem cells engraft and demonstrate site-specific differentiation after in utero transplantation. Nat Med. 6:1282–1286.

Lipshutz GS, Sarkar R, Flebbe-Rehwaldt L, Kazazian H, Gaensler KML. 1999. Short-term correction of factor VIII deficiency in a murine model of hemophilia A after delivery of adenovirus murine factor VIII in utero. Proc Natl Acad Sci USA. 96:13324–13329.

Lucarelli G, Clift RA, Galimberti M, Angelucci E, Giardini C, Baronciani D, Polchi P, Andreani M, Gaziev D, Erer B, Ciaroni A, D'Adamo F, Albertini F, Muretto P. 1999. Bone marrow transplantation in adult thalassemic patients. Blood. 93:1164–1167.

Luther-Wyrsch A, Costello E, Thali M, Buetti E, Nissen C, Surbek D, Holzgreve W, Gratwohl A, Tichelli A, Wodnar-Filipowicz A. 2001. Stable transduction with lentiviral vectors and amplification of immature hematopoietic progenitors from cord blood of preterm human fetuses. Hum Gene Ther. 12:377–389.

Lutzko C, Kruth S, Abrams-Ogg ACG, Lau K, Li L, Clark BR, Ruedy C, Nanji S, Foster R, Kohn D, Shull R, Dubé ID. 1999. Genetically corrected autologous stem cells engraft, but host immune responses limit their utility in canine α-L-iduronidase deficiency. Blood. 93:1895–1905.

Lutzko C, Omori F, Abrams-Ogg AC, Shull R, Li L, Ruedy C, Nanji S, Gartley C, Dobson H, Foster R, Kruth S, Dube ID. 1999. Gene therapy for canine alpha-L-iduronidase deficiency: In utero adoptive transfer of genetically corrected hematopoietic progenitors results in engraftment but not amelioration of disease. Hum Gene Ther. 10:1521–1532.

Manno CS, Chew AJ, Hutchison S, Larson PJ, Herzog RW, Arruda VR, Tai SJ, Ragni MV, Thompson A, Ozelo M, Couto LB, Leonard DG, Johnson FA, McClelland A, Scallan C, Skarsgard E, Flake AW, Kay MA, High KA, Glader B. 2003. AAV-mediated factor IX gene transfer to skeletal muscle in patients with severe hemophilia B. Blood. 101:2963–2972.

Mason CAE, Bigras J-L, O'Blenes SB, Zhou B, McIntyre B, Nakamura N, Kaneda Y, Rabinovitch M. 1999. Gene transfer in utero biologically engineers a patent ductud arteriosus im lambs by arresting fibronectin-dependent neointimal formation. Nat Med. 5:176–182.

May C, Rivella S, Callegari J, Heller G, Gaensler KM, Luzzato L, Sadelain M. 2000. Therapeutic haemoglobin synthesis in beta-thalassaemic mice expressing lentivirus-encoded human beta-globin. Nature (London). 406:82–86.

Milner R, Shaaban AF, Kim HB, Fichter C, Scully M, Asakura T, Flake AW. 1998. Long term high level expression of donor hemoglobin after in utero HSC transplantation in thalassemic knockout mice—selective advantage for donor red cells. Blood. 92:696a.

Milner R, Shaaban A, Kim HB, Fichter C, Flake AW. 1998a. Natural killer cells do not mediate allograft rejection after in utero hematopoietic stem cell transplantation. Blood. 92:288b.

Miyasaka M, Morris B. 1988. The ontogeny of the lymphoid system and immune responsiveness in sheep. Prog Vet Microbiol Immun. 4:21–55.

Miyoshi H, Smith KA, Mosier DE, Verma IM, Torbett BE. 1999. Transduction of human CD34+ cells that mediate long-term engraftment of NOD/SCID mice by HIV vectors. Science. 283:682–686.

Naldini L, Blomer U, Gallay P, Ory D, Mulligan R, Gage FH, Verma IM, Trono D. 1996. In vivo gene delivery and stable transduction of nondividing cells by a lentiviral vector. Science. 272:263–267.

Noia G et al. 2004. The intracoelomic route: A new approach for in utero human cord blood stem cell transplantation. Fetal Diagn Ther. 19:13–22.

Norbeck O, Tolfvenstam T, Shields LE, Westgren M, Broliden K. 2004 Nov. Parvovirus B19 capsid protein VP2 inhibits hematopoiesis in vitro and in vivo: Implications for therapeutic use. Exp Hematol. 32(11):1082–1087.

Omori F, Lutzko C, Abrams-Ogg A, Lau K, Gartley C, Dobson H, Nanji S, Ruedy C, Singaraja R, Li L, Stewart AK, Kruth S, Dube ID. 1999. Adoptive transfer of genetically modified human hematopoietic stem cells into preimmune canine fetuses. Exp Hematol. 27:242–249.

Peebles D et al. 2004. Widespread and efficient marker gene expression in the airway epithelia of fetal sheep after minimally invasive tracheal application of recombinant adenovirus in utero. Gene Therapy. 11:70–78.

Peranteau WH, Hayashi S, Hsieh M, Shaaban AF, Flake AW. 2002. High-level allogeneic chimerism achieved by prenatal tolerance induction and postnatal nonmyeloablative bone marrow transplantation. Blood. 100:2225–2234.

Peranteau WH, Endo M, Adibe OO, Merchant A, Zoltick PW, Flake AW. 2006. CD26 inhibition enhances allogeneic donor-cell homing and engraftment after in utero hematopoietic-cell transplantation. Blood. 108:4268–4274.

Peranteau WH, Endo M, Adibe OO, Flake AW. 2007. Evidence for an immune barrier after in utero hematopoietic-cell transplantation. Blood. 109:1331–1333.

Persons DA et al. 2001. Functional requirements for phenotypic correction of murine beta-thalassemia: Implications for human gene therapy. Blood. 97:3275–3282.

Petersen SM, Gendelman M, Murphy KM, Torbenson M, Jones RJ, Althaus JE, Stetten G, Bird C, Blakemore KJ. 2007. Use of T-cell antibodies for donor dosaging in a canine model of in utero hematopoietic stem cell transplantation. Fetal Diagn Ther. 22(3):175–179.

Piacibello W, Sanavio F, Garetto L, Severino A, Bergandi D, Ferrario J, Fagioli F, Berger M, Aglietta M. 1997. Extensive Extensive amplification and self-renewal of human primitive hematopoietic stem cells from cord blood. Blood. 89:2644–2653.

Pirovano S et al. 2004. Reconstitution of T-cell compartment after in utero stem cell transplantation: Analysis of T-cell repertoire and thymic output. Haematologica. 89:450–461.

Porada CD, Tran N, Eglitis M, Moen RC, Troutman L, Flake AW, Zhao Y, Anderson WF, Zanjani ED. 1998. In utero gene therapy: Transfer and longterm expression of the bacterial neo-r gene in sheep after direct injection of retroviral vectors into the preimmune fetus. Hum Gene Ther. 9:1571–1585.

Portmann-Lanz B, Schoeberlein A, Huber A, Sager R, Malek A, Holzgreve W, Surbek DV. 2006. Placental mesenchymal stem cells as potential autologous graft for pre- and perinatal neuroregeneration. Am J Obstet Gynecol. 194:664–673.

Portmann-Lanz B, Surbek D. 2007. Placental mesenchymal stem cells. Am J Obstet Gynecol. 196:e18–e9.

Qin L, Ding Y, Pahud DR, Chang E, Imperiale MJ, Bromberg JS. 1997. Promotor attenuation in gene therapy: Interferon-γ and tumor necrosis factor-α inhibit transgene expression. Hum Gene Ther. 6:1039.

Rucker M et al. 2004. Rescue of enzyme deficiency in embryonic diaphragm in a mouse model of metabolic myopathy: Pompe disease. Development. 131:3007–3019.

Sanna MA et al. 1999. In utero stem cell transplantation for beta-thalassemia: A case-report. Bone Marrow Transplant. 23:S109.

Santolaya-Forgas J, Galan I, Deleon-Luis J, Wolf R. 2007. A study to determine if human umbilical cord hematopoietic stem cells can survive in baboon extra-embryonic celomic fluid: A prerequisite for determining the feasibility of in-utero stem cell xeno-transplantation via celocentesis. Fetal Diagn Ther. 22(2):131–135.

Schachtner S, Buck C, Bergelson J, Baldwin H. 1999. Temporally regulated expression patterns following in utero adenovirus-mediated gene transfer. Gene Ther. 6:1249–1257.

Schneider H, Coutelle C. 1999. In utero gene therapy—The case for. Nat Med. 5:256–257.

Schoeberlein A, Holzgreve W, Dudler L, Hahn S, Surbek DV. 2004. In utero transplantation of autologous and allogeneic fetal liver stem cells in ovine fetuses. Am J Obstet Gynecol. 191:1030–1036.

Schoeberlein A, Schatt S, Troeger C, Surbek D, Holzgreve W, Hahn S. 2004. Engraftment kinetics of human cord blood and murine fetal liver stem cells following in utero transplantation into immunodeficient mice. Stem Cells Dev. 13:677–684.

Schoeberlein A, Holzgreve W, Dudler L, Hahn S, Surbek DV. 2005. Tissue-specific engraftment after in utero transplantation of mesenchymal stem cells into sheep fetuses. Am J Obstet Gynecol. 192:

Sekhon HS, Larson JE. 1995. In utero gene transfer into the pulmonary epithelium. Nat Med. 1:1201–1203.

Senut M-C, Suhr S, Gage FH. 1998. Gene transfer to the rodent placenta in situ. A new strategy for delivering gene products to the fetus. J Clin Invest. 101:1565–1571.

Seppen J et al. 2003. Long-term correction of bilirubin UDPglucuronyltransferase deficiency in rats by in utero lentiviral gene transfer. Mol Ther. 8:593–599.

Shaaban AF, Kim HB, Gaur L, Liechty KW, Flake AW. 2006. Prenatal transplantation of cytokine-stimulated marrow improves early chimerism in a resistant strain combination but results in poor long-term engraftment. Exp Hematol. 34:1278–1287.

Shapiro E et al. 2000. Long-term effect of bone-marrow transplantation for childhood-onset cerebral X-linked adreno-leukoldystrophy. Lancet. 356:713–718.

Shields LE, Bryant EM, Easterling TR, Andrews RG. 1995. Fetal liver cell transplantation for the creation of lymphohematopoietic chimerism in fetal baboons. Am J Obstet Gynecol. 173:1157–1160.

Shields LE, Gaur LK, Gough M, Potter J, Sieverkropp A, Andrews RG. 2003. In utero hematopoietic stem cell transplantation in nonhuman primates: The role of T cells. Stem Cells. 21:304–314.

Shields LE, Gaur L, Delio P, Potter J, Sieverkropp A, Andrews RG. 2004. Fetal immune suppression as adjunctive therapy for in utero hematopoietic stem cell transplantation in nonhuman primates. Stem Cells. 22:759–769.

Snyder RO, Miao C, Meuse L, Tubb J, Donahue BA, Lin H-F, Stafford DW, Patel S, Thompson AR, Nichols TN, Read MS, Bellinger DA, Brinkhous KM, Kay MA. 1999. Correction of hemophilia B in canine and murine models using recombinant adeno-associated viral vectors. Nat Med. 5:64–70.

Surbek DV, Gratwohl A, Holzgreve W. 1999. In utero hematopoietic stem cell transfer: Current status and future strategies. Eur J Obstet Gynecol Reprod Biol. 85:109–115.

Surbek DV, Tercanli S, Holzgreve W. 2000. Transabdominal first trimester embryofetoscopy as potential approach to early in utero stem cell transplantation and gene therapy. Ultrasound Obstet Gynecol. 15:302–307.

Surbek DV, Holzgreve W, Nicolaides KH. 2001. Hematopoietic stem cell transplantation and gene therapy in the fetus: Ready for clinical use? Hum Reprod Update. 7:85–91.

Surbek DV, Holzgreve W. Why do we need non-invasive prenatal diagnosis? In: Hahn S., Holzgreve, W. (Eds.) Fetal cells and fetal DNA in maternal blood. New developments for a new millenium, pp. 21–27, 2001, Karger, Basel-New York.

Surbek DV, Young A, Danzer E, Schoeberlein A, Holzgreve W. 2002. Ultrasound-guided stem cell sampling from the early ovine fetus for prenatal ex-vivo gene therapy. Am J Obstet Gynecol. 187:960–963.

Surbek DV, Schoeberlein A, Portmann-Lanz B. 2007. Perinatology. 8:249–262.

Sutton RE, Wu HTM, Rigg R, Böhnlein E, Brown P. 1998. Human immunodeficiency virus type 1 vectors efficiently transduce human hematopoietic stem cells. J Virol. 72:5781–5788.

Tavian M, Coulombel L, Luton D, San Clemente H, Dieterlen-Lièvre F, Péault B. 1996. Aorta-associated CD34+ hematopoietic cells in the early human embryo. Blood. 87:67–72.

Themis M, Schneider H, Kiserud T, Cook T, Adebakin S, Jezzard S, Forbes S, Hanson M, Pavirani A, Rodeck C, Coutelle C. 1999. Successful expression of beta-galactosidase and factor IX transgenes in fetal and neonatal sheep after ultrasound-guided percutaneous adenovirus vector administration into the umbilical vein. Gene Ther. 6:1239–1248.

Themis M et al. 2005. Oncogenesis following delivery of a non-primate lentiviral gene therapy vector to fetal mice. Mol Ther. 12:257–266.

Touraine JL et al. 2004. Reappraisal of in utero stem cell transplantation based on long-term results. Fetal Diagn Ther. 19:305–312.

Tran ND, Porada CD, Zhao Y, Almeida-Porada G, Anderson WF, Zanjani ED. 2000. In utero transfer and expression of exogenous genes in sheep. Exp Hematol. 28:17–30.

Troeger C, Surbek D, Schoeberlein A, Schatt S, Dudler L, Hahn S, Holzgreve W. 2007. In utero hematopoietic stem cell transplantation—experiences in mice, sheep and humans. Swiss Med Wkly. 136:498–503.

Tsukamoto M, Ochiya T, Yoshida S, Sugimura T, Terada M. 1995. Gene transfer and expression in progeny after intravenous DNA injection into pregnant mice. Nat Genet. 9:243–248.

Turner CW, Archer DR, Wong J, Yeager AM, Fleming WH. 1998. In utero transplantation of human fetal haemopoietic cells in NOD/SCID mice. Br J Haematol. 103:326–334.

Uchida N, Sutton RE, Friera AM, He D, Reitsma MJ, Chang WC, Veres G, Scollay R, Weissmann IL. 1998. HIV, but not murine leukemia virus, vectors mediate high efficiency gene transfer into freshly isolated G0/G1 human hematopoietic stem cells. Proc Natl Acad Sci USA. 95:11939–11944.

Verma IM, Somia N. 1997. Gene therapy—promises, problems and prospects. Nature (London). 389:239–242.

Waddington S et al. 2003. In utero gene transfer of human factor IX to fetal mice can induce tolerance of the exogenous clotting factor. Blood. 101:1359–1366.

Waddington S et al. 2004. Permanent phenotypic correction of haemophilia B in immunocompetent mice by prenatal gene therapy. Blood. 104:2714–2721.

Wang G, Williamson R, Mueller G, Thomas P, Davidson BL, McCray PB. 1998. Ultrasound-guided gene transfer to hepatocytes in utero. Fetal Diagn Ther. 13:197–205.

Westgren M et al. 1996. Lack of evidence of permanent engraftment after in utero fetal stem cell transplantation in congenital hemoglobinopathies. Transplantation. 61:1176–1179.

Winkler A, Kiem H-P, Shields L, Sun Q-H, Andrews RG. 1999. Gene transfer into fetal baboon hematopoietic progenitor cells. Hum Gene Ther. 10:667–677.

Woo YJ, Raju GP, Swain JL, Richmond ME, Gardner TJ, Balice-Gordon RJ. 1997. In utero cardiac gene transfer via intraplacental delivery of recombinant adenovirus. Circulation. 96:3561–3569.

Wu Y, Liu J, Woo S, Finegold MJ, Brandt ML. 1999. Prenatal orogastric gene delivery results in transduction of the small bowel in the fetal rabbit. Fetal Diagn Ther. 14:323–327.

Wyrsch A, Dalle Carbonare V, Jansen W, Chklovskaia E, Nissen C, Surbek D, Holzgreve W, Tichelli A, Wodnar-Filipowicz A. 1999. Preterm umbilical cord blood is enriched in early hematopoietic progenitors with expansion properties equal to term cord blood progenitors. Exp Hematol. 27:1338–1345.

Ye X, Gao GP, Pabin C, Raper SE, Wilson JM. 1998. Evaluating the potential of germ line transmission after intravenous administration of recombinant adenovirus in the C3H mouse. Hum Gene Ther. 9:2135–2142.

Young A et al. 2003. Engraftment of human cord blood-derived stem cells in preimmune ovine fetuses after ultrasound-guided in utero transplantation. Am J Obstet Gynecol. 189:698–701.

Zanjani ED, Flake AW, Rice H, Hedrick M, Tavassoli M. 1994. Long-term repopulating ability of xenogeneic transplanted human liver hematopoietic stem cells in sheep. J Clin Invest. 93:1051.

Zanjani ED, Anderson WF. 1999. Prospects for in utero human gene therapy. Science. 285:2084–2088.

Ziegler BL, Valtieri M, Porada GA, De Maria R, Muller R, Masella B, Gabbianelli M, Casella I, Pelosi E, Bock T, Zanjani ED, Peschle C. 1999. KDR receptor: A key marker defining hematopoietic stem cells. Science. 285:1553–1558.

Zufferey R, Nagy D, Mandel RJ, Naldini L, Trono D. 1997. Multiply attenuated lentiviral vector achieves efficient gene delivery in vivo. Nat Biotechnol. 15:871–875.

5

UMBILICAL CORD MESENCHYMAL STEM CELLS

Laurent Boissel, Monica Betancur, and Hans Klingemann

Tufts Medical Center, Molecular Oncology Research Institute, Division of Hematology and Oncology, Boston, MA, 02111

James Marchand

Tufts University, Department of Anesthesia Research and Pharmacology, Boston, MA, 02111

PREPARATION AND CULTURE OF UC–MSC

Various methods of isolation have been described to obtain mesenchymal stem cells (MSC) from umbilical cords [Can and Karahuseyinoglu, 2007]. The starting material differs between the reports, which implies that the isolated cells, although generally showing the properties and characteristics of MSC, are probably different cell subtypes. Most of the described methods share a common step, which is the manual separation of the tissue of interest within the umbilical cord, followed or not by enzymatic digestion of the extracellular matrix to release individual cells.

Whether obtained through cesarean section or natural delivery, care should be taken to remove contaminating blood from the umbilical cord. Cord blood itself can be a source of MSC, but the rate of recovery is highly variable and generally low [Secco, 2008]. The blood vessels present within the cord are another source of MSC [Romanov, 2003]. Sarugaser and collaborators [Sarugaser, 2005] described an isolation method where the vessels are stripped from the surrounding tissue (Wharton's Jelly) and incubated with collagenase to obtain "perivascular" cells. These cells have MSC characteristics when using criteria like plastic adherence, surface markers expression, and ability to transdifferentiate, but they may be considered different from the MSC present in the Wharton's Jelly as they express 3G5 [Sarugaser, 2005] and CD146 markers [Baksh, 2007]. In addition, even

Perinatal Stem Cells. Edited by C. L. Cetrulo, K. J. Cetrulo, and C. L. Cetrulo, Jr.

though the cord blood vessels can be removed from the Wharton's Jelly relatively easily, it is very likely that some of the Jelly "clings" to the vessels and thus "contaminates" the preparation with a different cell type. Conversely, Wang and collaborators [Wang, 2004] described a method where the blood vessels are removed before the Wharton's Jelly is scraped from the inside of the cord. The isolated Jelly is then incubated sequentially with collagenase and trypsin. Although this method presumably yields "pure" Jelly cells, one cannot rule out that the scraping of the inside of the amnionic membrane causes a contamination with subamnionic cells. Weiss and collaborators [Weiss, 2006] described a method for obtaining cells from the Wharton's Jelly without contaminating perivascular or endothelial cells. After removing the vessels, the cord is cut in segments that are incubated with a cocktail of hyaluronidase, collagenase, and trypsin.

Other laboratories report isolation protocols that do not involve enzymatic digestions. Mitchell et al. described a method where the umbilical cord is also cut in segments and vessels are removed, but the remaining tissue (Wharton's Jelly and amnionic membrane) is cut into small pieces and directly placed in the culture dish [Mitchell, 2003].

After trying several other methods, our group developed a simple isolation method that allows us to obtain MSC-like cells from the Wharton's Jelly without enzymatic treatment. The whole cord is cut in segments of $\sim$1 cm in length, which are further minced into submillimeter-sized particles and placed directly in medium [Friedman, 2007]. We did not notice any significant difference in the ability of various segments of the cord (proximal, median, and distal to the fetus) to yield MSC populations.

An important aspect of this method is that umbilical cord (UC) microparticles preparation can be frozen directly, in autologous cord plasma with 10% dimethyl sulfoxide (DMSO). For cord blood banking purposes, it is important to be able to process and freeze the incoming cord blood and UC derived MSC on the same day without the need of prior culture. If UC–MSC are cultured before freezing, the cells are considered manipulated and the process is regulated differently by the Food and Drug Administration (FDA). Further, the use of heat-inactivated *autologous* plasma in the cryoprotectant avoids exposure to allogeneic human serum, which can alter the antigen expression profile of MSC [Dimarakis and Levicar, 2006; Shahdadfar, 2005]. Furthermore, it avoids the requirement of fetal calf serum (FCS) or other animal-derived serum, which can be contaminated with viruses or prions. Umbilical cord samples frozen in this way readily yield colony-forming MSC when thawed and put in culture dishes [Friedman, 2007].

When small pieces of the cord are placed on a culture dish, the UC–MSC become plastic-adherent usually within 24 h, and form fibroblast-like colonies within 8–10 days. We observed that subsequent growth was enhanced when these first CFU-F were allowed to reach near-confluence (unpublished observation). Most of the available protocols describe culture of UC–MSC in DMEM medium supplemented with 10% FCS, with or without antibiotics or other additions [Lu, 2006; Mitchell, 2003; Sarugaser, 2005; Secco, 2008; Wang, 2004; Weiss, 2006]. Because of the potential clinical uses, we intended to test whether MSC grow (without changing their phenotype and activity) under conditions that comply with good manufacturing practices (GMP). Hence, we were able to substitute RPMI 1640 or DMEM with X-VIVO 10 medium, without changing the MSCs growth or characteristics. However, when trying to replace FCS with human serum (HS), we noticed that UC–MSC adopted different morphology and growth characteristics, spontaneously forming clumps of cells scattered in the dish after 2–3 days in culture (Fig. V.1). When these clumps were harvested and replated in culture media supplemented with FCS, the cells spread out and formed a monolayer again.

At this point FCS seems to be essential for the optimal culture expansion of UC–MSC. The dependence on FCS has also been reported for BM–MSC, which are already used

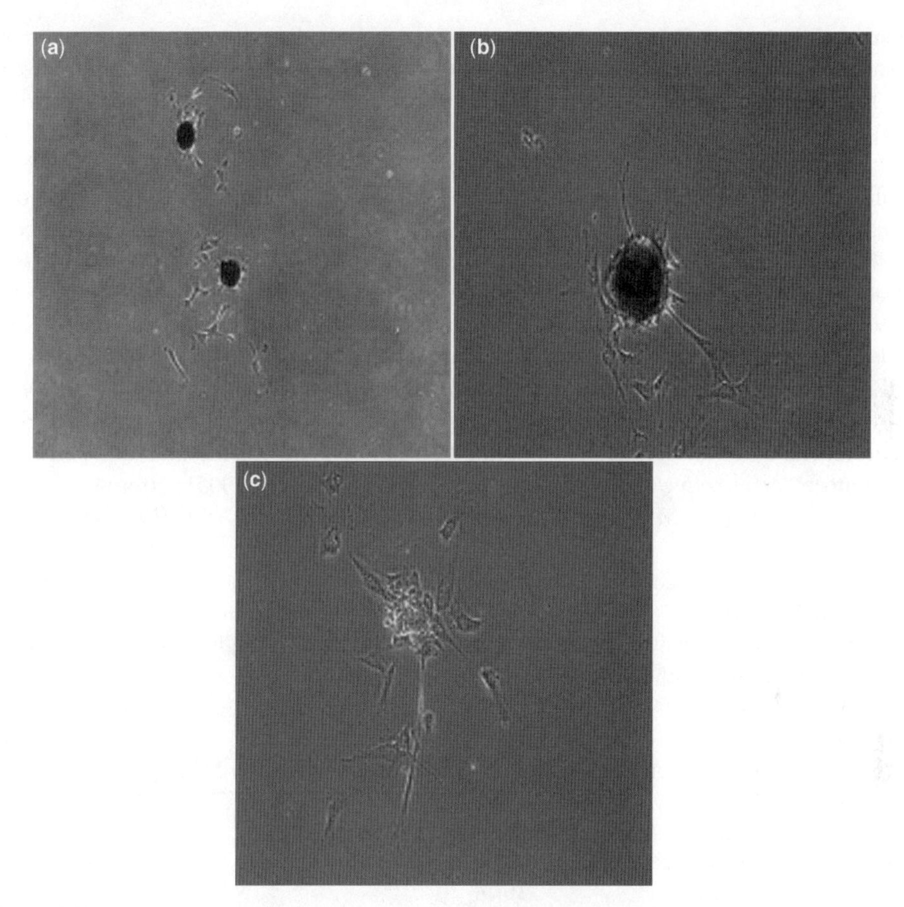

Figure V.1. Morphology of UC–MSC in various media. Morphology of UC–MSC cultured in 20% HS–RPMI 1640 medium. The UC–MSC tend to form scattered clumps with fewer plastic-adherent isolated cells. (a) Magnification ×20, (b) magnification ×100. Individual clumps spread out again when transferred into 20% FCS-RPMI 1640 medium (c) magnification ×100. (See color insert.)

clinically. Unfortunately, until equally effective serum replacements are available, FCS serum will have to be included in culture media. The FDA requires, however, that the final product before infusion into the patient contains only negligible amounts of FCS. Although 10% HS gave sufficient growth–expansion, 20% HS was clearly superior. In contrast, autologous cord plasma (obtained from the cord donor) did not support growth and expansion [Friedman, 2007].

Another important variable to consider for optimization of culture conditions is the source of the plastic container, as reported by Sotiropoulou et al. for BM–MSC [Sotiropoulou, 2006b]. We observed that UC–MSC have a similar dependence to specific plastic ware (unpublished observation).

CHARACTERISTICS–PROPERTIES AND BONE MARROW COMPARISON

As stated before, MSC can be found in different tissues and although they are believed to represent a homogeneous cell type, it is becoming clear that they differ when additional biomarkers are tested (e.g., cytokine–chemokine production or gene expression) [Panepucci, 2004]. When comparing UC–MSC with BM–MSC with respect to cytokine production, we have found that they have a very different secretion profile, as summarized in Table V.1. The UC–MSC produce significantly more of the G-CSF, GM–CSF, HGF, and LIF growth factors than BM–MSC, while BM–MSC produce more VEGF and SDF1β. Both cell types produce similar amounts of TGFβ and TPO. The UC–MSC secrete considerably more IL-8 and IL-11 than BM–MSC, as well as more IL-1a and IL-6. Secretion of IFNα, IFNγ, IL-7, IL-15, and IL-18 are similarly very low in the two populations. Note that both MSC types do not secrete any TNF-α, IL-2, IL-12, or IL-15.

Microarray analysis (Genechip–Affymetrix) provides relative quantitation of RNA expression of virtually the whole genome. It provides a powerful tool for comparing different cell types by simultaneously measuring expression of hundreds of intra- and extracellular markers. However, it can only provide a snapshot of the cell population state in a given environment, and special attention should be given to culture conditions if comparison of different cell types is intended. Wagner and co-workers [Wagner, 2005] performed a microarray analysis of three MSC types isolated from BM, UCB, and AT (adipose tissue) and found many differences between the profiles of each MSC type. However, MSC of all origins do share some common signature in transcription regulation in response to some changes in conditions.

Our group compared the transcription profiles of UC–MSC and BM–MSC cultured in the same conditions for the same amount of time. Expression profiles of two UC–MSC cell

TABLE V.1. Comparison of Cytokines Production by MSC[a]

Cytokine	UC–MSC	BM–MSC
GCSF	1844 ± 881	N.D.
GM–CSF	107 ± 20	N.D.
TPO	82.5 ± 4.5	77 ± 3
VEGF	120 ± 5	2731 ± 99
SDF1β	403 ± 53	2342 ± 178
TGFβ	3812 ± 585	2226 ± 31
HGF	12596 ± 7101	402 ± 203
LIF	152 ± 35	41 ± 2.5
TNF-α	N.D.	N.D.
IFN-α	3.9 ± 0.5	2.3 ± 0.2
IFN-γ	1.5 ± 0.3	0.2 ± 0.1
IL-1a	12 ± 4	N.D.
IL-2	N.D.	N.D.
IL-6	1570 ± 617	704 ± 51
IL-7	6.8 ± 1.0	7.9 ± 0.8
IL-8	10765 ± 5383	216 ± 27
IL-11	1472 ± 625	70 ± 1
IL-12	N.D.	N.D.
IL-15	4.4 ± 0.3	4.3 ± 0.1
IL-18	20 ± 1.2	17 ± 0.1

[a]Cytokine production by UC–MSC compared to BM–MSC as measured by enzyme-linked immunosorbent assay (ELISA). The MSC were allowed to grow to confluence for 7 days in 20% FBS/RPMI. Medium was changed 24 h prior to collection of the supernatant. All values represent the mean of four experiments ± SEM in pg/mL. N.D. = not detectable.

lines (passages 4 and 6) and two BM–MSC cell lines (passages 4 and 6) were compared, using the Affymetrix Human Gene 1.0 ST array. There were significant differences ($p < 0.05$, adjusted for multiple comparisons) in the expression profiles between UC–MSC and BM–MSC.

Of the top significantly upregulated genes in UC–MSC, the gene SLC7A2, a member of the solute carrier family 7, exhibited the greatest increase in expression versus BM–MSC, up 42-fold. This gene is a cationic amino acid transporter, also known as CAT-2A and functions to transport arginine into the cell. Numerous physiological processes are dependent on arginine transport into the cell, including protein synthesis, nitric oxide synthesis, and creatine and agmatine synthesis [Closs, 2006].

Another gene with increased expression in UC–MSC versus BM–MSC (up 34-fold) is aldehyde dehydrogenase (ALDH1A1), which catalyzes conversion of retinol to retinoic acid. Aldehyde dehydrogenase is highly enriched in hematopoietic stem cells (HSC) and, through its ability to synthesize retinoic acid, regulates differentiation of human HSC [Chute, 2006]. Two other upregulated genes, desmocollin 3 and desmoglein 2, are components of desmosomes, and more generally genes associated with cell–cell adhesion [Bonne, 2003].

In order to determine if these genomewide changes in UC–MSC gene expression are associated with global categories of gene function, an analysis of enrichment in Gene Ontology (GO) categories was performed using GOMiner (Genomics and Bioinformatics Group, National Cancer Institute [Zeeberg, 2003]). Gene ontology category cellular functions overrepresented in the significantly changed genes list include developmental processes (system development, anatomical structure development, organ development), cell signaling processes (response to external stimulus, response to chemical stimulus, cell–cell signaling, signal transduction), cell adhesion, cell migration and chemotaxis, and immune system functions (response to wounding, immune response, inflammatory response). Overall, these enriched categories give a picture of UC–MSC as preferentially involved in developmental processes, cellular motility, and immune functions.

UC–MSC AS FEEDER LAYER FOR EXPANSION OF HUMAN CELLS

One of the elusive goals in hematology is to recreate bone marrow microenvironment *in vitro* in order to allow the long-term *ex vivo* expansion of nondifferentiated, pluripotent hematopoietic stem cells (HSC) [Devine, 2003]. The BM–MSC secrete a number of cytokines and growth factors that support hematopoiesis, therefore investigators have used them as a feeder layer for HSC, alone or in combination with specific cytokines [Kogler, 2005; Sotiropoulou, 2006a]. Since UC–MSC produce many of the factors that are secreted by BM–MSC and also produce cytokines–chemokines that specifically support HSC, they may be considered as an alternative to BM–MSC. In addition, UC–MSC are more easily accessible than BM–MSC, are not subject to donor age-related senescence as BM [Stenderup, 2003], and the number of cells obtained from one cord is many times higher than from a BM sample. Lu et al. compared the hematopoiesis supportive functions of irradiated MSC from BM and UC and found similar results for both sources [Lu, 2006]. Friedman et al. also reported that a feeder layer of irradiated UC–MSC increases proliferation of CD34+ progenitor cells, and that it supports expansion of myeloid progenitor cells [Friedman, 2007]. Based on these observations, it was reasoned that a UC–MSC feeder layer could also support the expansion of immune cells [Condiotti, 2001; Gada, 2006]. Our group [Boissel, submitted for publication] explored the potential of UC–MSC for supporting proliferation of cord blood-derived NK cells. The rationale for this approach is that patients who relapse after cord blood transplant cannot be given donor lymphocyte infusion (DLI) because of the limited number of cells in a single unit. It was therefore desirable to develop an expansion system for NK cells, especially since it has been shown that cord blood is rich in NK cell progenitors [Dalle, 2005]. A feeder layer of irradiated UC–MSC alone increased proliferation of NK cells from cord blood ~sixfold, while the addition of a combination of the cytokines IL-2, IL-15, IL-3, and Flt-3 further increases proliferation to >60-fold. Contact between NK cells and feeder layer was necessary for their maximum expansion (Fig. V.2).

Interestingly, irradiated UC–MSC disappear during the first week of NK cells expansion, probably due to the cytolytic activity of the allogeneic NK cells [Boissel, submitted for publication; Spaggiari, 2006]. This raises the question of the immunogenicity of MSC and further studies are needed to determine whether MSC can also be the target of an NK response *in vivo* and whether this can affect the prospects of MSC therapies.

The types of secreted factors produced by UC–MSC make them useful for growing other, non-hematopoietic cell types. Saito and co-workers [Saito, 2002] reported that MSC derived from bovine UC were successful in helping the *ex vivo* expansion of equine embryonic stem (ES) cells, as well as keep them in an undifferentiated state in long-term cultures. The successful use of MSC as a supporting feeder layer was also reported for *ex vivo* growth of equine spermatogonia without differentiation [Marc Weiss, personal communication].

Although the use of MSC as feeder cells is a very promising field, the transition to actual clinical applications is still hindered by the fact that current culture conditions require the addition of serum (usually of animal origin) and/or cytokines, which are expensive and not always available in clinical grade quality, to the growth media. One way to overcome these problems is to genetically engineer the MSC, to make them produce the required factors. Several groups have shown that MSC can be successfully transduced [Baksh, 2007; Rachakatla, 2007; Studeny, 2004], and Friedman et al. reported that UC–MSC can be efficiently electroporated with a GFP bearing plasmid [Friedman, 2007]. This strategy could make UC–MSC customized feeders for the large-scale expansion of different types of human cells. It would also allow the differentiation of HSC into committed cells, such

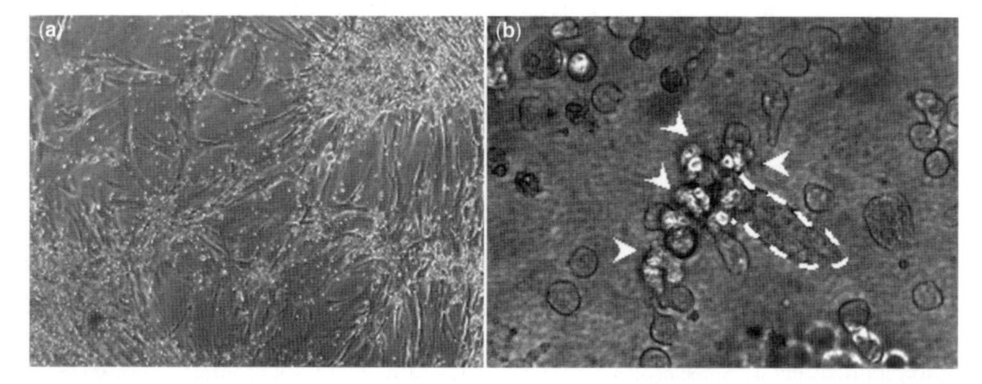

Figure V.2. The NK cells cluster around UC–MSC. The NK cells tend to grow in clusters in proximity with UC–MSC. (a) The UC–MSC (elongated fibroblast-like cells) with cord blood NK cells (small white dots), magnification ×100. (b) The UC–MSC (white outline) with NK-92 cells (white arrowheads), magnification ×400.

as platelets and erythrocytes, with important applications for blood transfusion. Methods are already available to enzymatically "clip" the AB antigens from red blood cells and obtain the universal red cell without the need for ABO matching.

CLINICAL INDICATIONS FOR UC–MSC

Bone marrow derived MSC (BM–MSC) is already used in many clinical trials (Osiris Inc., Columbia, MD—various press releases). The most advanced studies are performed in patients with acute graft versus host disease (GvHD) after allogeneic bone marrow–stem cell transplant for a hematological disease. This complication affects 50–70% of all transplanted patients and contributes to significant morbidity and mortality after a hematopoietic stem cell transplant. Affected organs include the skin, the gastrointestinal tract, and the liver. A phase 2 study using BM–MSC in advanced GvHD was just completed in Europe suggesting a significant benefit for patients who received treatment with MSC [Le Blanc, 2004, 2008]. Notably, the infusion of BM–MSC that were derived from unrelated, non-MHC matched volunteer donors did not cause any infusional (early) or late side effects.

Two multicenter randomized, placebo controlled trials are ongoing in the United States, Australia, and Europe: (1) for patients with severe GvHD who have failed steroid treatments and (2) for patients with newly diagnosed GvHD who will receive the cells in addition to steroids versus a group that will receive steroid therapy only. The rationale for these studies in GvHD is based on the preclinical observation that MSC can down regulate the immune response in addition to localizing preferentially to sites of inflammation [Khaldoyanidi, 2008]. The immunomodulatory effects of MSC are also studied in patients with Crohn's Disease and type I diabetes. Initial phase 1–2 studies in Crohn's disease showed a clear improvement in the activity index in patients who had failed previous treatments.

Making use of the ability of MSC to localize to sites of tissue damage and inflammation, BM–MSC were given to patients early after a myocardial infarction. The cells were injected into the coronary arteries and follow-up studies showed an improvement in the cardiac ejection fraction, as well as a lower incidences of arrhythmias in the patient group that received MSC. Studies are also underway in patients with type I diabetes, chronic lung disease (Chronic Obstructive Pulmonary Disease) and for local treatment of osteoarthritis.

Although UC–MSC have not yet entered clinical trials, there are clear advantages of UC–MSC over BM–MSC: Obtaining BM–MSC involves a painful bone marrow aspirate whereas the umbilical cord constitutes a waste product, and a relatively high number of cells can be obtained by simple preparation of the cord as described earlier in this chapter. The UC–MSC also represent "early" MSC, which are considered superior to cells obtained from more mature tissue like bone marrow, as they can undergo a significantly higher number of cell division before senescence. Considering all those advantages of UC–MSC, it is quite likely that UC–MSC will enter clinical trials soon and may be used for the same or similar indications as BM–MSC.

REFERENCES

Baksh D, Yao R, Tuan RS. 2007. Comparison of proliferative and multilineage differentiation potential of human mesenchymal stem cells derived from umbilical cord and bone marrow. Stem Cells. 25(6):1384–1392.

Boissel L, Tuncer H, Betancur M, Wolfberg A, Klingemann H. 2008. Umbilical Cord Mesenchymal. Stem Cells Increase Expansion of Cord Blood Natural Killer Cells. Biol Blood Marrow Transplant. 14:1031–1038.

Bonne S, Gilbert B, Hatzfeld M, Chen X, Green KJ, van Roy F. 2003. Defining desmosomal plakophilin-3 interactions. J Cell Biol. 161(2):403–416.

Can A, Karahuseyinoglu S. 2007. Concise review: Human umbilical cord stroma with regard to the source of fetus-derived stem cells. Stem Cells. 25(11):2886–2895.

Chute JP et al. 2006. Inhibition of aldehyde dehydrogenase and retinoid signaling induces the expansion of human hematopoietic stem cells. Proc Natl Acad Sci USA. 103(31):11707–11712.

Closs EI, Boissel JP, Habermeier A, Rotmann A. 2006. Structure and function of cationic amino acid transporters (CATs). J Membr Biol. 213(2):67–77.

Condiotti R, Zakai YB, Barak V, Nagler A. 2001. Ex vivo expansion of CD56+ cytotoxic cells from human umbilical cord blood. Exp Hematol. 29(1):104–113.

Dalle JH et al. 2005. Characterization of cord blood natural killer cells: Implications for transplantation and neonatal infections. Pediatr Res. 57(5 Pt 1):649–655.

Devine SM, Lazarus HM, Emerson SG. 2003. Clinical application of hematopoietic progenitor cell expansion: Current status and future prospects. Bone Marrow Transplant. 31(4):241–252.

Dimarakis I, Levicar N. 2006. Cell culture medium composition and translational adult bone marrow-derived stem cell research. Stem Cells. 24(5):1407–1408.

Friedman R, Betancur M, Boissel L, Tuncer H, Cetrulo C, Klingemann H. 2007. Umbilical cord mesenchymal stem cells: Adjuvants for human cell transplantation. Biol Blood Marrow Transplant. 13(12):1477–1486.

Gada P, Gleason M, McCullar V, McGlave PB, Miller JS. 2006. Optimal NK Cell Expansion Depends on Accessory Cells, Synergy between Physiologic Concentrations of IL-2 and IL-15, and Umbilical Cord Blood (UCB) NK Cell Precursors Expand Better Than Adult NK Cells. ASH Annual Meeting Abstracts. 108(11):1040a.

Khaldoyanidi S. 2008. Directing stem cell homing. Cell Stem Cell. 2(3):198–200.

Kogler G et al. 2005. Cytokine production and hematopoiesis supporting activity of cord blood-derived unrestricted somatic stem cells. Exp Hematol. 33(5):573–583.

Le Blanc K et al. 2004. Treatment of severe acute graft-versus-host disease with third party haploidentical mesenchymal stem cells. Lancet. 363(9419):1439–1441.

Le Blanc K et al. 2008. Mesenchymal stem cells for treatment of steroid-resistant, severe, acute graft-versus-host disease: A phase II study. Lancet. 371(9624):1579–1586.

Lu LL et al. 2006. Isolation and characterization of human umbilical cord mesenchymal stem cells with hematopoiesis-supportive function and other potentials. Haematologica. 91(8):1017–1026.

Mitchell KE et al. 2003. Matrix cells from Wharton's jelly form neurons and glia. Stem Cells. 21(1):50–60.

Panepucci RA et al. 2004. Comparison of gene expression of umbilical cord vein and bone marrow-derived mesenchymal stem cells. Stem Cells. 22(7):1263–1278.

Rachakatla RS, Marini F, Weiss ML, Tamura M, Troyer D. 2007. Development of human umbilical cord matrix stem cell-based gene therapy for experimental lung tumors. Cancer Gene Ther. 14(10):828–835.

Romanov YA, Svintsitskaya VA, Smirnov VN. 2003. Searching for alternative sources of postnatal human mesenchymal stem cells: Candidate MSC-like cells from umbilical cord. Stem Cells. 21(1):105–110.

Saito S et al. 2002. Isolation of embryonic stem-like cells from equine blastocysts and their differentiation in vitro. FEBS Lett. 531(3):389–396.

Sarugaser R, Lickorish D, Baksh D, Hosseini MM, Davies JE. 2005. Human umbilical cord perivascular (HUCPV) cells: A source of mesenchymal progenitors. Stem Cells. 23(2):220–229.

Secco M et al. 2008. Multipotent stem cells from umbilical cord: Cord is richer than blood! Stem Cells. 26(1):146–150.

Shahdadfar A, Fronsdal K, Haug T, Reinholt FP, Brinchmann JE. 2005. In vitro expansion of human mesenchymal stem cells: Choice of serum is a determinant of cell proliferation, differentiation, gene expression, and transcriptome stability. Stem Cells. 23(9):1357–1366.

Sotiropoulou PA, Perez SA, Salagianni M, Baxevanis CN, Papamichail M. 2006a. Characterization of the optimal culture conditions for clinical scale production of human mesenchymal stem cells. Stem Cells. 24(2):462–471.

Sotiropoulou PA, Perez SA, Gritzapis AD, Baxevanis CN, Papamichail M. 2006b. Interactions between human mesenchymal stem cells and natural killer cells. Stem Cells. 24(1):74–85.

Spaggiari GM, Capobianco A, Becchetti S, Mingari MC, Moretta L. 2006. Mesenchymal stem cell-natural killer cell interactions: Evidence that activated NK cells are capable of killing MSCs, whereas MSCs can inhibit IL-2-induced NK-cell proliferation. Blood. 107(4):1484–1490.

Stenderup K, Justesen J, Clausen C, Kassem M. 2003. Aging is associated with decreased maximal life span and accelerated senescence of bone marrow stromal cells. Bone. 33(6): 919–926.

Studeny M et al. 2004. Mesenchymal stem cells: Potential precursors for tumor stroma and targeted-delivery vehicles for anticancer agents. J Natl Cancer Inst. 96(21):1593–1603.

Wagner W et al. 2005. Comparative characteristics of mesenchymal stem cells from human bone marrow, adipose tissue, and umbilical cord blood. Exp Hematol. 33(11):1402–1416.

Wang HS et al. 2004. Mesenchymal stem cells in the Wharton's jelly of the human umbilical cord. Stem Cells. 22(7):1330–1337.

Weiss ML et al. 2006. Human umbilical cord matrix stem cells: Preliminary characterization and effect of transplantation in a rodent model of Parkinson's disease. Stem Cells. 24(3):781–792.

Zeeberg BR et al. 2003. GoMiner: A resource for biological interpretation of genomic and proteomic data. Genome Biol. 4(4):R28.

6

WHARTON'S JELLY-DERIVED MESENCHYMAL STROMAL CELLS

Barbara Lutjemeier, Deryl L. Troyer, and Mark L. Weiss

Department of Anatomy and Physiology, Kansas State University, Manhattan, KS 66506

MESENCHYMAL STROMAL CELLS

Friedenstein first demonstrated that plastic-adherent cells isolated from the stroma of bone marrow (BM) were capable of osteogenesis, and had a relatively rare population of cells that could form fibroblast colonies [Friedenstein, 1961]. Since then, bone marrow derived mesenchymal stromal cells (BMSCs) have been functionally defined as the non-hematopoietic cells from the BM cavity that are multipotential cells (e.g., cells that can differentiate into at least three connective tissue types: bone, cartilage, and adipocytes). The frequency of BM colony-forming units that were fibroblastic (CFU-F) has been reported to range from $1/10^4$ or 10^5 marrow mononuclear cells [Castro-Malaspina, 1980]. Efforts to enrich for the CFU-F population from the marrow cavity have produced enrichment on the order of $100\times$ using surface markers, such as Stro-1 [Simmons, 1991], or CD271 and CD140b [Buhring, 2007]. The possibility exists that the marrow cavity niche is not well-reproduced *in vitro*, and thus could affect surface phenotyping [see Jones, 2008]. Thus, a distinction is to be made between the cells expanded and characterized *in vitro*, and those results obtained following engraftment (*in vivo*). The scientific definition of MSCs is "soft" because the physical description of MSCs is general (a spindle shaped, fibroblastic cell), rather than specific. In fact, the acronym "MSC" has been used for (bone) marrow stromal cells, mesenchymal stem cells, and multipotent MSCs. To quote a 2008 review by Jones and McGonagle ... Indeed, the confusion to the *in vivo* identity of the BM MSC has lead to difficulty with terminology whereby the MSC acronym continues to signify both MSCs and marrow stromal stem cells ... [Jones, 2008] (Fig. VI.1).

Perinatal Stem Cells. Edited by C. L. Cetrulo, K. J. Cetrulo, and C. L. Cetrulo, Jr.

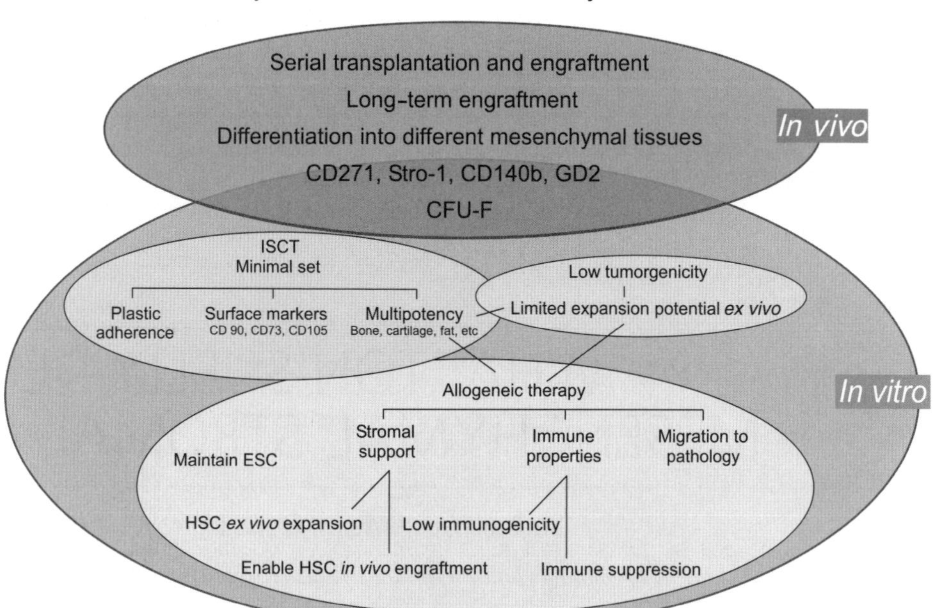

Figure VI.1. Venn diagram illustrating properties in common and difference between mesenchymal stem cells (purple oval) and mesenchymal stromal cells (MSC, blue oval). Mesenchymal stem cells are rare cells that have they ability to self-renewal and replace the MSC progenitor cells. Due to current technical limitations, the mesenchymal stem cells cannot be prospectively identified, and is characterized by *in vivo*, rather than *in vitro* properties shown. It is highly likely that the heterogenous population of cells isolated, for example, from bone marrow, contains a stem cell component because markers, such as CD271, Stro-1, CD140b, GD2, enrich the stem population as indicated by CFU-F assay. The MSCs were defined by International Society for Cellular Therapy working group as cells that are plastic adherent, possess specific surface markers and are capable of differentiating into multiple mesenchymal lineages (e.g., bone, cartilage, muscle, tendon, and adipose). As indicated by the overlap, the WJCs share these properties. The MSCs have other properties, such as stromal support, specific immune properties of low immunogenicity and immune suppression, and the ability to migrate to pathology. These properties are observed in WJCs, too. Adult MSCs have limited expansion capability *in vitro* before their multipotency is compromised. (See color insert.)

The International Society of Cellular Therapy (ISCT) recognized the need to define the MSC. This issue was taken up to distinguish between mesenchymal stromal cells and mesenchymal stem cells. The ISCT provided a clear starting place: MSCs are progenitor cells for the mesenchymal lineages, in contrast to mesenchymal stem cells that replenish and maintain the MSC population *in vivo* [Dominici, 2006].

The ISCT minimal definition of an MSC is a cell that adheres to tissue culture plastic (adherent cell), has a particular surface marker phenotype (expresses SH2, SH3, SH4, CD44, and HLA-Class I and does not express markers of the hematopoietic lineage CD34 and CD45 or HLA-Class II), and has the capacity to differentiate into three mesenchymal lineages (bone, cartilage, and adipocytes) *in vitro*.

An effort has been made, and will continue to be made, to prospectively identify or enrich true mesenchymal stem cells from the mixed population containing mostly stromal cells. Compounding the problem in identifying the stem cells are several factors, such as the donor's age [Rauscher, 2003; Zhang, 2005], limited expansion capability *in vitro*, and the inability to identify definitive surface markers that discriminate the stem cells from MSCs. To date, the gold standard for identifying mesenchymal stem cells is their engraftment and long-term reconstitution (Fig. VI.1).

IMMUNOPHENOTYPE OF MESENCHYMAL STROMAL CELLS

The ISCT in 2006 defined bone marrow derived MSCs in a clear, minimalist fashion as a plastic-adherent cell population isolated from the bone marrow cavity with the following surface markers: CD13, CD44, CD90, CD73, CD105, HLA-ABC positive, CD14, CD11b, CD79, CD34, CD45, and HLA-DR negative [Dominici, 2006].

Other markers, such as CD271 and CD140b [Buhring, 2007], ganglioside molecule, GD2 [Martinez, 2007], and stage specific antigen 4, CD349 (frizzled-9), Oct4, nanog-3, and nestin [Battula, 2007] may also be prospective markers for mesenchymal stem cells. More work is needed to confirm that these markers perform consistently, and across various putative MSC sources.

MSCs ARE MULTIPOTENT CELLS

The MSCs derived from BM and from adipose tissue self-renew and differentiate into specialized cells *in vitro*. The MSCs have been shown to differentiate into bone [Bruder, 1997; Jaiswal, 1997; Kadiyala, 1997], muscle [Ferrari, 1998], adipose tissue [Dennis, 1999], cartilage [Kadiyala, 1997], and tendon [Young, 1998]. There are also reports that MSCs differentiate into neural cells (e.g., neurons and glial cells). However, the differentiated cells do not possess all properties of mature neurons [Hermann, 2004].

MSC AND STROMAL SUPPORT OF THE STEM CELL NICHE

The MSCs provide a supportive role and a microenvironment that enables engraftment–culturing of hematopoietic stem cells (HSCs). For example, MSCs support HSC expansion *ex vivo* (most likely via release of diffusible factors) [de Silva, 2005; Ito, 2006] and MSCs support HSC engraftment *in vivo* when cografted [Anklesaria, 1987; Meida-Porada, 1999]. Nilsson et al. suggested [Nilsson, 2003] that hyaluronic acid (HA) may be an important factor for the HSC niche.

DEFINING PRIMITIVE STROMAL CELLS: DIFFERENCES BETWEEN FETAL AND ADULT MSCs

The MSCs may be harvested from BM, from the human fetus, or adults. While the literature is sparse, several differences between fetal and adult MSCs were noted. First, fetal MSCs appear to have greater expansion capacity *in vitro* and faster doubling times than adult MSCs. The reason may be because they have longer telomeres than adult MSCs [Campagnoli, 2001; Guillot, 2007]. Second, fetal MSCs appear to lack some of the immune suppression properties observed in adult MSCs [Gotherstrom, 2003]. Third, fetal

MSCs appear to lack human leukocyte antigen (HLA) class II, in contrast to adult MSCs [Gotherstrom, 2005]. Similarly, fetal MSCs appear to synthesize HLA-G, which is absent in adult MSCs [Gotherstrom, 2003]. Fourth, fetal MSCs express a slightly different cytokine profile than adult MSCs. In summary, primitive MSCs have a greater ability to expand in culture perhaps due to their relative youth and have different physiology that is likely due to their naïve status. These differences are similar to those observed between umbilical cord blood and adult peripheral blood.

UMBILICAL CORD MSCs

While MSCs have been isolated from several compartments of the umbilical cord; we will limit this discussion to MSCs isolated from Wharton's Jelly (WJ). Within WJ, MSCs have been isolated from three relatively indistinct regions: the perivascular, the intervascular and the subamnion zones; these zones were identified and investigated by Dr. Can's lab [Karahuseyinoglu, 2006]. It is unknown whether MSCs isolated from the different compartments of the umbilical cord represent different populations [see Karahuseyinoglu, 2006 for a review of differences]. Wharton's Jelly cells (WJCs) display MSC properties, suggesting that they are members of the MSC family. A broad comparison of adult-derived MSCs and those from the umbilical cord was recently reviewed (see a review by Troyer, 2008).

The WJCs have stromal support properties. For example, extraembryonic mesenchyme; that is, primitive WJ, surrounds the migrating embryonic blood island cells during their migration to the aorta-gonad mesonephros (AGM) from the yolk sac region prior to day E10.5 [Sadler, 2004]. The WJCs retain this property as demonstrated by their role in *ex vivo* hematopoeitic expansion [Lu, 2006] and *in vivo* engraftment of HSCs [Friedman, 2006]. Lu and co-workers reported that WJCs produced cytokines similar to those of BMSCs, and WJCs synthesized granulocyte macrophage and granulocyte colony stimulating factors (GM–CSF and G–CSF, respectively) that BMSCs did not [Lu, 2006]. The WJCs differ from bone marrow derived MSCs because WJCs are slower to differentiate to adipocytes [Karahuseyinoglu, 2006; Lu, 2006]. Since this feature and others listed below are shared with MSCs derived from umbilical cord blood (UCB), it is unclear whether the MSCs derived from UCB differ from those found in WJ.

Both UCB–MSCs and WJCs have several common properties (e.g., poor ability to differentiate to adipocytes [Bieback, 2004; Karahuseyinoglu, 2006; Kern, 2006], shorter doubling times than BMSCs, and greater number of passages to senescence) [Baksh, 2007; Bieback, 2004; Karahuseyinoglu, 2006; Kern, 2006; Lund, 2007]. Like WJCs, UCB–MSCs may make GM–CSF [Gao, 2006], although this has not been consistently found [Wang, 2004; Ye, 1994]. It is clear that HSCs can be expanded by MSCs from both UCB and WJ [Gao, 2006; Jang, 2006; Lu, 2006; Wang, 2004; Ye, 1994].

There are differences between BMSCs, UCB–MSCs, and WJCs. First, the isolation frequency of CFU-F from BM is estimated to be in the range of 1–10 CFU-F per 10^6 mononuclear cells (MNCs), whereas that of umbilical cord blood is reported to be $\sim$1 CFU-F clones per 10^8 MNCs [Bieback, 2004] to 1–3 CFU-F per 10^6 MNCs [Goodwin, 2001; Wang, 2004]. (*Note*: The isolation frequency from first semester fetal blood derived MSCs was 8.2 CFU-F per 10^6 MNCs, Campagnoli, 2001.) In contrast, cells derived from WJ have a higher frequency of CFU-F [Karahuseyinoglu, 2006; Sarugaser, 2005]. Thus, an order of magnitude more MSCs may be found in the initial isolation from WJ compared to BM or UCB. Second, coupled with the greater CFU-F frequency, the doubling time of WJCs and UCB–MSCs is shorter than adult BMSCs [Baksh, 2007; Karahuseyinoglu,

2006; Lund, 2007]. Faster doubling time is a common feature for MSCs derived from fetal blood [Campagnoli, 2001], cord blood and WJ and is thought to reflect their relatively "primitive" nature compared to adult BMSCs. An important difference between UCB–MSCs and WJCs is that WJCs can be isolated from close to 100% of the samples, even umbilical cords that are delayed in their processing up to 48 h [Weiss, 2006]. The UCB-MSCs have been more difficult to isolate, and with optimized procedures, the success rate has reached 63% for optimal samples [Bieback, 2004].

The role of embryonic fibroblasts to support embryonic stem cells is well established. The WJCs appear to support ESCs and ESC-like cells. Specifically, equine ESC-like cells may be derived and maintained using bovine fibroblasts, isolated from WJ as a feeder layer [Saito, 2002]. Histologically, cells in WJ are fibroblastic in appearance [Karahuseyinoglu, 2006]; this suggests that Saito's lab was isolating WJCs [Saito, 2002]. Similarly, human ESCs may be maintained using human WJCs in coculture [Hiroyama, 2007] or by using conditioned medium from human WJCs [A. Toujmade and J. Auerbach, GlobalStem, unpublished observations]. Further evidence for the supportive function of WJCs for primitive stem cell populations is suggested by (1) their proximity during embryonic germ cell migration and (2) the extracellular release of glial derived neurotrophic factor (GDNF), a factor important to maintain spermatogonial stem cells in the undifferentiated state [Kubota, 2004] by WJCs [Weiss, 2006].

CHARACTERIZATION OF WHARTON'S JELLY DERIVED CELLS

Wharton's Jelly cells are not derived from umbilical blood but from the cushioning matrix between the umbilical blood vessels [Mitchell, 2003]. Previously, we have termed these cells "umbilical cord matrix stem cells" to distinguish them from cells isolated from other umbilical vein endothelial cells or UCB. The WJCs meet the criteria for MSCs: they self-renew and can be induced to differentiate into various cell types *in vitro*. There are numerous descriptions and references to these stem cells in the literature since the original report [Mitchell, 2003], and they have been given different names by different groups (see review for further discussion [Troyer, 2008]). The term Wharton's Jelly cells is used here and includes cells derived from the perivascular space, the intravascular space and the subamnion [Karahuseyinoglu, 2006].

Wharton's Jelly allows the rapid initial isolation of large numbers of cells, with 10–15,000 cells/cm of human umbilical cord to 10 times that number of cells/cm reported [Baksh, 2007; Karahuseyinoglu, 2006; Lund, 2007; Sarugaser, 2005; Seshareddy, 2008; Wang, 2004], avoiding the necessity of extensive multiplication and potential epigenetic damage [Boquest, 2006; Noer, 2007]. This source has the advantage of cells that are isolated from fetal structure in the perinatal period, and perhaps like umbilical cord blood, may be better tolerated following transplantation with less incidence of graft versus host disease (GvHD). Recent work has indicated that WJCs have reduced immunogenicity and are immunosuppressive *in vitro* ([Weiss, 2006], accepted and [Ennis, 2008]), confirming and extending a report of immunogenicity *in vivo* by Cho et al. [Cho, 2007]. Thus, WJCs have the same immune properties of other MSCs.

OTHER MSC-LIKE CELLS FROM UMBILICAL CORD

Four MSC populations have been identified in the umbilical cord. First, MSC cells have been identified in WJ. Second, MSCs cells have been identified surrounding the umbilical vessels.

Third, MSCs have been isolated from UCB. Fourth, MSCs have been isolated from the subendothelium of umbilical vein.

Some authors have focused on a cell population that is isolated from around the umbilical vessels [Baksh, 2007; Sarugaser, 2005]. These have been termed human umbilical cord perivascular cells (HUCPV cells). While the non-vascular components of the umbilical cord may harbor more than one cell population, the WJ is a continuum from the subamnion to the perivascular region [Karahuseyinoglu, 2006]. Karahuseyinoglu compared the expansion ability of WJ stromal cells and the HUCPV cells and reported that the WJ stromal cells had greater expansion capability and faster doubling times than HUCPV cells. Further, the HUCPV cells stain for pan-cytokeratin more strongly than WJ stromal cells. This group suggested that HUCPV cells are more differentiated than WJ stromal cells and this explains why the HUCPVs may not differentiate to neuronal cells [Karahuseyinoglu, 2006].

Several laboratories have isolated MSCs from UCB. There are difficulties in isolating MSCs from UCB and high consistency has not yet been obtained. Recently, it was reported that MSC culture must be initiated <5 h after cord blood collection to increase the chances of MSC isolation. The clinical application of MSCs derived from cord blood may be limited due to the inconsistency initiating cultures and the required expansion time of the relatively small numbers of cells.

Human umbilical vein subendothelial cells are another source of MSC like cells. The isolation of MSCs from umbilical vein was first described by Romanov's lab [Romanov, 2003]. This source has been shown to be similar to BM derived MSCs, has osteogenic capability and multilineage potential [Covas, 2003; Kim, 2004; Panepucci, 2004; Romanov, 2003].

COMPARISON OF WJCs TO ADULT DERIVED MSCs

Wharton's Jelly cells are CD45, CD34, and HLA class II negative; CD73, CD90, CD105, and HLA-class I positive; are plastic adherent, and multipotent. Similar to early passage MSCs, WJCs grow robustly, can be frozen–thawed, and engineered to express exogenous proteins. Thus, the similarities of WJCs and MSCs can be summarized as follows: WJCs share the basic criteria used to define adult-derived MSCs. Additionally, WJCs express GD2 synthase, a marker that has been proposed to uniquely identify MSCs in a BM aspirate (Troyer, Ganta, Weiss, unpublished data).

In all reports, WJCs have faster proliferation and greater *ex vivo* expansion capabilities than BMSC. This finding may be due to the expression of telomerase by WJCs [Mitchell, 2003]. Karahuseyinoglu et al. were able to expand WJC numbers >300-fold over seven passages [Karahuseyinoglu, 2006]. Other groups have expanded WJC to $>10^{15}$ cells [Lund, 2007]. The MSCs are generally used within six passages. In contrast to MSCs, WJCs have a significantly higher CFU-F frequency [Lu, 2006]. A subset of WJCs express nestin, a marker for neural and other stem cells [Fu, 2004; Weiss, 2006; Yarygin, 2006]. Others have reported that WJCs may be superior to MSCs for repair of photoreceptor damage [Lund, 2007] or for tissue engineering [Bailey, 2007].

The WJCs may prove to be superior in other significant ways over BMSCs. A potential serious limitation with BMSCs entails age-related exhaustion of BM derived vascular repair cells [Rauscher, 2003]. Cardiovascular disease further compounds the effects of aging on BMS cells number and function [Dzau, 2005; Heeschen, 2004]. Thus, a further incentive to use a source of neonatal cells is elucidated by Heeschen et al. [2004], who identified the "significantly reduced migratory and colony-forming activity *in vitro* and a reduced neovascularization capacity *in vivo*" in BM derived stem cells isolated from patients with chronic ischemic cardiomyopathy (ICMP). The authors stated that, "this functional impairment of BMS cells from patients with ICMP may limit their therapeutic potential for clinical cell therapy." Thus, utilizing a source of neonatal cells that are not negatively impacted by aging or cardiovascular disease would be clinically relevant.

The WJCs appear to be similar to bone marrow stromal and other MSCs, since in addition to the three surface markers mentioned above, they express CD10, CD13, CD29, and CD44 [Fu, 2004; Karahuseyinoglu, 2006; Lu, 2006; Wang, 2004; Weiss, 2006]. Like BMSCs, WJCs express the stem cell factor gene [Lu, 2006]. However, some of these markers appear to be down regulated as passage number increases [Weiss, 2006] and unpublished data. The WJCs are negative for CD34, CD45, CD14, CD33, CD56, CD31, and HLA-DR [Wang, 2004; Weiss, 2006]. Karahuseyinoglu and co-workers divided WJCs into Type I cells that were more fusiform in appearance, but also expressed cytokeratins and other differentiation markers, and Type II cells which were more elongated but were more efficiently induced into neural cells [Karahuseyinoglu, 2006]. These authors reported that WJCs in chondrogenic media were strongly positive for type II collagen, whereas BMSCs in the same induction media only showed weak positive type II collagen immunofluorescence.

IN VITRO DIFFERENTIATION OF WJCs

The WJCs, such as BMSCs, can be induced *in vitro* to become cells with morphologic and biochemical characteristics of neural cells [Ma, 2005; Mitchell, 2003; Wang, 2004]. When they were cultured in media conditioned by primary rat brain neurons, WJCs could be invoked with glutamate to have an inward current and express neuronal proteins, such as NeuN, neurofiliament, glutamate decarboxylase, and subunits of the kainate receptor, and exposure to primary neuron-conditioned media upregulated the astrocyte protein GFAP [Fu, 2004]. Interestingly, this group also showed they could increase CD11b (microglial) cell numbers from 1.7 to 3.0% by subjecting WJCs to neuronal-conditioned media. Ma and co-workers induced WJCs to express neuronal markers β-tubulin III, neurofilament and GFAP by treatment with *Salvia miltorrhiza* [Ma, 2005].

Mesenchymal stromal cells isolated from WJ have been induced to form bone, cartilage, and adipose cells [Conconi, 2006; Fu, 2004; Karahuseyinoglu, 2006; Sarugaser, 2005; Wang, 2004]. The WJCs were spun into PLGA scaffolds and cultured in media containing chondrogenic growth factors and were compared to similarly treated cells isolated from condylar cartilage from the temporomandibular joint [Bailey, 2007]. Interestingly, these authors found that the WJCs outperformed the cartilage cells with regard to collagens I and II, glycosaminoglycans and cellularity of the constructs after 4 weeks. The WJCs can also be induced toward heart cells. After 5-azacytidine treatment for 3 weeks, they expressed the cardiomyocyte markers cardiac troponin I, connexin 43, and desmin, and exhibited cardiac myocyte morphology [Wang, 2004]. They have been used along with tissue engineering to generate artificial blood vessels and heart valves [Breymann, 2006; Hoerstrup, 2002; Schmidt, 2006]. The WJCs can be directed toward skeletal muscle cells; when they were placed in myogenic media they expressed Myf5 from day 7, Myo-D from day 11, and formed elongated, multinucleated cells [Conconi, 2006]. After undifferentiated WJCs were injected into rat muscle damaged by bupivacaine chloridrate, elongated cells expressing sarcomeric tropomyosin that were HLA immunopositive were identified within the muscle [Conconi, 2006]. Finally, it has been shown that human WJCs can be differentiated successfully into endothelial cells [Wu, 2007] after the addition of vascular endothelial growth factor (VEGF) and basic fibroblast growth factor (b-FGF) to cultures. Moreover, the authors reported that undifferentiated human WJCs injected into a murine ischemic model differentiated into endothelial cells.

COMPARISON OF WJCs TO FETAL MSCs

In contrast to what has been observed for adult MSCs, WJCs share several of the properties unique to fetal-derived MSCs. First, they have greater expansion potential *in vitro* than adult MSCs (reviewed above). Second, WJCs express some HLA-class I and the lack expression of HLA-class II [Lu, 2006; Sarugaser, 2005; Weiss, 2006]. In contrast to what has been published for fetal MSCs, WJCs are immune suppressive in mixed lymphocyte assays and inhibit T cell proliferation [Cho, 2007; Ennis, 2008] and [Weiss ML et al., *Stem Cells*, accepted]. The WJCs are tolerated following allogenic transplant and stimulate an immune response following multiple injections or injection of WJCs exposed to interferon [Cho, 2007]. Human WJCs express markers of primitive stem cells, such as leukemia inhibitory factor receptor (LIFR) pathway, ESG1, and TERT [Weiss, 2006]. Porcine WJCs express the ESC markers Oct-4, Sox-2, and Nanog, albeit at a lower level relative to embryonic stem cells, and are alkaline phosphatase positive [Carlin, 2006]. Flow cytometry indicates that a subpopulation of WJCs express the primitive stem cell markers SSEA4 and TRA-1-60 [Hoynowski, 2007]. When the Hoescht dye exclusion test is used to identify dye excluding, side population cells, $\sim$20% of human WJCs exclude dye. Flow-sorting to enrich the Hoescht-dim population resulted in cells that appeared morphologically smaller than the parent population, and many of the selected cells expressed CD44, the HA receptor [Weiss, 2006].

TRANSPLANTATION OF WJCs

Human WJCs ameliorate apomorphine-induced behavioral deficits in a hemiparkinsonian rat model [Weiss, 2006]. There was a significant decrease in apomorphine-induced rotations at 4 weeks continuing up to 12 weeks post-transplantation in PD rats that received human WJC transplants compared to the PD rats that received a sham transplant. The behavioral findings correlated with the numbers of TH positive cell bodies observed in the midbrain following sacrifice at 12 weeks, indicating a "rescue from a distance" phenomenon. One explanation for this effect may be that WJC synthesize GDNF, a potent survival factor for dopaminergic neurons, as well as other trophic factors such as VEGF and CNTF. In another report, WJCs were first induced toward dopaminergic neurons using neuron-conditioned media, sonic hedgehog, and fibroblast growth factor 8, and then transplanted into hemiparkinsonian rats [Fu, 2006]. Despite the fact that the rats were not immune suppressed, WJCs were identified 5 months later and prevented the progressive degeneration–behavioral deterioration seen in control rats with unilateral lesions. Similarly, rat WJCs transplanted into the brains of rats with global cerebral ischemia caused by cardiac arrest and resuscitation significantly reduced neuronal loss, apparently due to a rescue phenomenon [Jomura, 2007].

The WJCs have been shown to improved cardiac function in a rat myocardial infarction model with intramyocardial delivery of cells [Wu, 2007a, b]. Improvements in left ventricular function and attenuation of left ventricular dilation were observed. The authors report that the positive functional effects are probably related to differentiation of the cells within the infarcted myocardium, increased capillary and arteriole density, secretions of angiogenic factors, and prevention of apoptosis.

Whether MSCs differentiate into cardiac cells is a subject of debate [Kawada, 2004; Murry, 2004; Orlic, 2001]. Numerous studies have examined the use of MSCs as a therapeutic cell population for the repair of heart tissue. Many of these studies report conflicting results regarding whether MSCs have the capacity to differentiate into cardiomyocytes or, if they do have cardiomyocyte potential, whether they differentiate into cardiomyocytes at a high enough graft efficiency to have a significant effect on heart function [Caplan, 2006]. What is elucidated from these studies is that the administration of MSCs to the infarcted myocardium results in increased heart function compared to controls, as shown by hemodynamic measurements (systolic and end-diastolic pressure), echocardiography, and blood flow [Min, 2002]. Evidence for cellular fusion between MSC and recipient cardiomyocytes has also been reported [Noiseux, 2006]. The number of cellular fusion events was extremely low, but remained constant over time. The authors also reported rare differentiated cardiomyocytes. However, the frequency of those events was also very low. Thus, the role of early paracrine mechanisms mediated by MSCs is probably responsible for the therapeutic effects observed [Noiseux, 2006]. Tang et al. [Tang, 2004, 2006] have also reported the trophic effects of MSCs. They reported that MSCs implanted into ischemic myocardium significantly elevated VEGF, increased vascular density and regional blood flow in the infarcted area. The neovascularization resulted in a decrease in apoptosis of the hypertrophied myoctes.

The WJCs were administered into the eyes of a rodent model of retinal disease. Here, WJCs were compared to BMSCs and placental stem cells. They reported that the WJCs exhibited the best histological evidence of photoreceptor rescue [Lund, 2007]. In addition, they reported that WJCs significantly increased production of trophic factors, such as brain derived neurotrophic factor (BDNF) and fibroblast growth factor 2 (FGF2) compared to the other two cell types. Interestingly, WJCs transplanted into the vitreous demonstrated a rescue effect, indicating that they could enhance survival of photoreceptor cells without being in close proximity to them. This effect was presumably due to diffusible growth factors. These results, indicating a rescue phenomenon by WJCs, fit a model for the positive effects of MSC therapy in stroke [Borlongan, 2004] or myocardial infarction [Grinnemo, 2004].

The WJCs, like neural stem cells and mesenchymal stem cells [Aboody, 2000; Studeny, 2002, 2004], appear to migrate to areas of tumor growth. Human breast carcinoma (MDA 231) cells were intravenously injected into severe combined immunodeficient syndrome (SCID) mice, followed by intravenous transplantation of fluorescently labeled WJCs. One week after transplant WJCs were found near or within lung tumors and not in other tissues (see Fig. VI.2). WJCs were engineered to express human interferon beta and, were administered intravenously into SCID mice bearing MDA 231 tumors. This treatment significantly reduced the tumor burden [Rachakatla, 2007].

Recent work by Cho et al., evaluated the ability of allogeneic WJCs to stimulate the immune system in a swine model [Cho, 2007]. The WJCs were nonimmunogenic on the first injection into allogeneic recipients. However, repeated injection of WJCs produced an immunogenic response. Further, when WJCs were injected into inflamed skin, or when WJCs were exposed to interferon prior to injection, they were immunogenic. This work is the first to demonstrate immunogenicity of WJCs *in vivo* and has implications for their allogeneic use in disease tissues.

The migration or "homing" ability of umbilical cord and other MSCs is thought to be due to the expression of chemokine or other surface receptors. For example, a subpopulation

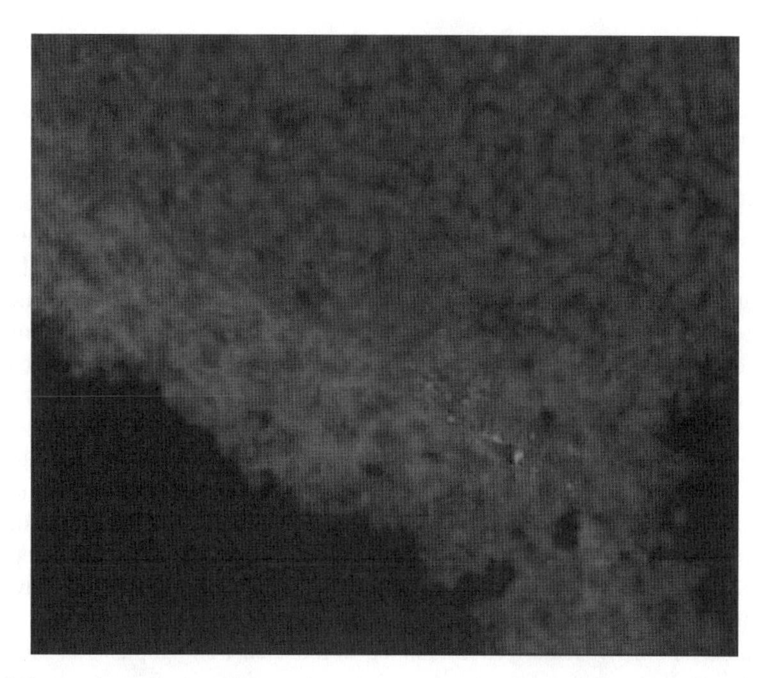

Figure VI.2. Homing of human WJCs to human cancer cells xenografted into SCID mice. The WJCs (labeled in red with SPDiI) were administered intravenously to SCID mice bearing lung metastases of human breast carcinoma (green). Nuclei are shown with DAPI staining in blue. (See color insert.)

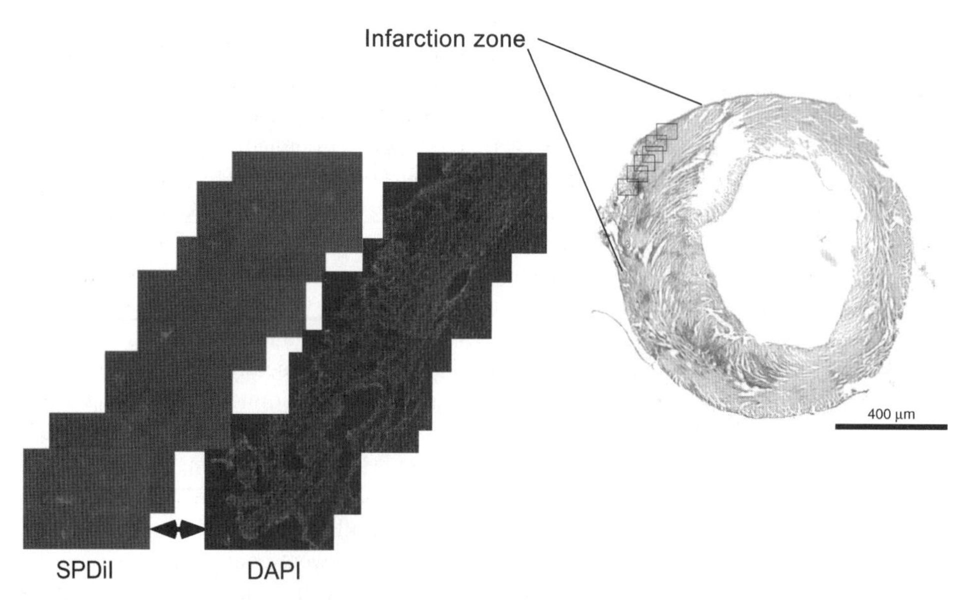

Figure VI.3. Homing of rat WJ derived cells to the ischemic region 48 h after myocardial infarction (MI) in rats. Right: Low power of left ventricle indicating the infracted zone 6 days after MI (and 4 days after injection of SpDiI labeled WJ derived mesenchymal stromal cells (WJCs). Note that the WJCs were derived from Fischer 344 rats and transplanted into Fischer 344 rats following MI. Far Left: SPDiI-labeled WJCs located in the peri-infarcted region. No WJCs were identified in the healthy myocardium. Middle: Nuclei stained by DAPI, same field as left. The plane of section shown is ~350–400 μm toward the base of the heart from ligation. (See color insert.)

of BMSCs was shown to express CXCR4, the receptor for the chemokine stromal-derived factor-1 (SDF-1) and CXCR3, the receptor for fractaline. The WJCs also express CXCR4 [Weiss, 2006]. Cell necrosis after an injury, such as myocardial infarction may cause the release of signals into the circulatory system, and these could induce the homing of stem cells to the myocardium [Jiang, 2006]. The SDF-1 plays an essential role in promoting the homing of cells toward the ischemic myocardium [Cheng, 2008]. Given that SDF-1 expression is up-regulated in the ischemic myocardium immediately following infarction [Abbott, 2004]. The WJCs expressing CXCR4 most likely play an important role in homing of the cells to infarcted myocardium following acute myocardial infarction. Another mobilizing factor associated with MSCs is granulocyte colony-stimulating factor (G-CSF), which has been shown to be involved in the mobilization of MSCs that have been implanted directly into the BM of mice that were later injected with G-CSF. This promoted the migration of the MSCs into the heart after myocardial infarction [Kawada, 2004]. Thus, G-CSF is also implicated as a general mobilization factor. Successful homing of intravenous injection of MSCs has been observed with concurrent increases in cardiac function [Boomsma, 2007; Jiang, 2006]. We have also observed homing of intravenous injected WJCs to the infarcted region of an acute myocardial infarcted rat model (see Fig. VI.3). Tumors also secrete factors that recruit cells from surrounding tissue as well as from the BM to provide support and nutrition.

WJCs AS PRIMITIVE STROMAL CELLS

Wharton's Jelly cells represent a unique, easily accessible and noncontroversial source of early stem cells that can be readily manipulated. In one species examined, the pig, WJCs resemble pluripotent cells, since they express Oct-4, Nanog, and Sox-2 and are alkaline phophatase positive [Carlin, 2006]. In agricultural species this is important, as there are few economically viable sources of primitive stem cells available for future transgenic or biotechnology applications. Our contention that umbilical matrix cells are primitive stromal cells is based on plastic adherence, immunophenotyping, real-time polymerase chain reaction (PCR), immunohistochemistry, enzyme-linked immunosorbent assay (ELISA), multipotency, and other results mentioned above. Since the cells are isolated at birth, there is less of a temporal separation from fetal cells than those isolated from adult tissues, and the isolates are consistent in terms of age at collection. Other authors also surmised that WJCs are earlier-stage cells than MSCs derived from adult fat or bone marrow [Wang, 2004; Wu, 2007]. This argument is based upon population doubling times and more extensive expansion prior to senescence. Another indication that WJCs are a primitive population is that an unusually high percentage of them express the ABCG2 transporter and efflux Hoescht dye [Carlin, 2006; Weiss, 2006], since these are markers of other primitive stem cells. When we used flow sorting to enrich for this population, the enriched cells appeared morphologically smaller than the Hoechst-bright population [Weiss, 2006].

Since the WJCs express some genes characteristic of primitive stem cells including embryonic stem cells (ESCs), and since ESCs sometimes form tumors after transplantation [Arnhold, 2004; Wakitani, 2003], we transplanted large numbers of WJCs into SCID mice. When the mice were examined 50 days later, there was no evidence of tumor formation or long-term engraftment [Rachakatla, 2007]. The fact the engraftment was not detected could be due to failure of the WJCs to engraft or lack of sensitivity in the detection method. Rat WJCs did not form tumors when transplanted into rats with retinal degeneration [Lund, 2007]. The WJCs are karyotypically stable over many passages [Karahuseyinoglu, 2006;

Lund, 2007; Weiss, 2006] and do not lose anchorage dependency, contact inhibition or serum dependence [Karahuseyinoglu, 2006; Weiss, 2006], as do cancer cells. A more definitive test of engraftment potential has yet to be performed. Such testing as transplantation following irradiation injury, is needed to evaluate whether WJCs, like MSCs, can engraft *in vivo* and thus be true stem cells.

SUMMARY

Wharton's Jelly cells, like bone marrow stromal cells, and other mesenchymal cells, are plastic-adherent, stained positively for markers of the mesenchymal cells, such as CD10, CD13, CD29, CD44, CD90, and CD105 and negatively for markers of the hematopoietic lineage. Moreover, WJCs morphologically resemble MSCs and can be expanded more than bone marrow derived MSCs in culture. Human WJCs express precursor cell markers, such as nestin. The WJCs can be induced to form adipose tissue, bone, cartilage, skeletal muscle cells, cardiomyocyte-like cells and neural cells, and are amenable to biomedical engineering applications. Therefore, these cells fit into the category of primitive stromal cells, and because WJ is a plentiful and inexpensive source of cells, it appears to have the potential to impact fields, such as regenerative medicine, biotechnology, and agriculture. Further work is needed to determine whether WJCs engraft long-term and display self-renewal and multipotency *in vivo*, and as such demonstrate that WJCs are a true stem cell population.

ACKNOWLEDGMENTS

The authors acknowledge support from the National Institute of Health (MLW NS34160), Midwest Institute for Comparative Stem Cell Biology, K-INBRE, Terry C Johnson Center for Basic Cancer Research, the KSU Developing Scholars Program, the KSU Targeted Excellence Program, and the State of Kansas. Thanks are given to: Drs. D. Davis, F. Marini, S. Medicetty, K. Mitchell, R. Rachakatla, C. Ganta, K. Seshareddy, J. Hong, and Y. Lopez for their assistance with this work. This chapter is a revision of a review that was published in Stem Cells: A Concise Review: Wharton's Jelly-derived cells are a primitive stromal population@ D.L. Troyer, M.L. Weiss, *Stem Cells*, 26(3):591–599, 2008 (Epub: PMID 18065397). Therefore, some of the ideas and a few sentences may have appeared first in that article. The editors of *Stem Cells* were informed.

REFERENCES

Abbott JD et al. 2004. Stromal cell-derived factor-1alpha plays a critical role in stem cell recruitment to the heart after myocardial infarction but is not sufficient to induce homing in the absence of injury. Circulation. 110:3300–3305.

Aboody KS et al. 2000. Neural stem cells display extensive tropism for pathology in adult brain: Evidence from intracranial gliomas. Proc Natl Acad Sci USA. 97:12846–12851.

Anklesaria P et al. 1987. Engraftment of a clonal bone marrow stromal cell line in vivo stimulates hematopoietic recovery from total body irradiation. Proc Natl Acad Sci USA. 84:7681–7685.

Arnhold S et al. 2004. Neurally selected embryonic stem cells induce tumor formation after long-term survival following engraftment into the subretinal space. Invest Ophthalmol Vis Sci. 45:4251–4255.

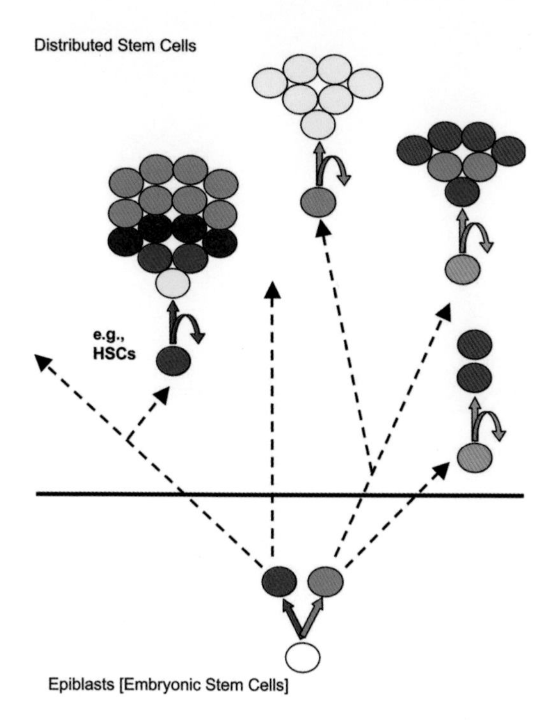

Figure II.1. The concept of distributed stem cells. Epiblasts in the early embryo divide asymmetrically to establish the complex cell differentiation plan of embryonic development. Their phenotypically asymmetric cell divisions do not preserve pluripotency during normal development. By preventing these divisions *in vitro*, embryonic stem cells are derived with the ability to preserve pluripotency. *In vivo* during embryonic development, the initiating embryonic pluripotency is "distributed" as multipotency and unipotency among postembryonic stem cells that become long lived in postnatal tissues [e.g., hematopietic stem cells (HSCs) found in umbilical cord blood (UCB)]. Unlike the phenotypically asymmetric divisions of epiblasts and other embryonic precursors, DSCs undergo asymmetric self-renewal divisions that preserve their specific potency even as they continuously renew tissue-specific differentiating cells.

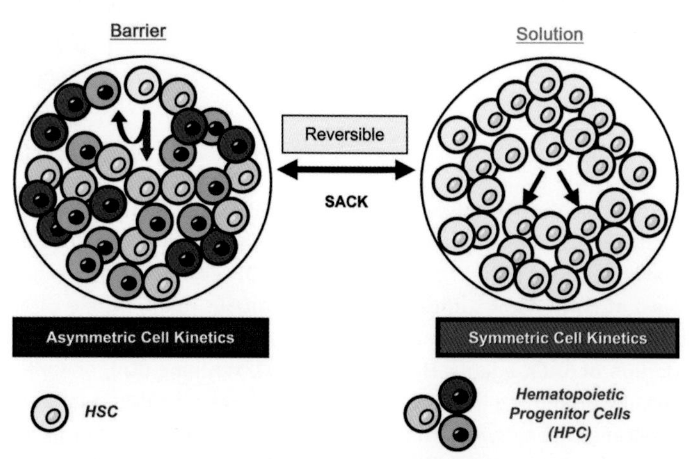

Figure II.2. Suppression of asymmetric cell kinetics: a solution to overcoming the cell kinetics barrier to hematopoietic stem cell expansion. Asymmetrically self-renewing HSCs divide with asymmetric cell kinetics to give rise to two daughter cells, a new HSC, and a hematopoietic progenitor cell (HPC). *Ex vivo*, the HPC divides to give rise to differentiated, maturing blood cells. As the culture continues, the HSC is predicted to become diluted and lost among the accumulating progeny from active HPCs and differentiated cell proliferation. The solution to overcome this cell kinetics barrier is SACK, which promotes symmetric self-renewal of HSCs. In this state, not only does the HSC number increase exponentially, but consequently the production of HPCs and differentiated cells is also reduced. In order for this solution to be ideally useful, it must be reversible; so that upon removal of SACK agents (see Fig. II.3), expanded HSCs can revert back to their asymmetrically self-renewing state to give rise to differentiated, functional blood cells, *in vitro* and *in vivo*.

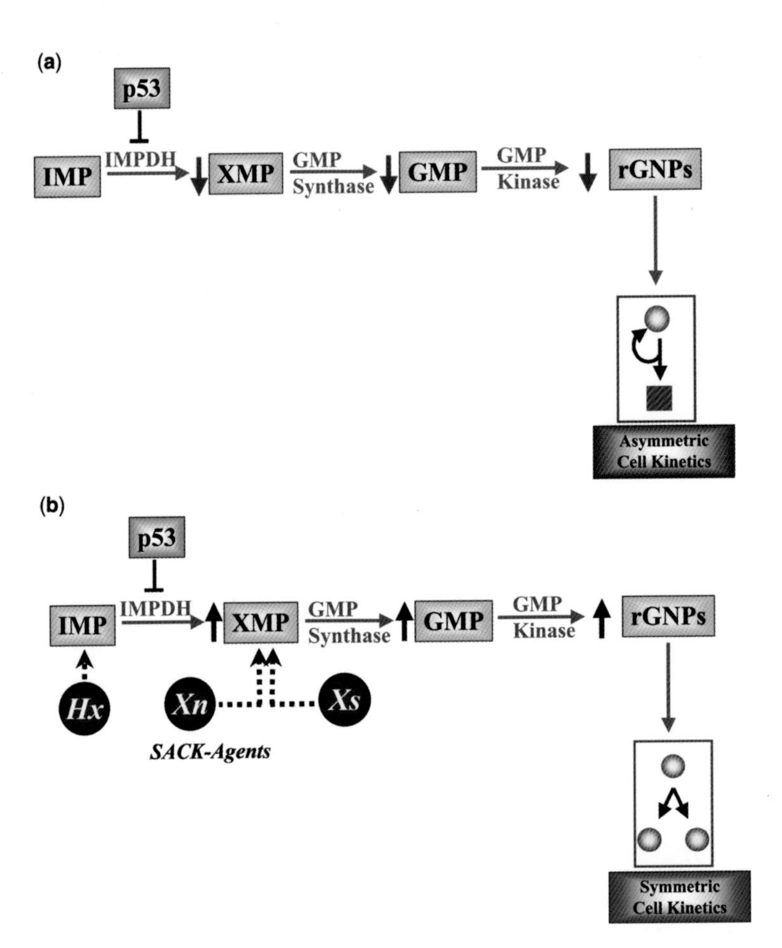

Figure II.3. A cellular pathway that controls the self-renewal pattern of DSCs. (a) The cellular pathway in its basal state in DSCs. The IMPDH is the rate-determining enzyme for the production of cellular guanine ribonucleotides (rGNPs). In normal DSCs, the p53 tumor suppressor protein downregulates IMPDH expression. This regulation keeps the rate of rGNP synthesis and the level of rGNP pools sufficiently low so that DSCs retain a state of asymmetric self-renewal. Inosine monophosphate (IMP); XMP, xanthosine monophosphate; GMP, guanosine monophosphate. (b) The cellular pathway's response to addition of SACK agents. Hypoxanthine (Hx), xanthine (Xn), or xanthosine (Xs) overcome or bypass the regulation of p53 by promoting increased synthesis and levels of rGNPs. Higher rGNP levels cause a shift of DSCs from asymmetric self-renewal to symmetric self-renewal. Mutations that lead to loss of p53 regulation or upregulation of cellular IMPDH activity are predicted to lock DSCs into a chronic state of symmetric exponential proliferation via this mechanism.

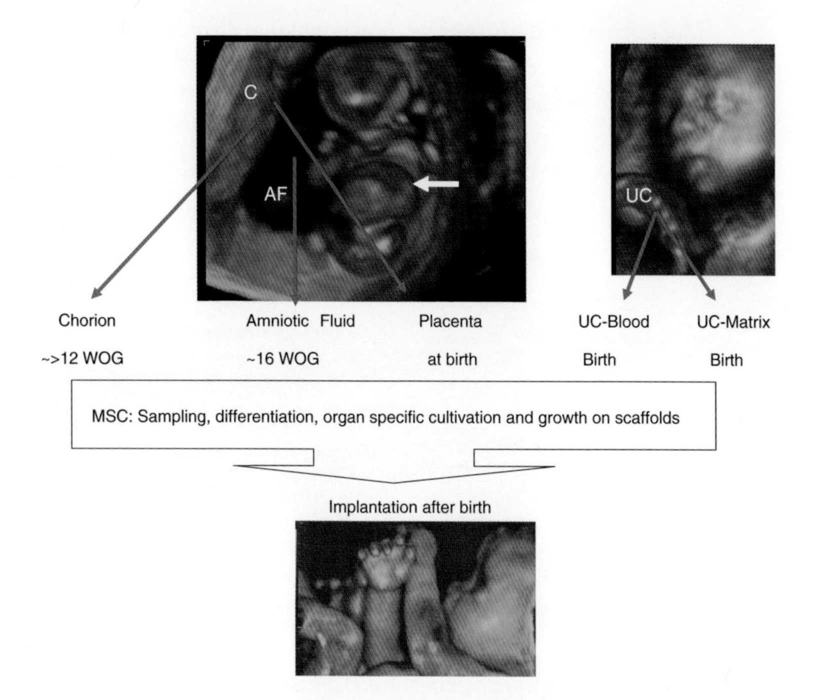

Figure III.1. Scenario of the use of fetal cells in regenerative medicine. This fetus shows an anterior body-wall defect (gastroschisis) that is, early detected (arrow = defect). Various cells sources can possibly be used for aquisition of progenitor cells. According to the needs for organ replacement (e.g., skin for abdominal wall closure after reposition of gastroschisis after birth) cells are then differentiated and used for regenerative organ growth.

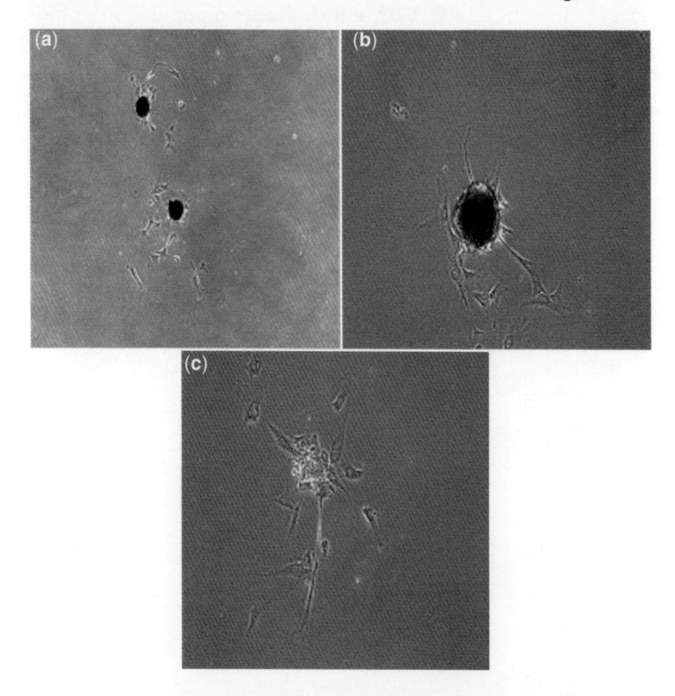

Figure V.1. Morphology of UC–MSC in various media. Morphology of UC–MSC cultured in 20% HS–RPMI 1640 medium. The UC–MSC tend to form scattered clumps with fewer plastic-adherent isolated cells. (a) Magnification ×20, (b) magnification ×100. Individual clumps spread out again when transferred into 20% FCS-RPMI 1640 medium (c) magnification ×100.

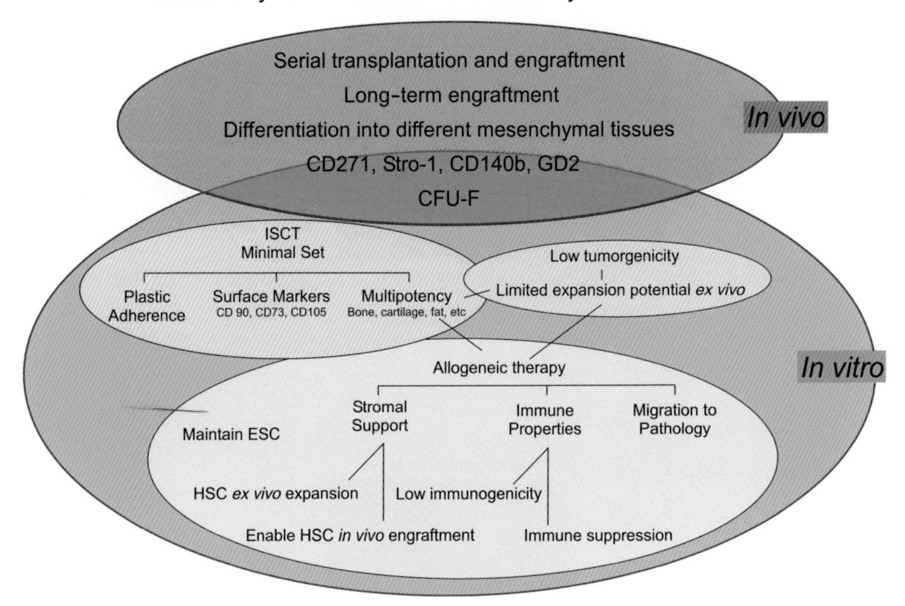

Figure VI.1. Venn diagram illustrating properties in common and difference between mesenchymal stem cells (purple oval) and mesenchymal stromal cells (MSC, blue oval). Mesenchymal stem cells are rare cells that have they ability to self-renewal and replace the MSC progenitor cells. Due to current technical limitations, the mesenchymal stem cells cannot be prospectively identified, and is characterized by *in vivo*, rather than *in vitro* properties shown. It is highly likely that the heterogenous population of cells isolated, for example, from bone marrow, contains a stem cell component because markers, such as CD271, Stro-1, CD140b, GD2, enrich the stem population as indicated by CFU-F assay. The MSCs were defined by International Society for Cellular Therapy working group as cells that are plastic adherent, possess specific surface markers and are capable of differentiating into multiple mesenchymal lineages (e.g., bone, cartilage, muscle, tendon, and adipose). As indicated by the overlap, the WJCs share these properties. The MSCs have other properties, such as stromal support, specific immune properties of low immunogenicity and immune suppression, and the ability to migrate to pathology. These properties are observed in WJCs, too. Adult MSCs have limited expansion capability *in vitro* before their multipotency is compromised.

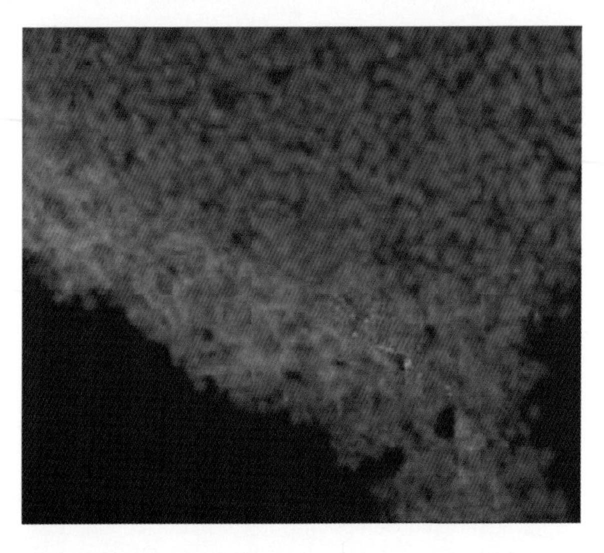

Figure VI.2. Homing of human WJCs to human cancer cells xenografted into SCID mice. The WJCs (labeled in red with SPDiI) were administered intravenously to SCID mice bearing lung metastases of human breast carcinoma (green). Nuclei are shown with DAPI staining in blue.

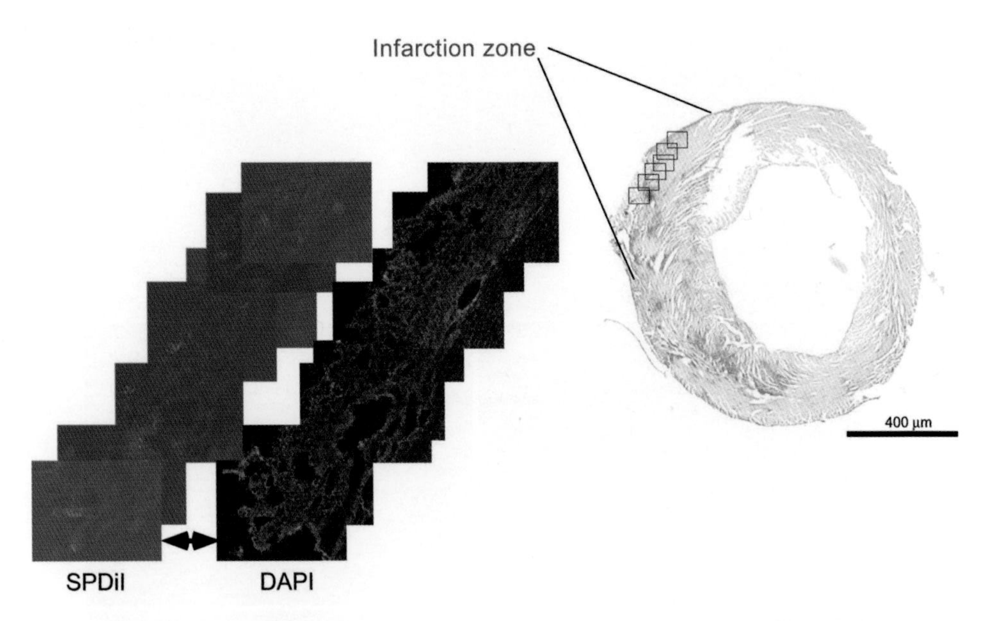

Figure VI.3. Homing of rat WJ derived cells to the ischemic region 48 h after myocardial infarction (MI) in rats. Right: Low power of left ventricle indicating the infracted zone 6 days after MI (and 4 days after injection of SpDil labeled WJ derived mesenchymal stromal cells (WJCs). Note that the WJCs were derived from Fischer 344 rats and transplanted into Fischer 344 rats following MI. Far Left: SPDil-labeled WJCs located in the peri-infarcted region. No WJCs were identified in the healthy myocardium. Middle: Nuclei stained by DAPI, same field as left. The plane of section shown is ~350–400 μm toward the base of the heart from ligation.

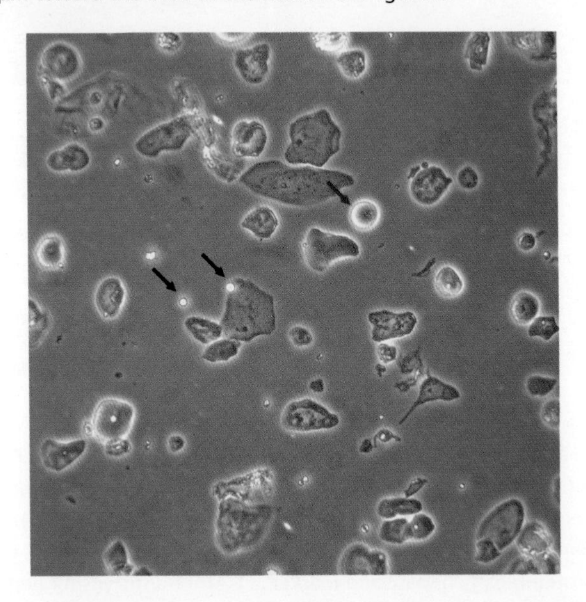

Figure IX.1. Amniotic fluid cells obtained from second trimester amniocentesis, two subpopulations, large cell (black arrow) and small cell (red arrow), of colony-formation cells are noted in fresh amniotic fluid.

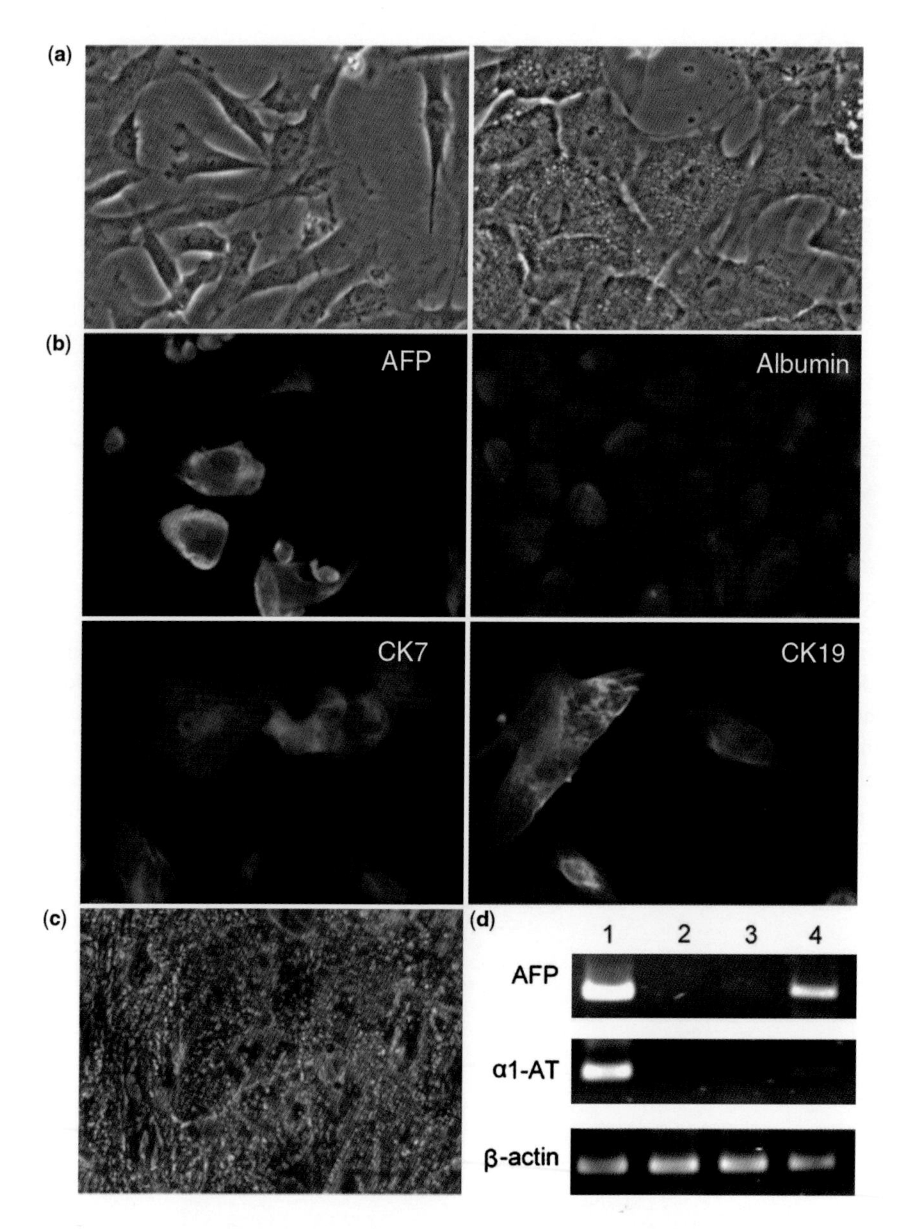

Figure IX.3. Differentiated amniotic fluid stem cells express markers for hepatocytes. Morphological of AFSCs before hepatic induction (a left) and at 4-weeks postinduction (a right). Differentiated cells at 6-weeks postinduction express immunofluorescence stains of AFP, Albumin, CK7 and CK19, markers for hepatic differentiation (b). Positive glycogen storage was noted at 4-weeks postinduction (c). The RT–PCR analyses are shown in (d). Lane 1: Hepa G2 for positive control, Lane 2: noninduction, Lane 3: at 2 weeks posthepatocyte-induction, Lane 4: at 3-weeks posthepatocyte induction. The AFP (α-fetoprotein): early hepatocyte marker, α1-AT(α1-antitrypsin): hepatocyte expression marker, a 52-kDa glycoprotein produced by hepatocytes and mononuclear phagocytes. β-Actin as a internal control.

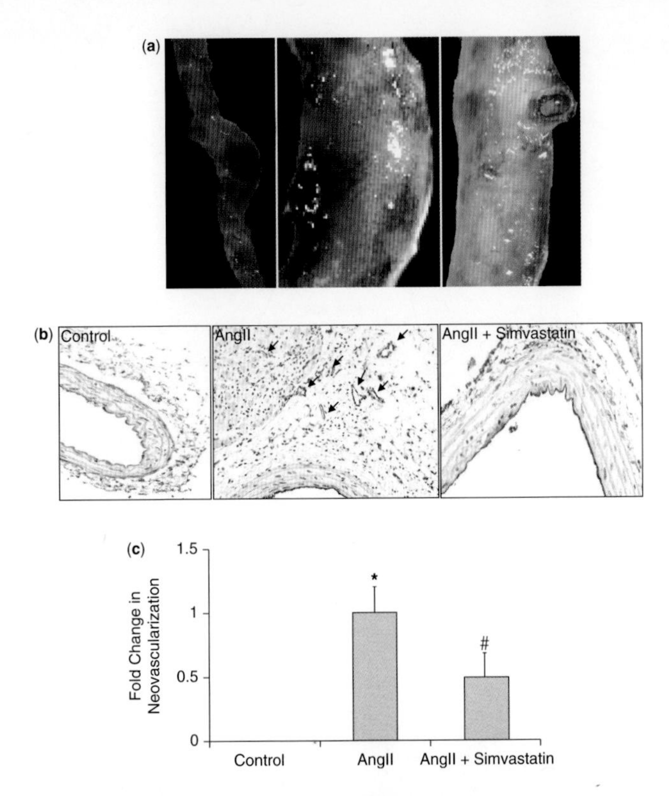

Figure XI.11. Simvastatin attenuates AngII-induced aneurysm formation and neovascularization in suprarenal aortic tissue. (a) Left panel: Representative whole-mount CD31 and Sudan staining of the aorta from an AngII treated mouse demonstrating multiple aneurysms (note the atheroma in the distal large aneurysm). Middle panels: high-power views of upper and lower regions shown in the left panel demonstrating extensive proliferation of VV. Right panel: high-power view of similar region from the aorta of a Simvastatin treated mouse. (b) CD31+ endothelial cells (brown color) in corresponding sections from aortas of control, AngII and AngII + Simvastatin treated mice. Arrows indicate new capillary formation. (c) Quantitative evaluation of CD31+ capillaries in media and adventitia normalized by area. $^{*}p < 0.05$ compared with control, $^{**}p < 0.05$ compared with AngII, data represent capillary counts in three fields in each of 15 sections.

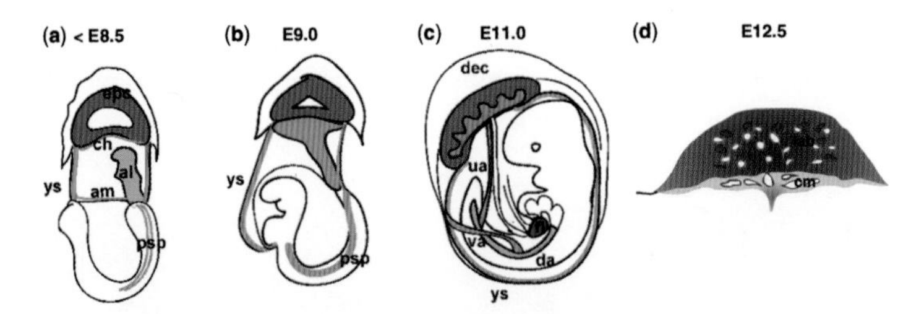

Figure XII.1. Development of the mouse placenta. (a) At E7.5–8.25 the allantois (red) has formed from mesodermal precursors from the primitive streak, and is growing toward the ectoplacental cone (brown). (b) Fusion of the allantois with the chorionic mesoderm occurs at E8.5, concomitant with the onset of heartbeat. Subsequently, chorioallantoic mesoderm interdigites with the trophoblasts and the placental vasculature starts to form. (c) By E10.5–11.0 large vessels that connect to the umbilical cord have formed in the chorioallantoic mesenchyme, and the feto-placental circulation is fully established. The placenta labyrinth is still developing and is therefore an active site of vasculogenesis–angiogenesis. (d) A E12.5 cross-section of the placenta displays the different regions of the placenta, namely, the chorioallantoic mesenchyme including the large vessels of the placenta (in red) and the placenta labyrinth, which is a unique region including trophoblast-lined maternal blood spaces (red spaces surrounded by brown trophoblasts) and fetal vessels lined by fetal endothelium (red vessels with lumens). al = allantois; ch = chorion; am = amnion; epc = ectoplacental cone; ys = yolk sac; psp = para-aortic splanchnopleura; dec = decidua; da = dorsal aorta; ua = umbilical artery; va = vitelline artery; fl = fetal liver; lab = labyrinth; cm = chorioallantoic mesenchyme.

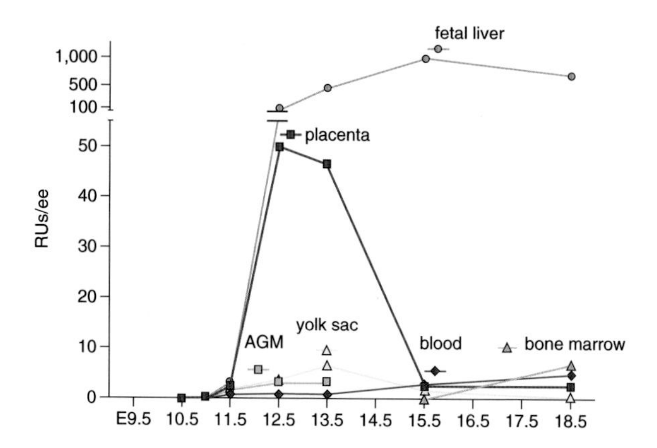

Figure XI.2. Kinetics of long-term reconstituting HSCs in the embryo and extraembryonic tissues. The graph depicts the number of HSCs with long-term reconstitution capability per tissue during development. The midgestation placenta harbors a large pool of HSCs, which diminishes toward the end of gestation while the fetal liver HSC pool is expanding. RU = reconstituting unit; ee = embryo equivalent.

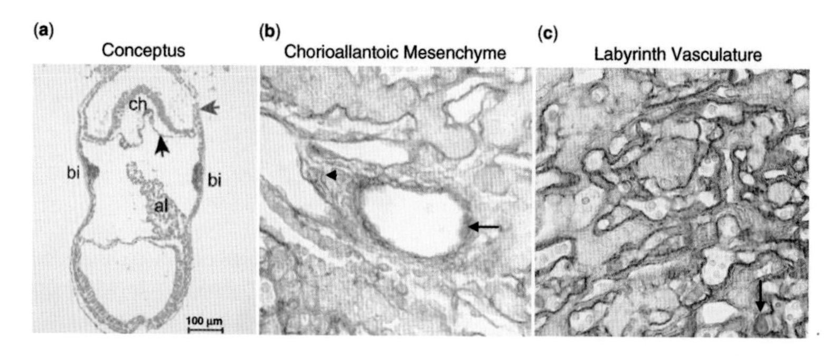

Figure XII.3. The Runx1–LacZ expression in the placenta. (a) A cross-section of a precirculation Runx1–LacZ conceptus documenting Runx1 expression in the chorionic mesoderm (black arrow) and the blood-islands of the yolk sac. The blue arrow denotes the ectoplacental cone. (b) The large vessels of the chorioallantoic mesenchyme harbor LacZ+ cells within the wall of the vessel (arrow) at E11.5, the time when HSCs emerge. The mesenchyme contains two distinct populations of LacZ+ cells, oblong-shaped cells that straddle within the stromal cells (arrowhead), and cuboidal LacZ+/cytokeratin+ cells derived from ectoplacental endoderm that have an organized structure (asterisk). (c) The LacZ+ definitive hematopoietic cells localize to the fetal labyrinth vasculature. (a–c) The Runx1–LacZ is in blue. (b,c) Laminin (pink) marks mesodermal tissues while cytokeratin (brown) marks trophoblasts and ectoplacental endoderm in the placenta. Bi = blood island; al = allantois.

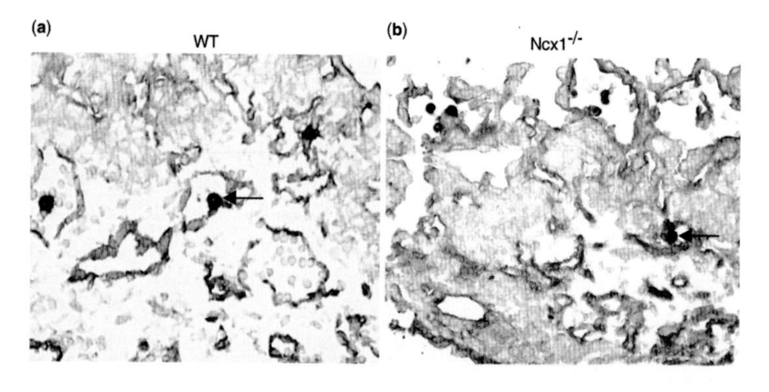

Figure XII.4. *De Novo* hematopoiesis in the placenta. (a) A wild-type placenta harboring CD41+ hematopoietic cells in the placental vasculature. (b) CD41+ hematopoietic cells emerge in the, vasculature in Ncx1$^{-/-}$ placenta in the absence of circulation. CD41 (blue) marks hematopoietic cells, whereas CD31 (red) identifies endothelium. Cytokeratin (brown) marks the trophoblasts.

Bailey MM et al. 2007. A comparison of human umbilical cord matrix stem cells and temporomandibular joint condylar chondrocytes for tissue engineering temporomandibular joint condylar cartilage. Tissue Eng. 13:2003–2010.

Baksh D, Yao R, Tuan RS. 2007. Comparison of proliferative and multilineage differentiation potential of human mesenchymal stem cells derived from umbilical cord and bone marrow. Stem Cells. 25:1384–1392.

Battula VL et al. 2007. Human placenta and bone marrow derived MSC cultured in serum-free, b-FGF-containing medium express cell surface frizzled-9 and SSEA-4 and give rise to multilineage differentiation. Differentiation. 75:279–291.

Bieback K et al. 2004. Critical parameters for the isolation of mesenchymal stem cells from umbilical cord blood. Stem Cells. 22:625–634.

Boomsma RA, Swaminathan PD, Geenen DL. 2007. Intravenously injected mesenchymal stem cells home to viable myocardium after coronary occlusion and preserve systolic function without altering infarct size. Int J Cardiol. 122:17–28.

Boquest AC, Noer A, Collas P. 2006. Epigenetic programming of mesenchymal stem cells from human adipose tissue. Stem Cell Rev. 2:319–329.

Borlongan CV et al. 2004. Central nervous system entry of peripherally injected umbilical cord blood cells is not required for neuroprotection in stroke. Stroke. 35:2385–2389.

Breymann C, Schmidt D, Hoerstrup SP. 2006. Umbilical cord cells as a source of cardiovascular tissue engineering. Stem Cell Rev. 2:87–92.

Bruder SP, Jaiswal N, Haynesworth SE. 1997. Growth kinetics, self-renewal, and the osteogenic potential of purified human mesenchymal stem cells during extensive subcultivation and following cryopreservation. J Cell Biochem. 64:278–294.

Buhring HJ et al. 2007. Novel markers for the prospective isolation of human MSC. Ann NY Acad Sci. 1106:262–271.

Campagnoli C et al. 2001. Identification of mesenchymal stem/progenitor cells in human first-trimester fetal blood, liver, and bone marrow. Blood. 98:2396–2402.

Caplan AI, Dennis JE. 2006. Mesenchymal stem cells as trophic mediators. J Cell Biochem. 98:1076–1084.

Carlin R et al. 2006. Expression of early transcription factors Oct4, Sox2, and Nanog by porcine umbilical cord (PUC) matrix cells. Reprod Biol Endocrinol. 4:8–21. DOI: 10.1186/1477-7827-4-8.

Castro-Malaspina H et al. 1980. Characterization of human bone marrow fibroblast colony-forming cells (CFU-F) and their progeny. Blood. 56:289–301.

Cheng Z et al. 2008. Targeted migration of mesenchymal stem cells modified with CXCR4 gene to infarcted myocardium improves cardiac performance. Mol Ther. 16:571–579.

Cho PS et al. 2007. Immunogenicity of umbilical cord tissue derived cells. Blood. 111:430–438. DOI: 10.1182/blood-2007-03-078774.

Conconi MT et al. 2006. CD105(+) cells from Wharton's jelly show in vitro and in vivo myogenic differentiative potential. Int. J Mol Med. 18:1089–1096.

Covas DT et al. 2003. Isolation and culture of umbilical vein mesenchymal stem cells. Braz J Med Biol Res. 36:1179–1183.

da Silva CL et al. 2005. A human stromal-based serum-free culture system supports the ex vivo expansion/maintenance of bone marrow and cord blood hematopoietic stem/progenitor cells. Exp Hematol. 33:828–835.

Dennis JE et al. 1999. A quadripotential mesenchymal progenitor cell isolated from the marrow of an adult mouse. J Bone Miner Res. 14:700–709.

Dominici M et al. 2006. Minimal criteria for defining multipotent mesenchymal stromal cells. The International Society for Cellular Therapy Position Statement Cytotherapy. 8:315–317.

Dzau VJ et al. 2005. Therapeutic Potential of Endothelial Progenitor Cells in Cardiovascular Diseases. Hypertension. 46:7–18.

Ennis J et al. 2008. In vitro immunologic properties of human umbilical cord perivascular cells. Cytotherapy. 10:174–181.

Ferrari G et al. 1998. Muscle regeneration by bone marrow-derived myogenic progenitors. Science. 279:1528–1530.

Friedenstein AJ. 1961. Osteogenetic activity of transplanted transitional epithelium. Acta Anat. (Basel). 45:31–59.

Friedman R et al. 2006. Co-Transplantation of Autologous Umbilical Cord Matrix Mesenchymal Stem Cells Improves Engraftment of Umbilical Cord Blood in NOD/SCID Mice ASH. Annual Meeting Abstracts. 108:2569.

Fu YS et al. 2004. Transformation of human umbilical mesenchymal cells into neurons in vitro. J Biomed Sci. 11:652–660.

Fu YS et al. 2006. Conversion of human umbilical cord mesenchymal stem cells in Wharton's jelly to dopaminergic neurons in vitro: Potential therapeutic application for Parkinsonism. Stem Cells. 24:115–124.

Gao L et al. 2006. Human umbilical cord blood-derived stromal cell, a new resource of feeder layer to expand human umbilical cord blood CD34+ cells in vitro. Blood Cells Mol Dis. 36:322–328.

Goodwin HS et al. 2001. Multilineage differentiation activity by cells isolated from umbilical cord blood: Expression of bone, fat, and neural markers. Biol Blood Marrow Transplant. 7:581–588.

Gotherstrom C et al. 2003. Immunomodulatory effects of human foetal liver-derived mesenchymal stem cells. Bone Marrow Transplant. 32:265–272.

Gotherstrom C et al. 2005. Difference in gene expression between human fetal liver and adult bone marrow mesenchymal stem cells. Haematologica. 90:1017–1026.

Grinnemo KH et al. 2004. Xenoreactivity and engraftment of human mesenchymal stem cells transplanted into infarcted rat myocardium. J Thorac Cardiovasc Surg. 127:1293–1300.

Guillot PV et al. 2007. Human first-trimester fetal. MSC express pluripotency markers and grow faster and have longer telomeres than adult. MSC Stem Cells. 25:646–654.

Heeschen C et al. 2004. Profoundly reduced neovascularization capacity of bone marrow mononuclear cells derived from patients with chronic ischemic heart disease 1. Circulation. 109:1615–1622.

Hermann A et al. 2004. Efficient generation of neural stem cell-like cells from adult human bone marrow stromal cells. J Cell Sci. 117:4411–4422.

Hiroyama T et al. 2007. Human umbilical cord-derived cells can often serve as feeder cells to maintain primate embryonic stem cells in a state capable of producing hematopoietic cells. Cell Biol Int. DOI: 10.1016/j.cellbi.2007.08.001.

Hoerstrup SP et al. 2002. Living, autologous pulmonary artery conduits tissue engineered from human umbilical cord cells. Ann Thorac Surg. 74:46–52.

Hoynowski SM et al. 2007. Characterization and differentiation of equine umbilical cord-derived matrix cells. Biochem Biophys Res Commun. 362:347–353.

Ito Y et al. 2006. Ex vivo expansion of human cord blood hematopoietic progenitor cells using glutaraldehyde-fixed human bone marrow stromal cells. J Biosci Bioeng. 102:467–469.

Jaiswal N et al. 1997. Osteogenic differentiation of purified, culture-expanded human mesenchymal stem cells in vitro. J Cell Biochem. 64:295–312.

Jang YK et al. 2006. Mesenchymal stem cells feeder layer from human umbilical cord blood for ex vivo expanded growth and proliferation of hematopoietic progenitor cells. Ann Hematol. 85:212–225.

Jiang W et al. 2006. Intravenous transplantation of mesenchymal stem cells improves cardiac performance after acute myocardial ischemia in female rats. Transpl Int. 19:570–580.

Jomura S et al. 2007. Potential treatment of cerebral global ischemia with Oct-4+ umbilical cord matrix cells. Stem Cells. 25:98–106.

Jones E, McGonagle D. 2008. Human bone marrow mesenchymal stem cells in vivo. Rheumatology. (Oxford). 47:126–131.

Kadiyala S et al. 1997. Culture expanded canine mesenchymal stem cells possess osteochondrogenic potential in vivo and in vitro. Cell Transplant. 6:125–134.

Karahuseyinoglu S et al. 2006. Biology of the Stem Cells in Human Umbilical Cord Stroma: In situ and in vitro. Surveys Stem Cells. 25:319–331.

Kawada H et al. 2004. Nonhematopoietic mesenchymal stem cells can be mobilized and differentiate into cardiomyocytes after myocardial infarction 11. Blood. 104:3581–3587.

Kern S et al. 2006. Comparative analysis of mesenchymal stem cells from bone marrow, umbilical cord blood, or adipose tissue. Stem Cells. 24:1294–1301.

Kim JW et al. 2004. Mesenchymal progenitor cells in the human umbilical cord. Ann Hematol. 83:733–738.

Kubota H, Avarbock MR, Brinster RL. 2004. Growth factors essential for self-renewal and expansion of mouse spermatogonial stem cells. Proc Natl Acad Sci USA. 101:16489–16494.

Lu LL et al. 2006. Isolation and characterization of human umbilical cord mesenchymal stem cells with hematopoiesis-supportive function and other potentials. Haematologica. 91:1017–1026.

Lund RD et al. 2007. Cells isolated from umbilical cord tissue rescue photoreceptors and visual functions in a rodent model of retinal disease. Stem Cells. 25:602–611.

Ma L et al. 2005. Human umbilical cord Wharton's Jelly-derived mesenchymal stem cells differentiation into nerve-like cells. Chin Med J (Engl.). 118:1987–1993.

Martinez C et al. 2007. Human bone marrow mesenchymal stromal cells express the neural ganglioside GD2: A novel surface marker for the identification of MSCs. Blood. 109:4245–4248.

Meida-Porada G et al. 1999. Cotransplantation of stroma results in enhancement of engraftment and early expression of donor hematopoietic stem cells in utero. Exp Hematol. 27:1569–1575.

Min JY et al. 2002. Significant improvement of heart function by cotransplantation of human mesenchymal stem cells and fetal cardiomyocytes in postinfarcted pigs. Ann Thorac Surg. 74:1568–1575.

Mitchell KE et al. 2003. Matrix cells from Wharton's jelly form neurons and glia. Stem Cells. 21:50–60.

Murry CE et al. 2004. Haematopoietic stem cells do not transdifferentiate into cardiac myocytes in myocardial infarcts. Nature (London). 428:664–668.

Nilsson SK et al. 2003. Hyaluronan is synthesized by primitive hemopoietic cells, participates in their lodgment at the endosteum following transplantation, and is involved in the regulation of their proliferation and differentiation in vitro. Blood. 101:856–862.

Noer A, Boquest AC, Collas P. 2007. Dynamics of adipogenic promoter DNA methylation during clonal culture of human adipose stem cells to senescence BMC. Cell Biol. 8:18–29.

Noiseux N et al. 2006. Mesenchymal stem cells overexpressing Akt dramatically repair infarcted myocardium and improve cardiac function despite infrequent cellular fusion or differentiation 1. Mol Ther. 14:840–850.

Orlic D et al. 2001. Bone marrow cells regenerate infarcted myocardium. Nature (London). 410:701–705.

Panepucci RA et al. 2004. Comparison of gene expression of umbilical cord vein and bone marrow-derived mesenchymal stem cells. Stem Cells. 22:1263–1278.

Rachakatla RS et al. 2007. Development of human umbilical cord matrix stem cell-based gene therapy for experimental lung tumors. Cancer Gene Ther. 14:828–835.

Rauscher FM et al. 2003. Aging, progenitor cell exhaustion, and atherosclerosis. Circulation. 108:457–463.

Romanov YA, Svintsitskaya VA, Smirnov VN. 2003. Searching for alternative sources of postnatal human mesenchymal stem cells: Candidate MSC-like cells from umbilical cord. Stem Cells. 21:105–110.

Sadler TW. disc2004. Second week of development: Bilaminar germ. 9:51–64.

Saito S et al. 2002. Isolation of embryonic stem-like cells from equine blastocysts and their differentiation in vitro. FEBS Lett. 531:389–396.

Sarugaser R et al. 2005. Human umbilical cord perivascular (HUCPV) cells: A source of mesenchymal progenitors. Stem Cells. 23:220–229.

Schmidt D et al. 2006. Engineering of biologically active living heart valve leaflets using human umbilical cord-derived progenitor cells. Tissue Eng. 12:3223–3232.

Seshareddy K, Troyer D, Weiss ML. 2008. Method to isolate mesenchymal-like cells from Wharton's Jelly of umbilical cord. Methods Cell Biol. 86:101–119.

Simmons PJ, Torok-Storb B. 1991. Identification of stromal cell precursors in human bone marrow by a novel monoclonal antibody, STRO-1. Blood. 78:55–62.

Studeny M et al. 2002. Bone marrow-derived mesenchymal stem cells as vehicles for interferon-beta delivery into tumors. Cancer Res. 62:3603–3608.

Studeny M et al. 2004. Mesenchymal stem cells: Potential precursors for tumor stroma and targeted-delivery vehicles for anticancer agents. J Natl Cancer Inst. 96:1593–1603.

Tang YL et al. 2004. Autologous mesenchymal stem cell transplantation induce VEGF and neovascularization in ischemic myocardium 1. Regul Pept. 117:3–10.

Tang J et al. 2006. Mesenchymal stem cells participate in angiogenesis and improve heart function in rat model of myocardial ischemia with reperfusion. Eur J Cardiothorac Surg. 30:353–361.

Troyer DL, Weiss ML. 2008. Wharton's jelly-derived cells are a primitive stromal cell population. Stem Cells. 26:591–599.

Wakitani S et al. 2003. Embryonic stem cells injected into the mouse knee joint form teratomas and subsequently destroy the joint. Rheumatology (Oxford). 42:162–165.

Wang HS et al. 2004. Mesenchymal Stem Cells in the Wharton's Jelly of the Human Umbilical Cord. Stem Cells. 22:1330–1337.

Wang JF et al. 2004. Mesenchymal stem/progenitor cells in human umbilical cord blood as support for ex vivo expansion of CD34(+) hematopoietic stem cells and for chondrogenic differentiation. Haematologica. 89:837–844.

Weiss ML et al. 2006. Human umbilical cord matrix stem cells: Preliminary characterization and effect of transplantation in a rodent model of Parkinson's disease. Stem Cells. 24:781–792.

Wu KH et al. 2007. In vitro and in vivo differentiation of human umbilical cord derived stem cells into endothelial cells. J Cell Biochem. 100:608–616.

Wu KH et al. 2007a. Therapeutic potential of human umbilical cord derived stem cells in a rat myocardial infarction model. Ann Thorac Surg. 83:1491–1498.

Wu KH et al. 2007b. Human umbilical cord derived stem cells for the injured heart. Med Hypotheses. 68:94–97.

Yarygin KN et al. 2006. Comparative study of adult human skin fibroblasts and umbilical fibroblast-like cells. Bull Exp Biol Med. 141:161–166.

Ye ZQ et al. 1944. Establishment of an adherent cell feeder layer from human umbilical cord blood for support of long-term hematopoietic progenitor cell growth. Proc Natl Acad Sci USA. 91:12140–12144.

Young RG et al. 1998. Use of mesenchymal stem cells in a collagen matrix for Achilles tendon repair. J Orthop Res. 16:406–413.

Zhang H et al. 2005. Increasing donor age adversely impacts beneficial effects of bone marrow but not smooth muscle myocardial cell therapy. Am J Physiol Heart Circ Physiol. 289:H2089–H2096.

7

PERINATAL ENDOTHELIAL PROGENITOR CELLS

Curtis L. Cetrulo, Jr. and Margaret J. Starnes

Division of Plastic and Reconstructive Surgery, Department of Surgery, Keck School of Medicine, University of Southern California, Los Angeles, CA 90069

THE ROLE OF VASCULAR STEM CELLS IN POSTNATAL BLOOD VESSEL FORMATION: ENDOTHELIAL PROGENITOR CELLS AND POSTNATAL VASCULOGENESIS

Hypotheses regarding the role of stem cells in disease and in the physiologic response to injury have exploded in the past decade with new insights into stem cell biology and the mysteries of stem cell quiescence and self-replication. A role for stem cells in many cancers has been postulated (with obvious tremendous therapeutic potential) and has led to a paradigm shift in cancer research. Similarly, basic scientists have begun to examine the role of stem cells in developmental processes and morphogenesis to gain insight to the body's response to injury by stem cells.

Neovascularization represents an additional biologic process that has recently been reexamined with regard to a role for stem cells, in particular, by revisiting the participation of vascular stem cells in development. Neovascularization is known to occur in the fetus by vasculogenesis (defined as blood vessel formation from "islands" of distinct vascular stem cell pools that form vessels *de novo*). In contrast, in the *postnatal* period, new vessel growth had previously been attributed solely to the process of angiogenesis, or the sprouting of microvessels from a preexisting capillary or arterial network.

In 1997, however, Asahara's description of the role of adult endothelial progenitor cells in new vessel formation suggested that perhaps vasculogenic processes occur postnatally, as well as in development [Asahara, 1997] and provided a paradigm shift by implicating vascular stem cells in postnatal neovascularization. This hypothesis has led to an explosion

Perinatal Stem Cells. Edited by C. L. Cetrulo, K. J. Cetrulo, and C. L. Cetrulo, Jr.

in vascular stem cell research and has enormous translational potential for regenerative medicine and tissue engineering applications [Asahara, 1999; Forrest, 2005; Hristov, 1999; Shaw, 2004; Shi, 1998].

SOURCES OF ENDOTHELIAL PROGENITOR CELLS

Endothelial progenitor cells (EPCs) were classically thought to be bone marrow derived cells circulating in the adult peripheral blood. Recently, however, EPCs have been identified in umbilical cord blood, the placenta and have been derived from the human umbilical vein endothelium [Au, 2008; Goessler, 2006; Hristov, 2003; Lin, 2000; Reyes, 2002; Tepper, 2003].

During embryonic development, three primary germ layers emerge: ectoderm, mesoderm, and endoderm. As embryogenesis proceeds, the mesoderm develops into various stem cell types, in particular hematopoietic and mesenchymal stem cells [Risau, 1988; Thomson, 1998]. Endothelial progenitor cells are most likely derived from either a hematopoietic stem cell or from a hemangioblast, a common precursor to both hematopoietic stem cells and endothelial progenitor cells [Loges, 2004]. Some disagreement exists regarding the definitive phenotype of an EPC with regard to cell-surface marker profile, as any such definition represents only developmental "snapshot" dependent on the degree of differentiation [Eguchi, 2007; Peichev, 2000].

Most investigators agree, however, that EPCs are non-leukocyte, bone marrow derived, multipotent stem cells that circulate in the blood and have the potential to differentiate into endothelial cells and participate in vascular tubule formation. In addition, most vascular biologists agree that, at minimum, EPCs express the marker CD133, which is not found on mature endothelial cells, and CD34, which is not found on the pluripotent stem cells from which EPCs are derived (i.e., hematopoietic stem cells) [Crosby, 2000; Dong, 2007; Eguchi, 2007; Murayama, 2002; Peichev, 2000; Urbich, 2004].

Early studies suggested that adult EPCs are normally recruited from the bone marrow to assist in various normal and pathological processes where new vessel growth is needed, including endometrial remodeling, colon cancer growth, and myocardial infarction [Tepper, 2005]. Other work demonstrated that these cells home to areas of active vascularization and express the vascular endothelial growth factor receptor 2 (VEGF-R2) [Crosby, 2000; Dong, 2007; Murayama, 2002; Peichev, 2000; Urbich, 2004]. In addition, it has been shown that fibronectin and VEGF promote the migration and differentiation of EPCs [Wijelath, 2004]. The EPCs are thought to promote healing of the endothelial monolayer and the vessel wall primarily through the processes of neoendothelialization and neovascularization [Ward, 2007].

In healthy adult individuals, circulating EPCs comprise approximately 0.01% of peripheral blood cells [Hristov, 2003]. Negative regulators of EPCs include aging, hypercholesteremia, diabetes, and smoking [Fadini, 2005; Heiss, 2005; Kondo, 2004; Rauscher, 2003]. Positive regulators include exercise, hyperbaric oxygen treatment, and exogenous drug therapy including VEGF, granulocyte colony stimulating factor (G-CSF), and statins [Kalka, 2000a, b; Steiner, 2005; Walter, 2002].

EPCs, VASCULOGENESIS AND ANGIOGENESIS

Angiogenesis (the sprouting of new vessels from preexisting ones) is a dynamic process involving many cell types and growth factors, and is induced by hypoxia-mediated vascular permeability and orchestrated by cytokine and growth factor-mediated migration of differentiated endothelial cells into the extracellular matrix. Vasculogenesis, however, is a less well-understood process in which EPCs appear to play a critical role. The exact ratio of angiogenesis to vasculogenesis in the body is uncertain, although the discovery of EPCs has dramatically increased investigation in this area [Folkman, 1971; Magri, 2007].

Early studies suggest that EPCs are most active in areas that have undergone large vascular insults or trauma, rather than in normal vessel cell turnover. Early work by Tepper and others found that EPCs can form tubules in extracellular matrices *in vitro* and *in vivo* and are able to induce vascular invasion by host tissue if implanted [Kaushal, 2001; Tepper, 2002]. In a recent 2006 study by Capla et al., replacement of donor graft vasculature by EPCs from the recipient represented the predominant mechanism for skin graft revascularization, suggesting at minimum a concomitant relationship between angiogenesis and vasculogenesis in graft revascularization [Capla, 2006]. Kung and co-workers have recently published date suggesting that *in vivo* perfusion of human skin substitutes could be achieved with microvessels formed by adult circulating EPCs [Kung, 2008].

The EPCs and vasculogenesis have also been implicated in the surgical technique of "vascular delay." Using this technique, surgeons render a tissue ischemic by incising the skin around a flap before transfer to induce an increase in vascularity in the unincised area. Increased vascularity allowed a greater area of the unincised tissue to be transferred. Based on early work, this increased vascularity was believed to be a result of ischemia, sympathetic nerve exhaustion, and dilation of "watershed choke vessels," which allowed reorientation of vessels within the graft, secretion of growth factors, and finally neovascularization. Many early clinical applications demonstrated that this technique was effective in improving the survival of soft tissue flaps used in surgical reconstruction, as well as in the size of the flap available for transfer, after a "vascular delay" of 8–10 days. However, until recently, a role for EPCs in this process had not yet been postulated. Tepper and co-workers reexamined the vascular delay phenomenon in the mouse flap model and demonstrated bone marrow derived endothelial progenitor cell recruitment to the ischemic "delayed" tissue within 72, in proportion to rising levels of VEGF [Ghali, 2007]. By day 14, the EPCs had coalesced into vascular cords and at by day 21 had developed into functional vessels. Although low numbers of EPCs were found in flap tissue at 3 and 7 days postoperatively, the contribution of EPCs was maximal at 2 weeks. Taken with the observance of angiogenic mechnanisms in early revascularization, these results suggested that perhaps angiogenesis is an "early" response to hypoxemia and vasculogenesis is a "late" response initiated by EPCs.

Note that several of these aforementioned studies have been conducted with EPCs from different anatomic sources. Some studies have used bone marrow derived/peripheral blood cells and others have used umbilical cord blood cells. More recent studies are also using cells induced from mesenchymal progenitor cells or myeloid cells. It is unclear yet what differences may lay between these EPCs and the impact this may have on tissue engineering and neovascularization [Lineaweaver, 2004; Milton, 1965; Myers, 1969; Wang, 1996; Wong, 2004].

Techniques to induce pluripotency and "stemness" have pervaded stem cell research directives in recent years and EPCs have been investigated in this regard. The "EPC-like" cells have been induced from other stem cell types, including myeloid progenitor cells, mesenchymal progenitor cells, and cardiac stem cells [Baal, 2004].

EPCs AND TISSUE ENGINEERING

Tissue engineering is most often accomplished by seeding a cell type of interest on a biocompatible scaffold. The construct then requires subsequent vascular ingrowth for survival.

The amount of tissue grown by this model is limited by the extent of this vascular ingrowth. Therefore, creating a vascularized scaffold first, on which to subsequently seed a selected cell type, is a promising alternative for the engineering of complex tissue. With new understanding of endothelial stem cells, "therapeutic vasculogenesis" with EPCs has become a rapidly expanding area of interest for the vascularization of engineered tissue [Finkenzeller, 2007].

Therapeutic vasculogenesis offers several advantages to the tissue engineer. For one, EPCs can deliver both the substrate and the growth factors–cytokines involved in neovascularization [Suuronen, 2006]. Furthermore, in tissue engineering, EPCs may be critical in the patency of vascular grafts. Kaushal et al. demonstrated that the duration of vascular patency in grafts seeded with EPCs was longer than that of nonseeded grafts [Amiel, 2006]. Tepper also observed that EPC supplementation in synthetic grafts enhanced thrombus regression [Tepper, 2003]. Finkenzeller and co-workers recently investigated the utility of EPCs in the *ex vivo* vascularization of tissue-engineered grafts as compared to human umbilical vein endothelial cells (HUVECs). Their study found that both EPCs and HUVECs formed vascular structures in both two- and three-dimensional (2D and 3D) models, however, EPCs showed a significantly decreased level of apoptosis in serum-deprived conditions. The increased survival capacity of EPCs and their ability to perform well in various matrices make them an ideal autologous cell source for vascularization of *ex vivo* engineered tissues [Finkenzeller, 2007].

Other researchers have explored the augmentation of vasculogenesis on different biocompatible scaffolds and with various growth factors. Some studies have shown that fibrin and collagen based matrices were effective in delivering EPCs to ischemic tissue *in vivo*, thereby allowing these cells to incorporate into vascular structures [Sales, 2007]. Other studies have focused on the use of biodegradable–biocompatible elastomers as scaffolds for soft tissue engineering with EPCs. Polyglycerol sebacate (PGS) and poly-1, 8-co-citrate (POC) are two elastomers in particular that have been shown to be an appropriate scaffold material for blood vessel tissue engineering. Other work has focused on nanocomposites, such as polyhedral oligomeric silsesquioxane (POSS) with incorporated bioactive peptides [Gao, 2007]. VEGF, basic fibroblast growth factor, and fibronectin are some of the many cytokines and growth factors implicated in the mobilization and differentiation of EPCs [Conklin, 2004]. Recently, researchers found elastomer scaffold precoating with such extracellular matrix proteins as these allowed more precise "engineering" of cellular behavior in the development of engineered heart valves with EPCs. Scaffold precoating led to increased production of the desired cell phenotype [Alobaid, 2006]. Another group of researchers found that stimulation of EPC seeded tissue engineering scaffolds with TGF-beta1 *in vitro* resulted in a more organized cellular architecture with glycoprotein, collagen, and elastic synthesis, than scaffolds not stimulated with TGF-beta1 [Sales, 2006].

EPCs AND CARDIOVASCULAR TISSUE ENGINEERING

The EPCs have been studied in regard to therapies for peripheral artery disease, cardiac remodeling after myocardial infarction, artificial heart valves, and vascular healing [Fang, 2007; Mendelson, 2007]. There is evidence that EPCs play a role in the repair of damaged blood vessels following myocardial infarction. In fact, higher levels of circulating EPCs detected in the bloodstream predict for better outcomes and fewer repeat heart attacks [Werner, 2005]. In another study, it was demonstrated that EPCs isolated from human umbilical cord blood could differentiate into endothelial cells *in vitro* and form a functional endothelium atop decellularized heart valve scaffolds [Schmidt, 2006]. One group implanted pulmonary heart valves engineered with autologous endothelial progenitor cells in an 11- and a 13-year old child and followed the children 3.5 years postimplantation. They found over the 3 years that the heart valve annulus increased, the postoperative regurgitation decreased, the transvalvular gradient decreased in one patient and increased in the other, the end-diastolic ventricle diameter decreased or remained the same, and the heart's body surface area increased. These findings suggested that tissue engineering of heart valves using autologous EPCs is a feasible and safe method for pulmonary valve replacement [Cebotari, 2006].

In other recent studies involving urologic tissue engineering, investigators found that decellularized porcine small bowel could be recellularized with smooth muscle cells, urothelial cells, and EPCs for function as a neobladder. Upon reimplantation into a porcine model, the reseeded neobladder remained thrombus free with intact perfusion for 1–3 h, while unseeded bowel showed perfusion stagnation and thrombus formation after 30 min [Schultheiss, 2005]. Other studies have also studied the experimental generation of functional and vascularized trachea [Walles, 2004]. Capla and co-workers recently conducted a study that suggested EPC dysfunction might contribute to altered wound healing in diabetic patients [Capla, 2007].

THE FUTURE: VASCULARIZED ENGINEERED TISSUE

Findings from many of these studies will prove essential in the developing understanding of EPCs from both perinatal and adult sources and will impact "vascularized" tissue engineering for numerous applications. For example, in a recent study by Mukai, the authors investigated the mechanism by which EPCs promote the formation of new vessels. By using early EPCs from human peripheral blood, late EPCs from umbilical cord blood, and HUVECs, the authors discovered that late EPCs and HUVECs formed tubular structures, whereas early EPCs randomly migrated and failed to form such structures. Late EPCs appeared to participate in tubule formation with HUVECs and migrated toward preexisting structures constructed by HUVECs. In contrast, early EPCs promoted the sprouting of HUVECs from tubular structures. These observations suggested that early EPCs may cause the disorganization of preexisting vessels (or possibly angiogenesis) and late EPCs constitute and orchestrate vascular tube formation. Furthermore, there appears an inextricable link between angiogenic endothelial cells and vasculogenic adult endothelial progenitor cells in neovascularization [Mukai, 2008].

REFERENCES

Allen J et al. 2008. Characterization of porcine circulating progenitor cells: Toward a functional endothelium. Tissue Eng. 14(1):183–194.

Alobaid N et al. 2006. Nanocomposite containing bioactive peptides promote endothelialisation by circulating progenitor cells: An in vitro evaluation. Eur J Vascular Endovascular Surg. 32(1):76–83.

Amiel GE et al. 2006. Engineering of blood vessels from acellular collagen matrices coated with human endothelial cells. Tissue Eng. Aug; 12(8):2355–2365.

Asahara T et al. 1997. Isolation of putative progenitor endothelial cells for angiogenesis. Science. 275:964–967.

Asahara T et al. 1999. Bone marrow origin of endothelial progenitor cells responsible for postnatal vasculogenesis in physiologic and pathologic neovascularization. Circulation Res. 85:221.

Au P et al. 2008. Differential in vivo potential of endothelial progenitor cells from human umbilical cord blood and adult peripheral blood to form functional long-lasting vessels. Blood. 111(3):1302–1305.

Baal N et al. 2004. Expression of transcription factor Oct-4 and other embryonic genes in CD133 positive cells from human umbilical cord blood. Thromb Haemost. Oct; 92(4):767–775.

Capla JM et al. 2006. Skin graft vascularization involves precisely regulated regression and replacement of endothelial cells through both angiogenesis and vasculogenesis. Plastic Reconstructive Surg. 117(3):836–844.

Capla JM et al. 2007. Diabetes impairs endothelial progenitor cell-mediated blood vessel formation in response to hypoxia. Plast Reconstr Surg. 2007 Jan; 119(1):59–70.

Cebotari S et al. 2006. Clinical application of tissue engineered human heart valves using autologous progenitor cells. Circulation. 114(1 Suppl):1132–1137.

Conklin BS et al. 2004. Basic fibroblast growth factor coating and endothelial cell seeding of a decellularized heparin-coated vascular graft. Artificial Organs. 28(7):668–675.

Crosby JR et al. 2000. Endothelial cells of hematopoietic origin make a significant contribution to adult blood vessel formation. Circulation Res. 87:728.

Dong C et al. 2007. Endothelial progenitor cells: A promising therapeutic alternative for cardiovascular disease. Journal of Interventional Cardiology. 20:93–99.

Eguchi M et al. 2007. Endothelial progenitor cells for postnatal vasculogenesis. Clinics Exp Nephro. 11:18–25.

Fadini GP et al. 2005. Circulating endothelial progenitor cells are reduced in peripheral vascular complications of type 2 diabetes mellitus. J Am College Cardiol. 45:1149–1157.

Fang NT et al. 2007. Construction of tissue-engineered heart valves by using decellularized scaffolds and endothelial progenitor cells. Chinese Med J. 120(8):696–702.

Finkenzeller G et al. 2007. In vitro angiogenesis properties of endothelial progenitor cells: A promising tool for vascularization of ex vivo engineered tissues. Tissue Eng. 13(7):1413–1420.

Folkman J et al. 1971. Tumor angiogenesis: Therapeutic implications. New England J Med. 285:1182.

Forrest CR. 2005. What's new in plastic and maxillofacial surgery. J Ame College Surg. 200(3):399–408.

Gao J et al. 2007. Poly(glycerol sebacate) supports the proliferation and phenotypic protein expression of primary baboon vascular cells. J Biomed Mater Res. 83(4):1070–1075.

Ghali S et al. 2007. Vascular delay revisited. Plastic Reconstructive Surg. 119(6):1735–1744.

Goessler UR et al. 2006. Perspectives of gene therapy in stem cell tissue engineering. Cells Tissues Organs. 183:169–179.

Heiss C et al. 2005. Impaired progenitor cells activity in age-related endothelial dysfunction. Am College Cardiol. 45:1141–1148.

Hristov M et al. 1999. Endothelial progenitor cells: Isolation and characterization. Trends Cardiovascular Med. 13(5):201–206.

Hristov M et al. 2003. Endothelial progenitor cells: Mobilization, differentiation, and homing. Arteriosclerotic Thrombotic Vascular Biol. 23:1185–1189.

Kalka C et al. 2000a. Vascular endothelial growth factor(165) gene augments circulating endothelial progenitor cells in human subjects. Circulation Res. 86:1198.

Kalka C et al. 2000b. VEGF gene transfer mobilizes endothelial progenitor cells in patients with inoperable coronary disease. Ann Thoracic Surg. 70:829.

Kaushal S et al. 2001. Functional small-diameter neovessels created using endothelial progenitor cells expanded ex vivo. Nat. Med. 7:1035.

Kondo T et al. 2004. Smoking cessation rapidly increases circulating endothelial progenitor cells in peripheral blood in chronic smokers. Arteriosclerotic Thrombotic Vascular Biol. 24:1442–1447.

Kung EF et al. 2008. In vivo perfusion of human skin substitutes with microvessels formed by adult circulating endothelial progenitor cells. Dermatologic Surg. 34(2):137–146.

Lin Y et al. 2000. Origins of circulating endothelial cells and endothelial outgrowth from blood. J Clin Inves. 105:71–77.

Lineaweaver WC. 2004. Vascular endothelium growth factor, surgical delay, and skin flap survival. Ann Surg. 239:866.

Loges S et al. 2004. Identification of the adult human hemangioblast. Stem Cells Devel. 13:229–242.

Magri D et al. 2007. Endothelial progenitor cells: A primer for vascular surgeons. Vascular. 15(6):382–394.

Mendelson K et al. 2007. Healing and remodeling of bioengineered pulmonary artery patches implanted in sheep. Cardiovascular Pathol. 16(5):277–282.

Milton S et al. 1965. The effects of delay on the survival of experimental studies on pedicled skin flaps. Br J Plastic Surg. 22:244.

Mukai N et al. 2008. A comparison of the tube forming potentials of early and late endothelial progenitor cells. Experimental Cell Research. 314(3):430–440.

Murayama T et al. 2002. Determination of bone marrow-derived endothelial progenitor cell significance to angiogenic growth factor-induced neovascularization in vivo. Exp Hematol. 30:967.

Myers MB. 1969. Mechanism of the delay phenomenon. Plastic Reconstructive Surg. 44:52.

Peichev M et al. 2000. Expression of VEGFR-2 and AC133 by circulating human CD34(+) cells indentifies a population of functional endothelial precursors. Blood. 95:952–958.

Rauscher FM et al. 2003. Aging, progenitor cell exhaustion, and atherosclerosis. Circulation. 108:457–463.

Reyes M et al. 2002. Origin of endothelial progenitors in human postnatal bone marrow. Clini Inves. 109:337–346.

Risau W et al. 1988. Vasculogenesis and angiogenesis in embryonic-stem-cell-derived embryoid bodies. Development. 102:471.

Sales VL et al. 2006. Transforming growth factor-beta1 modulates extracellular matrix production, proliferation, and apoptosis of endothelial progenitor cells in tissue-engineering scaffolds. Circulation. 114(1 Suppl):1193–1199.

Sales VL et al. 2007. Protein precoating of elastomeric tissue-engineering scaffolds increased cellularity, enhanced extracellular matrix protein production, and differentially regulated the phenotypes of circulating endothelial progenitor cells. Circulation. 116(11 Suppl):155–163.

Schmidt D et al. 2005. Living patches engineered from human umbilical cord derived fibroblasts and endothelial progenitor cells. Eur J Cardiothoracic Surg. 27(5):795–800.

Schmidt D et al. 2006. Engineering of biologically active living heart valve leaflets using human umbilical cord-derived progenitor cells. Tissue Eng. 12(11):3323–3332.

Schultheiss D et al. 2005. Biological vascularized matrix for bladder tissue engineering: Matrix preparation, reseeding technique, and short-term implantation in a porcine model. J Urol. 173(1):276–280.

Shaw J et al. 2004. Hematopoietic stem cells and endothelial cell precursors express Tie-2, CD31, and CD45. Blood Cells, Molecules, and Diseases. 32(1):168–175.

45. Shi Q et al. 1998. Evidence for circulating bone marrow-dervied endothelial cells. Blood. 92:362.

Steiner S et al. 2005. Endurance training increases the number of endothelial progenitor cells in patients with cardiovascular risk and coronary artery disease. Atherosclerosis. 181:305–310.

Suuronen EJ et al. 2006. Promotion of angiogenesis in tissue engineering: Developing multicellular matrices with multiple capacities. Inter J Artificial Org. 29(12):1148–1157.

Tepper OM et al. 2002. Endothelial progenitor cells isolated from human type-II diabetics exhibit impaired adhesion, proliferation, and incorporation into vascular structures. Circulation. 106:2781.

Tepper OM et al. 2003. Endothelial progenitor cells: The promise of vascular stem cells for plastic surgery. Plast Reconstr Surg. Feb; 111(2):846–854.

Tepper OM et al. 2003. Endothelial progenitor cells: The promise of vascular stem cells for plastic surgery. Plastic Reconstructive Surg. 111(2):846–854.

Tepper OM et al. 2005. Adult vasculogenesis occurs through in situ recruitment, proliferation, and tubulization of circulating bone marrow-derived cells. Blood. 105:1068.

Thomson JA et al. 1998. Embryonic stem cell lines derived from human blastocysts. Science. 282:1145.

Urbich C et al. 2004. Endothelial progenitor cells: Characterization and role in vascular biology. Circulation Res. 95:343.

Walles T et al. 2004. Experimental generation of a tissue-engineered functional and vascularized trachea. J Thoracic Cardiothoracic Surg. 128(6):900–906.

Walter DH et al. 2002. Statin therapy accelerates reendothelialization: A novel effect involvling mobilization and incorporation of bone marrow-derived endothelial progenitor cells. Circulation. 105:3017–3024.

Wang WZ et al. 1996. Ischemic preconditioning versus intermittent reperfusion to improve blood flow to a vascular isolated skeletal muscle flap of rats. J Trauma. 45:953.

Ward MR et al. 2007. Endothelial progenitor cell therapy for treatment of coronary disease, acute MI, and pulmonary arterial hypertension: Current perspectives. Catheterization Cardiovascular Interventions. 70(7):983–998.

Werner et al. 2005. Circulating endothelial progenitor cells and cardiovascular outcomes. N Engl J Med. Sep. 8; 353(10):999–1007.

Wijelath ES et al. 2004. Fibronectin promotes VEGF-induced CD34 cell differentiation into endothelial cells. J Vascular Surg. 39(3):655–660.

Wong MS et al. 2004. Basic fibroblast growth factor expression following surgical delay of rat tranverse abdominis myocutaneous flaps. Plastic and Reconstructive Surg. 113:2030.

Xia L et al. 2007. Decrease and dysfunction of endothelial progenitor cells in umbilical cord blood with maternal pre-eclampsia. J Obstetrics Gynaecology Res. 33(4):465–474.

8

UMBILICAL CORD DERIVED MAST CELLS AS MODELS FOR THE STUDY OF INFLAMMATORY DISEASES

Taxiarchis Kourelis, Duraisamy Kempuraj, and Akrivi Manola

Departments of Pharmacology and Experimental Therapeutics, Tufts University School of Medicine, Tufts Medical Center, Boston, MA 02111

Theoharis C. Theoharides

Laboratory of Molecular Immunopharmacology and Drug Discovery; Department of Pharmacology and Experimental Therapeutics; and Departments of Internal Medicine and Biochemistry, Tufts University School of Medicine, Tufts Medical Center, Boston, MA 02111

INTRODUCTION

Mast cells develop from CD34+, c-kit+ progenitor cells that arise from hematopoietic stem cells in the bone marrow [Kirshenbaum, 1999]. These progenitor cells express the receptor c-kit for Stem Cell Factor (SCF), which has growth, differentiative, and chemoattractant properties for mast cells (MCs).

Mast cell progenitors have also been described in the peripheral blood, where they circulate as mononuclear leukocytes lacking secretory granules and expressing CD13, CD33, CD38, CD34, and Kit [Czarnetzki, 1984; Valent, 1992]. Mast cells mature in tissues depending on microenvironmental conditions. In addition to SCF, mast cell chemoattractants include, nerve growth factor (NGF) [Aloe and Levi-Montalcini, 1977], RANTES (CCL5), and MCP 1 (CCL2) [Conti, 1997, 1998]. Several cytokines play an important role in mast cell differentiation and proliferation. The IL-3 directly stimulates proliferation of uncommitted progenitors and directly promotes granule assembly in rodent mast cells, while IL-9 may have similar effects on human MCs [Matsuzawa, 2003; Tsuji, 1990]. The IL-4 possesses mast cell growth factor activity [Bischoff, 1999], can promote phenotype

Perinatal Stem Cells. Edited by C. L. Cetrulo, K. J. Cetrulo, and C. L. Cetrulo, Jr.

switching to connective tissue type MCs [Toru, 1998], and enhances expression of neuropeptide receptors [van der Kleij, 2003]. However some studies showed that IL-4 inhibits SCF depended differentiation of human mast cells [Sillaber, 1994], suggesting therefore, that the *in vitro* response of mast cells to IL-4 could be modulated by the microenvironment. Thrombopoietin also has been shown to stimulate SCF depended MC development [Sawai, 1999].

However, the phenotypic expression of MCs does not appear to be fixed [Bischoff, 1999; Levi-Schaffer, 1986]. Mature mast cells vary considerably [Tainsh and Pearce, 1992] in their cytokine [Bradding, 1995] and proteolytic enzyme content. Differences in the concentration of SCF and certain other cytokines in the various tissues, and perhaps within sites of the same tissue, can determine the number, distribution, and phenotype of MCs at that site [Metcalfe, 1997]; this characteristic is referred to as MC heterogeneity and makes MCs capable of responding to the diverse stimuli they are likely to come across in various tissues. Mast cells differ also in their staining properties, response to stimuli, and expression of mediators and receptors. For example, there is a great variability in the expression of chymase among mast cells located in different tissues. Connective tissue-type MCs (CTMC), predominantly located in normal skin and in intestinal submucosa, contain tryptase, chymase, MC carboxypeptidase, and cathepsin G in their secretory granules. They are therefore named MC_{TC}. Mucosal-type MCs (MMC), the main type of MCs in normal alveolar wall and in small intestinal mucosa, contain tryptase in their secretory granules, but lack the other proteases. These are named MC_T [Irani, 1986]. In addition, MCs of the skin have an increased expression of FcγRII [Ghannadan, 1998] and human lung MCs of CD50 (ICAM-3) [Babina, 1999], activated MCs of various sources overexpress ICAM-1 antigen and b2 intergrins, complement receptors, as well as various cytokines receptors [Valent, 1994; Weber, 1997]. In fact, brain MCs in normal rodents lack c-kit receptor [Shanas, 1998] and FcεRI proteins [Dimitriadou, 1990]. The MC_T do not seem to respond to neuropeptides [Goetzl, 1985] while MC_{TC} do [Shanahan, 1985]. The MC_T contain mostly chondroitin sulfate in their granules, while MC_{TC} contain heparin [Scully, 1986; Serafin, 1986]. The FcepsilonRI aggregation promotes survival of connective tissue-like MCs, but not mucosal-like MCs [Ekoff, 2007]. Several studies also showed the phenotypic heterogeneity of MCs in various diseases. For example, in atheromatous plaques, the most abundant type found were MC_{TC} cells [Jeziorska, 1997], in rheumatoid arthritis synovial MCs were found to express C5aR [Kiener, 1998]. In addition, the phenotypic profile of MCs in mastocytosis seemed to differ greatly than that of normal patients [Valent, 2001]. It has been shown also that there is a phenotypic variation in MC responsiveness to nitric oxide [Koranteng, 2000].

Mast cells are important for allergic reactions [Metcalfe, 1981], but also in immunity [Galli, 2005a; Puxeddu, 2003; Wedemeyer, 2000], and inflammatory conditions [Galli, 2005b; Theoharides and Cochrane, 2004; Theoharides and Kalogeromitros, 2006]. Animal models of bacterial and parasitic infections have shown that MCs have essential roles in host defense, and in some models the presence of MCs is crucial for host survival. Mast cells respond to pathogen signals, such as bacteria, parasites, viruses, and toxins; depending on the extend of the trigger, the tissue and other factors mast cell mediators may deactivate noxious molecules or respond to them with massive degranulation. They can also degrade endogenous as well as exogenous (i.e., snake venoms) toxins [Metz and Maurer, 2007]. It is known that mast cells can phagocytose pathogens, interestingly, recent results indicate that MCs can also kill bacteria extracellularly, using reactive oxygen species (ROS), by structures composed of DNA, histones, tryptase, and the antimicrobial peptide LL-37 [von Kockritz-Blickwede, 2008].

Another fact indicating that MCs are involved in innate immune responses is that they express a number of toll-like receprtors (TLRs). These receptors were shown to be important in recognition of ligands associated with bacterial or viral infections [Aderem and Ulevitch, 2000; Rock, 1998]. Ten human TLRs have been identified so far [Aderem and Ulevitch, 2000; Akira, 2001; Heine and Lien, 2003]. Rodent MCs express bacterial TLR 2 and 4 [McCurdy, 2003; Varadaradjalou, 2003]. Human MCs express viral TLR-9 [Ikeda, 2003], activation of which produced IL-6 [Ikeda, 2003], while TLR-3 activation produced IFN [Kulka, 2004]. It has been shown that expression of MHC class II molecules by MCs is enhanced after TLR ligation, and that MCs can directly present antigen to T cells [Marshall, 2004]. They cross-talk with epithelial cells causing loosening of the tight junctions leading to edema, as well as inducing expression of adhesive molecules on their surface leading to accumulation of inflammatory cells to the site of inflammation; However, nonpathogenic *Escherichia coli* strains were found to inhibit degranulation of murine MCs under all activator conditions suggesting, thereby a possible basis for future antiallergic treatment or preventive measures [Magerl, 2008].

Mast cells are also known to contribute to wound repair. In recent studies [Weller, 2006] MC deficient (Kit^{W}/Kit^{W-v}) mice had reduced neutrophil recruitment, as well as impaired wound healing, especially during the early phase. In another study, mast cells were shown to degranulate during the early inflammatory phase, but not the second proliferative stage [Weller, 2006]; however, during the latter, mast cells can release angiogenic and growth promoting factors selectively without degranulation (see below).

The perivascular localization of MCs [Robinson-White and Beaven, 1982] in close proximity to neurons [Bienenstock, 1987; Dvorak, 1992b; Letourneau, 1996; Newson, 1983; Pang, 1995; Stead, 1987; Theoharides, 1996], specifically CRH-positive neurons [Theoharides, 1995b], makes the mast cells prime candidates to participate in neuroinflammatory diseases. However, as discussed above, mast cells in such diseases and the respective affected tissues appear to have different mediator content and secretory characteristics. Consequently, one needs to either isolate mast cells from those tissues or attempt to develop *in vitro* models. However, there are few tissues accessible to biopsy (e.g., skin and lung) from which MCs may be disaggregated and purified [Levi-Schaffer, 1995]. Moreover, the starting material is limited, the yield is low, trypsinization often affects mast cell viability, primary cells do not proliferate, and most importantly biopsy donors have been on numerous drug regimens (e.g., immunomodulators or cancer treatments) making any results very difficult to interpret and reproduce by other investigators. A number of human leukemic mast cells also have been used; these include the HMC-1 line [Alexandrakis, 2003a, b] that do not require SCF, do not respond to IgE or neuropeptides, and the LAD2 cells, which are grown using SCF and are more closely resemble CTMCs [Drexler and MacLeod, 2003; Kirshenbaum, 2003] since they respond to neuropeptides [Kulka, 2007]. However, LAD2 are still leukemic mast cells that are making it difficult to ascertain their relevance to normal mast cells or other MC related diseases.

The most reasonable solution to the problem of human mast cell culture is the growth of mast cells from progenitors in which case not only primary cultures are possible, but such cells may be induced to "mature" toward distinct phenotypes given addition of cytokines [Ochi, 2000] and growth factors relevant to the specific tissues–diseases in question. Such MCs have been grown using umbilical cord blood progenitor cells to generate human cord blood cultured mast cells (hCBMCs). The cells mostly have characteristics of MMC, unless IL-4 is added as described below.

The hCBMCs are derived by the culture of CD34+ progenitor cells as previously described [Kempuraj, 1999] with minor modifications. Briefly, mononuclear cells are isolated by pipeting heparin-treated cord blood onto lymphocyte separation medium (INC Biomedical, Aurora, OH). CD34+ progenitor cells are isolated from mononuclear cells by selection of cells positive for the AC133 antigen (CD133+/CD34+) by magnetic cell sorting (Miltenyi Biotec, Auburn, CA). For the first 4 weeks, cells are cultured in IMDM (GIBCO BRL, Long Island, NY) supplemented with 0.55 μM 2-mercaptoethanol, 200 mg/L insulin–transferin–selenium supplement (ITS) from GIBCO, 0.1% bovine serum albumin (BSA; Sigma, St. Louis, MO), penicillin/streptomycin, 200 ng/mL SCF (Amgen), and 50 ng/mL IL-6 (Chemicon) at 37°C in 5% CO_2 balanced air. The IL-6 modulates SCF dependent MC development directly via an IL-6R-gp130 system [Kinoshita, 1999]. After 4 weeks of culture, BSA and ITS in the culture medium are substituted with 10% FBS (GIBCO) as fetal bovine serum (FBS) inhibits cell proliferation during the first few weeks of culture [Ishida, 2003; Kirshenbaum and Metcalfe, 2006]. By 6 weeks, >99% of the cells in the culture are identified as mast cells by immunostaining for tryptase. The IL-4 (50 ng/mL) is added in the last 3 weeks to increase the chymase content of mast cells [Toru, 1998], an index of their maturation [Ahn, 2000; Toru, 1998]. The IL-4 upregulates the expression of functional high-affinity IgE receptor (FceRI) [Kinoshita, 1999], as well as functional expression of neurokinin-1 (NK-1) receptors [van der Kleij, 2003], and increases mediator release [Bischoff, 1999]. Mast cells cultured in the presence of SCF and IL-6 show increased cell size, frequency of chymase-positive cells, and intracellular histamine level compared with the cells cultured only with SCF alone [Kinoshita, 1999].

Unfortunately, hCBMCs still do not have the characteristics of mature skin MCs, as they do not readily respond to neuropeptides, such as neurotensin (NT) or substance P (SP) [Amano, 2000; Foreman, 1977]. One alternative would be to grow MCs from a more undifferentiated precursor, such as CD34-cells found in the cord blood matrix, which derive from CD34+ stem cells. These hCMMCs are 8-day embryonic stem cells, have the unique ability of being embryonic cells without the explicit creation of an embryo to obtain them, and this potential has only recently been appreciated. The hCMMCs are found in the Wharton's jelly of the umbilical cord, have capabilities and characteristics that are very close to embryonic stem cells [Romanov, 2003; Sarugaser, 2005; Wang, 2004]. The hCMMCs have been preliminarily characterized by some investigators, and shown to express receptors generally found on mesenchymal stem cells (CD44, CD105), but not hematopoietic lineage markers (CD34, CD45). The hCMMCs can give rise to differentiated cell types found in embryonic germ layers including nerve cells, bone, cartilage, fat, tendon, muscle, and marrow stroma. The hCMMCs have been reported to be c-kit positive [Mitchell, 2003], which means they can develop into cells of neuroectoderm origin, such as brain MCs [Hirota, 1993].

MAST CELL TRIGGERS

Mast cells are well known to be stimulated by IgE and antigen through aggregation of their specific receptors (FcεR1) [Blank and Rivera, 2004; Kraft, 2004], as well as immunoglobulin free light chains [Kraneveld, 2005; Redegeld, 2002; Redegeld and Nijkamp, 2005], anaphylatoxins, cytokines, hormones, toxins, and a number of drugs [Church, 1989; Foreman, 1987b; Goetzl, 1990; Janiszewski, 1994; Mousli, 1994; Swieter, 1993; Theoharides, 1996; Theoharides and Cochrane, 2004] (Table VIII.1). Mast cells are frequently found in close proximity to mucosal nerves and vagal nerves have been reported to influence mast cells [Stead, 2006]. Very early studies have shown that vagal stimulation causes mast cells to

TABLE VIII.1. Mast Cell Triggers

I. Natural	
A. Immunologic	D. Hormones
Anaphylatoxins (C3a, C5a)	CRH
CGRP	Ucn
IgG 1 + IgE	
IL-1	E. Infectious
IL-33	LPS (TLR 4)
Immunoglobulin—free light chains	Peptidoglycan (TLR 2)
PAF	Viral antigens (TLR 3,5,7,9)
PGE_2	
Superallergens	F. Toxins
MCP -1,-2,-3	Fire ants
MIP-1	Jelly fish
B. Neuropeptides–Neurotransmitters	Snake venoms
Acetylcholine	Wasps
Adrenomedullin	
Bombesin	G. Vascular
PACAP	Adenosine
Somatostatin	Endothelin
SP	Oxidized LDL
VIP	Reactive oxygen species
	Thrombin
C. Growth Factors	
NGF	
SCF	
Lymphopoietin	
II. Drugs	
Adenosine	
Contrast media	
Curare	
Ibuprofen (high doses)	
Morphine	

CGRP, calcitonin gene related peptide; CRH, corticotropin-releasing hormone; MCP, monocyte chemotactic protein; MIP, monocyte inflammatory peptide; NGF, nerve growth factor; PACAP, Pituitary adenylate cyclase activating peptide; SCF, Stem cell factor.

degranulate [Blandina, 1980; Fantozzi, 1978] in the gastrointestinal mucosa of rats and guinea pigs [Bani-Sacchi, 1986; Cho and Ogle, 1977]. Mediators released from mast cells, such as histamine, can then cause vagal stimulation [Casale and Marom, 1983; Christian, 1989]. Therefore, it has been hypothesized that there is cross-talk between mast cells and nerves *in vitro* [Blennerhassett, 1991; Suzuki, 1999] and in the central nervous system (CNS) [Rozniecki, 1999]. In fact, a number of neuropeptide triggers have been shown to act on mast cells. These include SP [Ali, 1986; Matsuda, 1989; Theoharides, 1987], NT [Carraway, 1982], nerve growth factor (NGF) [Bienenstock, 1987; Tal and Liberman, 1997], which is released under stress [De Simone, 1990], hemokinin [Camarda, 2002], and pituitary adenylate cyclase activating polypeptide (PACAP)

[Odum, 1998; Seebeck, 1998]. A number of diseases in which mast cells are known to play a key role are worsened by stress. Corticotropin-releasing hormone (CRH) [Theoharides, 1998] and it structurally related peptide, urocortin (Ucn) [Singh, 1999a] can activate skin mast cells and induce mast cell dependent vascular permeability in rodents. The CRH also increases vascular permeability in human skin [Clifton, 2002], a process dependent on mast cells. CRH-2 receptor expression was shown to be upregulated in stress-induced alopecia in humans [Katsarou-Katsari, 2001], while CRHR-1 expression was increased in chronic urticaria [Papadopoulou, 2005]. Acute restraint stress induces rat skin vascular permeability [Singh, 1999b], an effect inhibited by a CRH receptor antagonist and absent in mast cell deficient mice [Singh, 1999a; Theoharides, 1998]. Histamine is a major regulator of the hypothalamus [Roberts and Calcutt, 1983] and can increase its CRH mRNA expression [Kjaer, 1998]. Moreover, human mast cells can synthesize and secrete large amounts of CRH, as well as IL-1 and IL-6, which are independent activators of the HPA axis [Bethin, 2000].

Other triggers include SCF and IL-6 [Kandere-Grzybowska, 2003b], which have been shown to induce IL-6 release without degranulation. The IL-33 is a recently identified member of the IL-1 family of molecules that can induce cytokine production in human mast cells even in the absence of FceRI aggregation [Iikura, 2007]. The IL-33 has a number of effects on hCBMCs. These include increasing their survival, promoting their adhesion to fibronectin, inducing IL-8 and IL-13 production and enhancing production of these cytokines after IgE/anti-IgE-stimulation [Iikura, 2007]. The IL-33 also acts both alone and in concert with thymic stromal lymphopoietin (TSL) to accelerate the *in vitro* maturation of CD34(+) mast cell precursors and induce the secretion of Th2 cytokines and Th2-attracting chemokines [Allakhverdi, 2007b]. The TSL, a recently described potent mast cell activator [Allakhverdi, 2007a], is considered a "master switch" in allergic inflammation [Liu, 2006] and in airway inflammation.

Adrenomedullin (AM) is a peptide that has been found to be involved in carcinogenesis [Cuttitta, 2002], in the regulation of macrophage [Wong, 2005] and mast cell activation [Yoshida, 2001], as well as in complement fixation [Pio, 2001]. Mast cells in the vicinity of tumors were shown to produce AM [Zudaire, 2006]. In the same study, AM was found to be chemotactic for human MCs and stimulated mRNA expression of vascular endothelial growth factor (VEGF), MCP-1, and basic fibroblast growth factor (FGF). Differentiated mast cells in a hypoxic environment, such as a tumor, elevate mRNA and protein expression of AM. These results suggest a possible mechanism by which mast cells may contribute to neoagiogenesis, supporting tumor growth.

The OxLDL has been implicated in the pathogenesis of atherosclerosis and has been shown to induce microvascular dysfunction through degranulation of MCs [Liao and Granger, 1996]. The OxLDL increases mRNA and protein levels of interleukin IL-8 [Kelley, 2006].

Superallergens are proteins of various origins able to activate FcεRI+ cells (mast cells and basophils) through interaction with membrane-bound IgE. Several superallergens have been reported. Protein Fv, a sialoprotein produced in the human liver and released in biological fluids during viral hepatitis, protein L, a bacterial cell-wall component as well as antigens of the HIV virus and Staphylococcus aureus [Marone, 2007].

Mast cells have been found to synthesize and secrete endothelins (ET) and also have ET receptors, therefore suggesting an autocrine path of action [Ehrenreich, 1992]. Endothelins are cytokine-like agents that can be synthesized and bound by a large number of different cell types. Their role in the immune system and in the pathophysiology of different diseases is thus complicated [Kedzierski and Yanagisawa, 2001]. There are three different ET peptides,

ET1, ET2 and ET3, as well as two different receptors ET-A and ET-B. ET-1 caused degranulation, TNF-alpha, IL-6, TGF beta1, and VEGF production by fetal skin mast cell (FSMC), but not bone marrow-derived mast cells (BMMC). In addition, FSMC also produced ET-1 in response to Toll-like receptor ligands suggesting an autocrine–paracrine pathway in these cases [Matsushima, 2004]. A recent study [Maurer, 2004] showed that mast cells may be implicated in preventing ET-1 toxicity since mast cell secreted chymase is able to degrade ET-1 further complicating the interactions between mast cells and endothelin.

A number of drugs can also stimulate mast cells [Table VIII.1], best known of which is morphine [Barke and Hough, 1993].

MAST CELL MEDIATORS

Mast cells can secrete a multitude of biologically potent mediators (Table VIII.2) that permit them to participate in innate or acquired immunity [Galli, 2005a, b; Rottem and Mekori, 2005]. There are various vasodilatory and proinflammatory mediators, such as the preformed histamine, heparin, kinins, proteases, as well as the newly synthesized leukotrienes, prostaglandins, nitric oxide (NO), and cytokines [Dvorak, 1997; Hogan and Schwartz, 1997; Lagunoff, 1983]. Mast cells are the only cells known to have prestored TNF, which they release at the early stages of an inflammatory response [Gordon and Galli, 1990].

A number of growth factors are also secreted, such as VEGF [Boesiger, 1998; Galli, 2005b; Grutzkau, 1998], which has been shown to be released selectively in response to PGE2 [Abdel-Majid and Marshall, 2004], adenosine [Feoktistov, 2003]. Ultraviolet (UV) radiation, and CRH [Kempuraj, 2004]. The VEGF was specifically shown to induce dilation of microvessels [Laham, 2003] and was also chemottactic for MCs suggesting an autocrine–paracrine vicious cycle in which MCs, well as other types of cells located in the vicinity of a tumor secrete VEGF, which in turn attracts more mast cells supporting tumor neoangiogenesis and growth. Other angiogenic factors released from mast cells include fibrolast growth factor 2, TGF-beta, TNF, IL-8, angiogenin, hepatocyte growth factor, and the angiopoeitins-1 and -2.

Proteases released from mast cells could act on plasma albumin to generate histamine-releasing peptides [Carraway, 1989; Cochrane, 1993] that could further propagate mast cell activation and inflammation. Proteases could also stimulate protease activated receptors (PAR) inducing microleakage and widespread inflammation [Molino, 1997; Schmidlin and Bunnett, 2001]. Proteases can activate fibroblasts [Gruber, 1997], thereby promoting collagen deposition and fibrosis, and also directly stimulate angiogenesis [Blair, 1997].

Reactive oxygen species are produced in cells by a variety of enzymatic and nonenzymatic mechanisms. Degranulation, as well as an increase in intracellular ROS production, was observed in rat peritoneal mast cells (RPMCs) following stimulation with antigen [Brooks, 1999; Swindle, 2004; Tsinkalovsky and Laerum, 1994], compound 48/80 [Brooks, 1999; Swindle, 2004; Tsinkalovsky and Laerum, 1994] the calcium ionophore A23187 [Tsinkalovsky and Laerum, 1994], SP [Brooks, 1999], and NGF [Brooks, 1999]. In contrast, D-mannitol increased ROS production in stimulated mast cells without a marked effect on degranulation or histamine secretion [Tsinkalovsky and Laerum, 1994].

In addition, several chemicals [Brown, 2007; Cho, 2004; Kawamura, 2006; Kim and Ro, 2005; Wolfreys and Oliveira, 1997; Yoshimaru, 2006], as well as bacteria [Malaviya, 1994] and fungal components [Niide, 2006], can induce ROS production in mast cells.

TABLE VIII.2. Mast Cell Mediators and Their Actions

Mediators	Main Pathophysiologic Effects
Prestored	
Biogenic Amines	
Histamine	Vasodilation, angiogenesis, pain
5-Hydroxytryptamine (5-HT, serotonin)	Vasoconstriction, pain
Chemokines	
IL-8 (CXCL8), MCP-1 (CCL2), MCP-3 (CCL7), MCP-4 (CXCL13), RANTES (CCL5)	Chemoattraction and tissue infiltration of leukocytes
Enzymes	
Arylsulfatases	Lipid–proteoglycan hydrolysis
Carboxypeptidase A	Peptide processing
Chymase	Tissue damage, pain, angiotensin
II synthesis	
Kinogenases	Synthesis of vasodilatory kinins, pain
Metalloproteinases	Tissue damage
Phospholipases	Arachidonic acid generation
Tryptase	Tissue damage, activation of PAR, inflammation, pain
Peptides/Proteins	
Corticotropin-releasing hormone (CRH)	Inflammation, vasodilation
Endorphins	Analgesia
Endothelin	Sepsis
Kinins (bradykinin)	Inflammation, pain, vasodilation
Neurotensin	Inflammation, pain
Renin (CTMC only)	vascular contriction
Somatostatin (SRIF)	Anti-inflammatory (?)
Substance P (SP)	Inflammation, pain
Vasoactive intestinal peptide (VIP)	Vasodilation
Urocortin	Inflammation, vasodilation
Vascular endothelial growth factor (VGEF)	Neovascularization, vasodilation
Proteoglycans	
Chondroitin sulfate	Cartilage synthesis, anti-inflammatory
Heparin	Angiogenesis, NGF stabilization
Hyaluronic acid	Connective tissue
Angiogenic Factors	
Adrenomedullin	
Angiogenin	
Angiopoietin	
EGF	
FGF α	
basic FGF	
IL-8	
Neuropillin	
PDGF	
TGF β	
VEGF	

(*Continued*)

TABLE VIII.2. *Continued*

Mediators	Main Pathophysiologic Effects
De novo Synthesized	
Cytokines	
Interleukins (IL)-1,2,3,4,5,6,8,10,13,16,17,32	Inflammation, leukocyte migration, pain
IFN-γ; MIF; TNF-α,TGF-beta	Inflammation, leucocyte proliferation, activation
Growth Factors	
SCF, GM-CSF, b-FGF, NGF, VEGF	Growth of a variety of cells
Phospholipid metabolites	
Leukotriene B_4 LTB_4	Leukocyte chemotaxis
Leukotriene C_4 (LTC_4)	Vasoconstriction, pain
Platelet activating factor (PAF)	Platelet activation, vasodilation
Prostaglandin D_2 (PGD_2)	Bronchonstriction, pain
Others	
Nitric oxide (NO)	Vasodilation
ROS	Inflammation

TABLE VIII.3. Mechanisms of Mast Cell Secretion

Degranulation	Anaphylaxis
Selective release	Infection, inflammation
Piecemeal degranulation	MS, scleroderma
Exosomes	Infection
Transgranulation	Brain

Recently it was reported that 5-Lipoxygenase (5-LO) was the primary enzyme involved in ROS production by human mast cells and mouse BMMC following FcεRI aggregation [Swindle, 2007]. Finally, it has been shown that ROS play an essential role in the activation of MCs in ischemia reperfusion (IR) injury [Boros, 1992; Kurose, 1997].

SELECTIVE RELEASE OF MAST CELL MEDIATORS

Mast cell have been reported to secrete without degranulation, but through ultrastructural alterations of the cell electron dense granular core indicative of secretion. This process has been termed "activation" [Dimitriadou, 1990, 1991; Theoharides, 1995a], "intragranular activation" [Letourneau, 1996], or "piecemeal" degranulation [Dvorak, 1992a] (Table VIII.3). During this process, mast cells can release many mediators differentially or selectively (Table VIII.4) [Kops, 1984, 1990; Van Loveren, 1984], such as serotonin [Theoharides, 1982], eicosanoids [Benyon, 1989; Levi-Schaffer and Shalit, 1989; van Haaster, 1995], IL-6 in response to IL-1 [Kandere-Grzybowska, 2003b], VEGF in response to CRH [Cao, 2006b], VEGF and IL-8 in response to adenosine or VEGF in response to PGE2 [Feoktistov, 2003]. In fact, MCs could communicate with neurons through transgranulation [Wilhelm, 2005]. Mast cells appear to have unique ultrastructural changes indicative of selective secretion in diseases, such as interstitial cystitis [Theoharides, 1995a], or experimental allergic encephalomyelitis (EAE) [Letourneau, 2003]. Such activation may be associated with the ability of mast cells to release some mediators selectively [Kops,

TABLE VIII.4. Natural Mast Cell Inhibitors

Chondroitin sulfate
Flavonoids (luteolin, quercetin)
Heparin
Nitric oxide
Somatostatin (mucosal must cells only)
Vitamin A (retinoic acid)
Vitamin E (tocopherol)

1984, 1990; Van Loveren, 1984] as shown for serotonin [Theoharides, 1982], eicosanoids [Benyon, 1989; Levi-Schaffer and Shalit, 1989; van Haaster, 1995], or IL-6 [Gagari, 1997; Leal-Berumen, 1994]. Recently, we showed that IL-1 can stimulate selective release of IL-6 [Kandere-Grzybowska, 2003b], while CRH can induce selective release of VEGF [Kulka and Metcalfe, 2005], as can PGE_2 [Abdel-Majid and Marshall, 2004]. It was recently shown that the newly discovered IL-33 can induce selective release of IL-13 without degranulation [Ho, 2007; Iikura, 2007]. This unique ability of mast cells extends to many other pathophysiological trigger as reviewed recently [Theoharides, 2007b].

Exosomes containing MHC II antigens have been shown to be released by mast cells during degranulation [Raposo, 1997]. Exosomes are small, membrane vesicles, which are released extracellularly. They have emerged as a novel mechanism of genetic exchange since it has been shown that they are involved in transferring mRNAs and microRNAs between cells [Valadi, 2007]. It was also demonstrated that mast cell associated exosomes can induce immature dendritic cells (DCs) to acquire potent Ag-presenting capacity to T cells [Skokos, 2003]. Exosomes have been implicated in the induction of tolerance in the gut epithelium toward food antigens [Karlsson, 2001], and in the maintenance of pregnancy [Taylor, 2006] by several mechanisms.

NATURAL MAST CELL SECRETION INHIBITORS

There are no effective clinically available mast cell inhibitors. While disodium cromoglycate (cromolyn) has been shown to inhibit rodent MC histamine secretion [Theoharides, 1980] it is a very weak inhibitor of human MCs [Theoharides and Kalogeromitros, 2006].

Vitamin E inhibits protein kinase C (PKC), which is involved in degranulation and cytokine production [Ricciarelli, 1998]. In many studies, vitamin E reduced MC degranulation by scavenging free radicals by activating several drug-metabolizing enzymes (cytochrome P450 CYP3A and CYP4F2) [Brigelius-Flohe, 2005], and by modulating signal transduction and gene expression [Azzi, 2004], suggesting that it may have additional modes of action.

Retinoic acid was also shown to inhibit proliferation of leukemic MCs [Arici, 1996; Ishida, 2003].

Chondroitin sulfate, a major natural constituent of connective tissues, especially cartilage, and of MC secretory granules, had a dose-dependent inhibitory effect on rat peritoneal MC release of histamine induced by compound 48/80 [Theoharides, 2000]. Heparin, the structure of which is quite similar to that of chondroitin sulfate and is also stored in mast cells, had been shown to inhibit *in vivo* release of histamine [Inase, 1993].

The IL-10 inhibits the release of TNF-α, IL-8, and histamine by activated hCBMCs. These findings suggest that IL-10 might have anti-inflammatory effects on IgE/anti-IgE challenged human mast cell by inhibiting their release of TNF-alpha, IL-8, and histamine [Phillips, 1975]. However, other studies have reported lack of an inhibitory action on human mast cell tryptase and IL-6 release [Conti, 2003], suggesting there may be species differences.

The TGF-1b acts as a novel potent inhibitor and modulator of human intestinal mast cell effector functions [Gebhardt, 2005]. It also has been found to downregulate FcεRI expression [Gomez, 2005] in mouse mast cells. However, human mast cells can produce TGF β1 [Kanbe, 1999]. Both TGFβ and IL-6 are necessary for the proliferation and maturation of TH17, T cells secreting IL-17, which recently has been shown to be a cytokine critical for the development of autoimmune diseases [Hurst, 2002].

Somatostatin is an important regulator in the neuroendocrine-immune network. Several studies have shown that somatostatin inhibits MC activity and could prevent the intestinal responses to MC hyperplasia [Saavedra and Vergara, 2003; Tang, 2005]. In contrast, somatostatin was shown to stimulate connective tissue MCs *in vitro* [Benyon, 1989; Kassessinoff and Pearce, 1988].

Nitric oxide is a gas and free radical that has important physiological roles including defense against microorganisms, but also regulation of cellular activation [Grisham, 1999] and gene expression [Ray, 2007]. It has been demonstrated that MCs express NO synthases (NOS) and produce NO themselves [Bidri, 2001]. Importantly, it has been shown that NO directly inhibits IgE depended MC degranulation [Deschoolmeester, 1999; Eastmond, 1997]. Endogenous and exogenous NO protects against IR injury in isolated guinea pig hearts; endogenous and exogenous NO inhibits mast cell degranulation and protease release [Jorens, 1993], adherence to fibronectin [Wills, 1999], leukotriene [Gilchrist, 2004], cytokine, and chemokine production [Coleman, 2002; Sekar, 2005]. Recent data suggest that NO protects MCs from activation-induced cell death [Inoue, 2008].

Flavonoids are polyphenolic compounds present in fruit, vegetables, nuts, seeds, wine, tea, and coffee with antioxidant and anti-inflammatory actions [Middleton, Jr., 2000]. Kimata [2000] reported that luteolin, quercetin, and baicalein inhibited the release of histamine, leukotrienes, and prostaglandin D2, as well as the secretion of granulocyte macrophage-colony stimulating factor by human cultured mast cells in response to cross-linkage of FcεRI and subsequently showed that these compounds also inhibited IgE-mediated TNF-α and IL-6 production by bone marrow derived cultured murine mast cells. It was shown also that quercetin and other flavonoids could inhibit histamine, IL-6, IL-8, TNF-α, and tryptase from IgE-stimulated human MCs [Kempuraj, 2005]. The same flavonoids could also inhibit leukemic mast cell proliferation [Alexandrakis, 2003b].

MAST–T CELL INTERACTION

Mast cell activation could contribute to immune [Ashwood, 2006] and neuroinflammatory [Cohly and Panja, 2005; Pardo, 2005] abnormalities. Increasing evidence indicates that mast cells can interact with T cells [Bachelet and Levi-Schaffer, 2007; Mekori and Metcalfe, 1999; Nakae, 2006; Salamon, 2005], at least through TNF, and superactivate them [Theoharides, 2007a].

INFLAMMATORY SKIN DISEASES

Eczema is divided into atopic and non-atopic, but they both involve skin MCs [Brown and Reynolds, 2006]. Mast cells are located close to sensory nerve endings [Wiesner-Menzel,

1981] and can be triggered by neuropeptides [Church, 1989; Foreman, 1987a; Goetzl, 1985, 1990], such as NT [Carraway, 1982], NGF [Watt, 1991], SP [Fewtrell, 1982], and pituitary adenylate cyclase activating polypeptide (PACAP), all of which can be released from dermal neurons [Odum, 1998]. In fact, skin mast cells contain SP [Toyoda, 2000], while cultured mouse and human MCs contain and secrete NGF [Xiang and Nilsson, 2000]. Proteases released from MCs could act on plasma albumin to generate histamine releasing peptides [Carraway, 1989; Cochrane, 1993] that could further propagate mast cell activation and inflammation. Proteases could also stimulate protease-activated receptors (PAR) inducing microleakage and widespread inflammation [Molino, 1997; Schmidlin and Bunnett, 2001].

Many dermatoses, such as atopic dermatitis (AD), chronic urticaria and psoriasis, are triggered or exacerbated by stress [Katsarou-Katsari, 1999], which also worsens eczema [Graham and Wolf, 1953] and acne vulgaris [Murphy, 2001]. Computer-induced stress enhanced allergen specific responses, with concomitant increase in plasma SP levels, in patients with atopic dermatitis [Kimata, 2003a]. Similar findings with increased plasma levels of SP, VIP, and NGF, along with a switch to a TH2 cytokine pattern, were reported in patients with atopic dermatitis playing video games [Kimata, 2003b]. Exercise was also shown to increase the responsiveness of skin MCs to morphine only in patients with exercise-induced asthma [Choi, 2004].

Psoriasis is a chronic inflammatory skin disorder characterized by keratinocyte hyperproliferation and increased epidermal vascularization [Fortune, 2005; Remrod, 2007]. Psoriasis is well known to worsen by stress [Fortune, 2005; Katsarou-Katsari, 1999] and both mast cells and cutaneous nerves are involved [Harvima, 1993]. Both SP [Fortune, 2005] and corticotropin-releasing hormone (CRH) [Dewald, 2004] have been shown to be involved in stress-induced skin dermatoses. Stress alone has been shown to increase SP positive nerve fibers and MC contacts in mice [Peters, 2005]. Skin cells are activated by acute stress in rodents [Singh, 1999b] and SP is involved in stress-induced skin mast cell degranulation [Kawana, 2006]. Dermal mast cells are located in close proximity to SP containing sensory nerves [Dimitriadou, 1997; Rozniecki, 1999; Stead, 1989]. These SP positive nerve fibers are more dense in psoriatic lesions and have an increased number of MC contacts compared to normal skin [Al'Abadie, 1995; Chan, 1997; Jiang, 1998; Naukkarinen, 1996]. Recently, it was reported that there is increased expression of SP and NK-1 in lesional psoriatic skin with the latter primarily associated with MCs [Remrod, 2007].

Other neuropeptides are also involved in psoriasis [Saraceno, 2006]. They include CRH [Dewald, 2004], which induces VEGF release from MCs by activation of the CRH-1 receptor and p38 [Cao, 2005], as well as calcitonin gene-related peptide (CGRP) [Chan, 1997; He, 2000], which induces VEGF expression in keratinocytes through ERK activation [Yu, 2006]. Similar to SP, CRH, and CGRP are upregulated in psoriatic skin [Chan, 1997; He, 2000; Kim, 2007] and CRH is also increased in the plasma of psoriasis patients [Tagen, 2007]. Both SP and CRH are released locally in the skin during stress [Lytinas, 2003; Kawana, 2006].

Skin has its equivalent of the hypothalamic–pituitary–adrenal (HPA) axis [Slominski, 2000; Slominski and Wortsman, 2000]. Its main regulator, CRH and its receptors are present in the skin [Slominski, 2001] from where CRH is released in response to stress [Lytinas, 2003]. Acute stress also exacerbates skin-delayed hypersensitivity reactions [Dhabhar and McEwen, 1999] and chronic contact dermatitis in rats, an effect dependent on mast cells and CRH-1 receptors; [Kaneko, 2003] acute stress also induces redistribution of leukocytes from the systemic circulation to the skin [Dhabhar and McEwen, 1996]. Acute restraint stress induces rat skin vascular permeability [Singh, 1999b], an effect inhibited by a CRH receptor antagonist and is absent in MC deficient mice [Singh, 1999a; Theoharides,

1998], CRH [Theoharides, 1998], and its structurally related peptide, urocortin (Ucn) [Singh, 1999a], can activate skin MCs and induce MC dependent vascular permeability in rodents. Corticotropin-releasing hormone also increases vascular permeability in human skin [Clifton, 2002], a process dependent on MCs. The CRHR-2 receptor expression was shown to be upregulated in stress-induced alopecia in humans, [Katsarou-Katsari, 2001], while CRHR-1 expression was increased in chronic urticaria [Papadopoulou, 2005]. The immunoendocrine responses to stress in chronic skin inflammatory diseases have been reviewed [Buske-Kirschbaum and Hellhammer, 2003; Theoharides and Cochrane, 2004] and it was proposed that mast cells constitute the "sensor" of a "brain–skin" connection [Paus, 2006].

FIBROMYALGIA SYNDROME

Fibromyalgia Syndrome (FMS) is a complex disease characterized by fatigue, malaise, and muscle aches [Lucas, 2006; Mease, 2005; Nielsen and Henriksson, 2007; Stauder and Kovacs, 2003]. Many FMS patients demonstrate abnormal activity of the HPA axis, with elevated activity of CRH and SP [Mease, 2005] that may not only affect the PHA axis, but other endocrine and immune processes [Neeck, 2002; Neeck and Crofford, 2000; Riedel, 2002; Torpy, 2000].

Acute stress precipitates or intensifies symptoms and CRH has been shown to be increased in the urine of FMS patients [Lund, 2006]. Fibromyalgia Syndrome is often comorbid with other disorders [Aaron, 2001; Young and Redmond, 2007] that include interstitial cystitis–painful bladder syndrome (IC–PBS), migraines, irritable bowel syndrome (IBS), post-traumatic stress disorder migraines, and chronic fatigue syndrome (CFS) [Young and Redmond, 2007], all of which worsen by stress [Aaron and Buchwald, 2001b, 2003; Bennett, 1998; Hudson, 1992; Peres, 2002; Raphael, 2006; Smith, 2003]. A number of genetic polymorphisms have been identified, but these are not unique to FMS, since they are also present in the other comorbid conditions [Buskila, 2007]. The pathophysiology of FMS is still unknown [Abeles, 2007]. Fibromyalgia syndrome appears to be particularly vulnerable to stress, which can be triggered by acute emotional *upheavals*, infections, cytokines, and toxins, all of which affect brain function [Krueger, 2003; Theoharides, 2004b]. CRH and IL-1 can be increased locally in the skin or muscles [Donelan, 2006; Lytinas, 2003; Slominski, 2006; Solinas, 2006]. Moreover, IL-1 receptors have been identified in skeletal muscle in patients with idiopathic inflammatory myopathies [Grundtman, 2007].

Fibromyalgia syndrome may be another inflammatory disorder exacerbated by stress [Theoharides and Cochrane, 2004]. Recent publications have reported elevated levels of cytokines in the serum of FMS patients [Mullington, 2001]. In one study, serum IL-1 and IL-6 levels were not different from controls, but IL-8 was significantly elevated [Gur, 2002a]. Nevertheless, IL-6 was elevated in the supernatants from peripheral blood mononuclear cells from FMS patients [Wallace, 2001]. Moreover, injection of IL-6 produced excessive heart rate responses in FMS patients [Torpy, 2000]. Interestingly, young FMS patients with milder symptoms had significantly increased serum IL-8 levels [Gur, 2002b].

Migraines and IC–PBS are two of the most common comorbid conditions with FMS [Aaron, 2001; Aaron and Buchwald, 2001a; Clauw, 1997; Hudson, 1992; Korszun, 1998; Peres, 2002; Smith, 2003] and both have been associated with an increased number of activated mast cells [Theoharides, 2001, 2005]. In fact, family linkage studies have shown an association of panic disorder and IC–PBS on chromosome 13 [Weissman, 2004], with mast cells being considered a common underlying pathogenetic link [Theoharides, 2004]. We showed that acute stress increases serum histamine and IL-6 levels, both of which are absent in mast cell deficient W/W^v mice [Huang, 2002, 2003]. Moreover, stress increased ^{99}Technetium permeability in the dura of mice, but not in those deficient in mast cells (W/W^v) or NK-1 $-/-$ mice [Kandere-Grzybowska, 2003a] implying that mast cells and tachykinins are involved. Similarly, stress exacerbates interstitial cystitis symptoms [Rothrock, 2001; Theoharides, 2007; Theoharides and Cochrane, 2004; Theoharides and Kalogeromitros, 2006] and recently we showed that intravesical CRH induces VEGF release from bladder explants [Cao, 2006a]. We also showed that CRH released under stress can induce VEGF release from mast cells [Cao, 2005] and so can SP [Tagen, 2007], which was shown to be increased in the CSF of FMS patients [Russell, 1994]. Neuropeptides, released from dorsal root ganglia (DRG) or sensory nerve endings, could stimulate mast cells locally in the muscles or skin because mast cell–neuron interactions have previously have been described in the intestines [Rijnierse, 2007; Stead, 1989], the meninges [Rozniecki, 1999] and other tissues [Suzuki, 1999].

Once inflammation occurs, IL-1 could stimulate MCs to release IL-6 selectively [Kandere-Grzybowska, 2003b]. Increased activated MCs have been reported in association to IgG deposits in skin biopsies from FMS patients [Enestrom, 1997]. Moreover, skin biopsies from FMS patients had high IL-6 expression by RT-PCR [Salemi, 2003] and trapezius muscle biopsies had more SP immunoreactivity [De Stefano, 2000]. Corticotropin-releasing hormone content [Lytinas, 2003] and vascular permeability [Singh, 1999b], increased in the skin or rats in response to stress a process mimicked by intradermal administration of CRH [Theoharides, 1998]. The CRH could be acting together with SP or other neurokinin-1 receptor agonists [Kandere-Grzybowska, 2003a].

The TNF-α can be released along with histamine from rat brain MCs [Cocchiara, 1998] and was involved in both brain inflammation [Klinkert, 1997; Probert, 1997] and increased vascular permeability [Kim, 1992]. Mast cells also release VEGF [Boesiger, 1998], an isoform of which is particularly vasodilatory [Grutzkau, 1998]. For these reasons, mast cells have been considered important in brain pathophysiology [Silver, 1996; Theoharides, 1980]. TNF-α can be released along with histamine from rat brain mast cells [Cocchiara, 1998] and was involved in both brain inflammation [Klinkert, 1997; Probert, 1997] and increased vascular permeability [Kim, 1992]. Mast cells appear to have unique ultrastructural changes indicative of selective secretion in diseases, such as IC [Theoharides, 1995a] or experimental allergic encephalomyelitis (EAE) [Letourneau, 2003]. Such activation may be associated with the ability of mast cells to release some mediators selectively [Kops, 1984, 1990; Van Loveren, 1984], as shown for serotonin [Theoharides, 1982], eicosanoids [Benyon, 1989; Levi-Schaffer and Shalit, 1989; van Haaster, 1995], or IL-6 [Gagari, 1997; Leal-Berumen, 1994].

A recent study showed that 24-h urine levels of CRH correlated with FMS symptoms and CRH levels decreased along with symptoms after successful treatment [Lund, 2006]. There appears to be hypothalamic, but not pituitary dysregulation [Jones, 2007]. However, functional magnetic resonance imaging (MRI) studies do not show any specific lesions [Williams and Gracely, 2007]. Mast cells and their mediators have been implicated in all diseases that are comorbid with FMS [Lucas, 2006; Theoharides, 2005]. Mast cells are abundant in connective tissues and their activation, especially by triggers, such as CRH, could release molecules that participate in FMS pathogenesis. For example, IL-1, IL-6, and IL-8 appear to be dysregulated in FMS [Wallace, 2006]. Local tissue oxidative stress may also play a role in FMS pathogenesis [Ozgocmen, 2006]. Histamine [Kjaer, 1998] and IL-6 or IL-1 [Bethin, 2000; Krüger-Krasagakes, 1996; Leal-Berumen, 1994; Mastorakos, 1993; Navarra, 1991; Spinedi, 1992], all of which are released from MCs [Gordon, 1990; Grabbe, 1994], could then stimulate the HPA axis. Mast cells have been considered the key skin cell in "brain–skin connection" [Paus, 2006], as well as in the pathogenesis of FMS [Lucas, 2006]. Dermal IgG deposits and increased skin mast cells have been reported in FMS [Enestrom, 1990, 1997], along with neurogenic inflammation [Kim, 2007].

ASTHMA

Asthma is one of the most common chronic illnesses, affecting roughly, 300 million people worldwide [Papiris, 2002; Weiss, 2001]. The morbidity and mortality due to asthma continues to increase despite advances in both our scientific knowledge, as well as in hygiene and improved drugs for this disease [Papiris, 2002]. The World Health Organization (WHO) has estimated that 1/250 deaths worldwide is due to asthma, which highlights the need for an improved understanding of the cellular and molecular mechanisms that contribute to the pathogenesis of asthma.

The use of animal "models" has provided useful information about the mechanisms of airway inflammation and hyperreactivity characterizing asthma [Kips, 2003; Lloyd and Gutierrez-Ramos, 2004]. Chronic exposure to aerosolized ovalbumin leads to airway inflammation, airway hyperresponsiveness (AHR) [Wilson, 2000], as well as microvascular leakage in rodent airways [Olivenstein, 1997; Van Rensen, 2002; Wilson, 2000]. Microvascular leakage in the airway wall may also be important for the airway wall remodeling that characterizes most asthmatics [Johnson, 2004; Naureckas, 1999]. More recently, the house dust mite allergen model was shown to effectively induce chronic airway inflammation and AHR [Johnson, 2004; Sadakane, 2002].

The role of mast cells in asthma is undisputed, but their role may have been underestimated, especially in viral infections [Bradding, 2003; Brightling, 2003; Cho, 2002]. Viral infections commonly exacerbate asthma and contribute to >50% of asthma-associated deaths; moreover >80% of childhood asthma exacerbations are associated with viral airway infections [O'sullivan, 2005]. A number of studies have shown that viral infections increase AHR and antigen sensitization [Dakhama, 2005], as well as recruitment of inflammatory cells [Van Rijt, 2005]. Viruses that have been implicated in the pathogenesis of asthma are rhinovirus, adenovirus, as well as influenza and parainfluenza [Matsuse, 2005; Pelaia, 2006; Sackesen, 2005; Williams, 2005]. In fact, rhinovirus infections during infancy appear to predict childhood wheezing [Lemanske, Jr., 2005], while respiratory syncytial virus during the first 3 months of life was shown to promote a TH2 response, especially significantly high levels of IL-4 [Kristjansson, 2005]. Such early infancy viral respiratory infections may also induce metalloproteinases that are involved in airway remodeling in asthma [Gualano, 2006].

The TLRs were shown to be important in recognition of ligands associated with bacterial or viral infections, and play a key role in the development of adaptive immune responses [Aderem and Ulevitch, 2000; Rock, 1998], especially in asthma [Cristofaro and Opal, 2006]. Ten human TLRs have been identified so far [Aderem and Ulevitch, 2000; Akira, 2001; Heine and Lien, 2003]. Rodent mast cells express bacterial TLR 2 and 4 [McCurdy, 2003; Varadaradjalou, 2003]. Human MCs express viral TLR-9 [Ikeda, 2003], activation of which produced IL-6, [Ikeda, 2003] while TLR-3 activation produces IFN [Kulka, 2004]. Lipopolysaccharide (LPS) induces TNF release through TLR-4, while peptidoglycan induces histamine release through TLR-2 from rodent MCs. Fetal rat skin derived MCs express TLR 3, 7, and 9 and activation by CPG oligodeoxynucleotide induces release of TNF and IL-6, as well as RANTES and MIP, but without degranulation [Anthony and

Lance, 1971; Cairns and Walls, 1996]. Lipopolysaccharides could not induce release of GM–CSF, IL-1 or LTC_4 [McCurdy, 2003]. However, LPS did induce secretion of TH2 cytokines, IL-5, IL-10, and IL-13 and increased their production by FcεRI cross-linking [Masuda, 2002]. Elsewhere, it was shown that TLR-2 activation produced IL-4, IL-6, and IL-13, but not IL-1, [Supajatura, 2002], while LPS produced TNF, IL-1, IL-6, and IL-13, but not IL-4 or IL-5, without degranulation [Supajatura, 2002]. Finaly, human epithelial cells of atopic subjects (atopic dermatitis and asthma patients) after TLR stimulation produced thymic stromal lymphopoietin (TSL), a potent activator of mast cells; TSL in concert with IL-1, was shown to stimulate MCs to secrete selective Th2 cytokines (IL 5, IL-13, IL-6, IL-10, GM–SCF) [Holgate, 2007]. These data suggest that innate immune responses, in part through MC involvement, may promote asthma development, as well as exacerbation.

Stress has long been postulated to negatively impact asthma, but the mechanisms by which this occurs remains unknown [Gordon and Rifkind, 1989; Joachim, 2003; Kilpelainen, 2002; Laube, 2002; Lawrence, 2002; Liu, 2002]. Stress can induce asthma exacerbations [Bienenstock, 2002; Joachim, 2003; Kilpelainen, 2002; Laube, 2002; Lawrence, 2002; Schmaling, 2002], including the possibility that maternal stress may contribute to subsequent cellular responses in childhood asthma [von Hertzen, 2002]. It has been postulated that stress associated with urban living promotes asthma [Dendorfer, 1994]. One study showed that adolescents with asthma in a low socioeconomic group, with more stressful life events had more asthma exacerbations and higher serum Th-2 cytokines than those in higher socioeconomic status [Chen, 2003]. The Inner City Asthma Study showed a correlation between community violence and asthma morbidity [Wright, 2004]. Post-traumatic psychological stress following the 9/11 attacks on the World Trade Center correlated with increased symptom severity in subjects with moderate to severe asthma and with utilization of urgent care in New York City [Centers for Disease Control and Prevention (CDC), 2002; Fagan, 2003]. In an epidemiological study carried out among 10,667 Finnish first-year university students (18–25 years old), excess of stressful events, such as concomitant severe disease or death of immediate family members or family conflicts, were associated with exacerbations of asthma [Kilpelainen, 2002]. Moreover, college students with mild asthma had increased sputum eosinophil counts, as well as eosinophil-derived neurotoxin and IL-5 [Liu, 2002]. It was suggested that a shift in cytokine generation to that of a Th2 type may be the defining parameter [Bienenstock, 2002]. In one longitudinal study (12 months) of 92 adults with asthma, it was determined that subjects who reported more negative life events and had low levels of social support had more episodes of asthma exacerbations induced by upper respiratory tract infections [Smith and Nicholson, 2001]. A prospective long-term follow-up community-based cohort study of young adults ($n = 591$, 19–40 years old) showed a dose-response relationship between panic and asthma [Hasler, 2005]. In fact, one study showed that the greater the levels of caregiver-perceived stress at 2–3 months the higher the risk of subsequent repeated wheezing among children during the first 14 months of life [Wright, 2002]. Even though the HPA axis apparently functions normally in asthmatic adult patients in response to stress [Kapoor, 2003], in one study, there was a significantly lower cortisol response to stress in asthmatic children [Buske-Kirschbaum, 2003], suggesting there may be age differences.

The timing of stress application appears to lead to different results. Short-term (3 days) stress before allergen challenge decreased the number of inflammatory cells, but increased IL-6, while long-term (7 days) stress evidently increased the number of inflammatory cells, but did not alter IL-6 levels [Forsythe, 2004].

Several mast cell mediators, as well as surface molecules, have been reported to play a key role in the pathogenesis of asthma. FcepsilonRI was expressed on mast cells during the development of airway hyperresponsiveness after allergen exposure [Taube, 2004] and MC secreted LTB4, which acted as a chemoattractant for CD8+ cells *in vitro* [Ott, 2003]. In addition, histamine receptor 1 has been shown to be important for allergic airway disease development as it can attract Th2 cells in sites of allergen exposure. Finally, recent studies demonstrate that mast cell produced TNF plays a role in airway hyperreactivity and induces cytokine production from Th2 cells during the challenge phase [Nakae, 2007]. In fact, anti-TNF antibodies seemed to have promising effects in reducing the severity of clinical symptoms in asthma patients [Howarth, 2005]. Moreover, IgE was found to promote human lung mast cell survival through the autocrine production of IL-6, suggesting a possible mechanism for the ongoing activation of mast cells in the absence of antigen [Cruse, 2008].

INFLAMMATORY ARTHRITIS

The inflammatory and subsequent lytic processes in inflammatory arthritis are not well understood, especially the initial triggering events. Until recently, discussion of the cellular elements involved did not even mention mast cells, even though a number of papers had reported the presence of mast cells in joints [Crisp, 1984; de Paulis, 1996; Gotis-Graham, 1998; Tetlow and Woolley, 1995a; Wasserman, 1984]. Osteoarthritis (OA) is a degenerative joint disease and is characterized by the breakdown of the joint's cartilage and loss of joint space, especially in hands and weight-bearing joints, such as knees, hips, feet, and the back [Malone, 2002]. Cartilage breakdown allows bones to rub against each other, causing inflammation, pain, and loss of movement.

Rheumatoid arthritis (RA) involves chronic inflammation and destruction of the joints. Rheumatoid arthritis is a systemic disease that affects the entire body and is characterized by inflammation and reactive hyperplasia of the synovium that causes redness, swelling, stiffness, and pain. It typically affects many different joints; it is noted for remissions and flares, often precipitated or exacerbated by stress [Herrmann, 2000; Thomason, 1992]. Articular damage in RA involves cartilage erosion, inflammatory cell accumulation, as well as bone destruction and reactive bone spur formation. The involved joints lose their shape and alignment, resulting in pain.

While mast cells constitute a small fraction of the cells in a normal human synovium, they increase to almost 10% in RA [Nigrovic and Lee, 2005]. Mast cells, cytokines, and metalloproteinases were also colocalized in the RA lesion [Tetlow and Woolley, 1995b; Woolley and Tetlow, 2000]. Activation of synovial MCs was shown to be autoregulated by their own mediators, histamine and tryptase [He, 2001]. The accumulation of MCs in the joints may be due to the local release of chemoattractants, such as RANTES and MCP-1 [Conti, 1997], found in joints of patients with arthrosynovitis [Conti, 2002].

It was convincingly argued that MCs may be involved in inflammatory arthritis; however, this presentation was still limited primarily to the role of histamine and TNF-α [Woolley, 2003]. For example, it did not discuss potential triggers, such as C3a and C5a, or the involvement of T-cells [Askenase, 2005]. One reason why MCs have been neglected comes from the persistence of associating them only with allergies in spite of mounting evidence to the contrary. Another reason is the fact that MCs are not commonly seen to degranulate in tissues, such as joints that do not get allergic reactions, they were, therefore, considered an "innocent bystander."

Animal Models of Inflammation

Mast cells are required in autoimmune arthritis [Lee, 2002a] and inflammatory arthritis [Mattheos, 2003], as knee involvement was absent in the joints of W/W^v mast cell deficient mice as compared to their +/+ controls.

In spite of the fact that TNF blockers are used clinically with considerable success in adult RA [Taylor, 2003], increasing evidence indicates that IL-6 may be equally or more important [de Hooge, 2000; Kaneko, 2000]. In fact, we showed that experimental inflammatory arthritis could not develop in TNF −/− mice [Mattheos, 2003]. Instead, the IL-6 −/− mice were resistant to inflammatory arthritis [Boe, 1999], and humanized antibody to human IL-6 receptor inhibited collagen-induced arthritis [Mihara, 2001]. These findings are of particular significance as stress was shown to increase IL-6 release in juvenile RA [Roupe van der Voort, 2000] and we showed that MCs are the sole source of serum IL-6 elevations induced by acute stress in mice [Huang, 2003a]. It is quite interesting that stress increased serum IL-6 levels in care givers of chronically ill patients [Kiecolt-Glaser, 2003], but it was decreased in those who went to church regularly [Lutgendorf, 2004], indicating that reduction of stress could lead to decrease in a key proinflammatory cytokine production.

Mast cells in the joints of RA patients express CRH receptors [McEvoy, 2001]. Moreover, both CRH [Lowry, 1996; McEvoy, 2001], and Ucn [Kohno, 2001; Uzuki, 2001], as well as CRH receptors are increased in the joints of RA patients, the symptoms of which worsen by stress [Herrmann, 2000; Thomason, 1992].

Mast cell deficient and CRH −/− mice were shown be resistant to autoimmune [Lee, 2002a] and inflammatory arthritis. These results are of interest because CRH is increased in the joints of RA patients and CRH receptors (CRHR) are present on articular mast cells [McEvoy, 2001], implying that CRH could trigger or regulate synovial mast cell activation. Corticotropin-releasing hormone was recently shown to also induce mast cell dependent vascular permeability in humans [Crompton, 2003]. The pathophysiological implication of such finding is that CRH could be released locally under stress [Lytinas, 2003] and exacerbate inflammatory diseases [Theoharides, 2004a; Theoharides and Cochrane, 2004].

The K/BxN mouse has been used as a model for human inflammatory arthritis [Ditzel, 2005] and was recently shown that the production of autoantibodies against glucose-G phosphate isomerase dependent on mast cells [Wipke, 2005].

CORONARY INFLAMMATION

The presence of mast cells has been established in human heart tissue from very early studies [Kovanen, 2007; Patella, 1995]. Their location, in conjunction with certain cytokines that mast cells produce, suggested that they may play an important role in cardiac diseases [Frangogiannis, 2007]. Cardiac MCs can participate in the development of atherosclerosis, coronary inflammation and cardiac ischemia [Constantinides, 1995; Kaartinen, 1994; Laine, 1999; Patella, 1995]. In fact, mast cells are particularly prominent in coronary arteries during spasm [Forman, 1985], they accumulate in the shoulder region of human coronary plaque rupture [Kovanen, 2007; Parikh and Singh, 2001; Patella, 1995], and were found to be decreased in normal coronary intima as opposed to their increased numbers in fatty streaks [Kaartinen, 1994].

Mast cell mediators released are increased in blood or urine of patients suffering from acute coronary syndromes. Histamine is elevated in the coronary circulation of patients with variant angina [Sakata, 1996], can be released under acute stress [Huang, 2002] and induces tissue factor expression [Steffel, 2005]. In addition, IL-6 is released in the coronary sinus of patients with acute coronary syndromes [Deliargyris, 2000; Raymond, 2001], and is also exclusively released from mast cells during acute stress in mice [Huang, 2003]. Another observation stressing the importance of mast cells in coronary inflammation and pathology is that IL-6 is increased in ischemia-reperfusion in mice, but not in MC deficient mice [Bhattacharya, 2007].

Many mediators released from cardiac MCs could influence cardiovascular pathophysiology. Mast cell proteases participate in plaque rupture [Lee-Rueckert and Kovanen, 2006]. In fact, cardiac MC derived histamine [Gristwood, 1981] can cause coronary constriction [Genovese and Spadaro, 1997] and can sensitize nerve endings [Christian, 1989]. Tryptase can degrade HDL [Lee, 2002c] and has been implicated in acute coronary syndromes [Deliargyris, 2005; Kervinen, 2005]. Moreover chymase can induce the removal of cholesterol from HDL particles and increase uptake by macrophages that become "foam" cells, major components of coronary atheromas [Kovanen, 1996; Lee, 2002b, 2003; Lindstedt, 1996]. This cholesterol removal may be due to impairment of HDL to act as a high affinity cholesterol receptor [von Eckardstein, 2001] and subsequent destabilization of the LDL bound to the heparin proteoglycan component of the exocytosed granules of MCs, thus leading to LDL retention in the arterial intima [von Eckardstein, 2001]. Chymase may also play a role in plaque destabilizaton by inhibiting collagen synthesis. Notably, chymase is responsible for the ACE independent coversion of angiotensin-I (ANG I) to angiotensin-II (ANG II), after vascular injury [Doggrell and Wanstall, 2004; Lee, 2002b].

Mast cells were recently shown to be a novel source of renin in human and rodent myocardium [Silver, 2004]. In myocardial IR, mast cells degranulate and release renin, which activates the rennin–angiotensin (ANG) pathway leading eventually to release of norepinephrine (NE) from sympathetic nerve terminals. Recent studies [Dell'Italia, 1997] indicate that ANG I found in the heart is synthesized *in situ*, and most of cardiac ANG II derives from the conversion of locally produced, rather than blood-derived, ANG I. It has also been shown that pharmacological mast cell stabilization, renin inhibition, or AT_1-receptor blockade prevent reperfusion arrhythmias, that are mainly mediated by ANG II and NE [Mackins, 2006].

Acute psychological stress affects cardiac mast cells and this has been hypothesized to play a significant role in coronary heart disease (CHD) [Deanfield, 1984; Deedwania, 1995; Freeman, 1987; Rozanski, 1988]. Acute stress induces rat cardiac MCs activation, an effect blocked by cromolyn [Pang, 1998]. It is known that histamine [Clejan, 2002] and IL-6 [Suzuki, 2003] are independent factors of CHD morbility and mortality. Acute stress induced histamine release from mouse heart [Huang, 2002], as well as increased histamine and IL-6 in the serum [Huang, 2002, 2003a]. These effects are dependent on MCs and are greater in apolipoprotein E (ApoE) knockout mice that develop atherosclerosis [Huang, 2002, 2003a]. Serum IL-6 elevations in patients with acute CHD were documented to derive primarily from the coronary sinus [Deliargyris, 2000]. Finally, there are reports of anaphylactic CHD in patients with drug eluting coronary stents that has been termed the "Kounis" syndrome, underlying the significance of the mast cell in cardiac pathophysiology [Kounis, 1999, 2007; Kounis and Zavras, 1996]. Hypersensitivity reactions that do not involve IgE appear to involve activation of the complement (C) system and are sometimes called "C activation-related pseudoallergy" (CARPA) [Szebeni, 2005]. Drugs and agents causing CARPA include radiocontrast media, liposomal drugs and amphiphilic lipids, such as the vehicle for certain drugs used in drug eluting stents [Kounis, 2007].

Mast cells could also have a useful role through adrenomedullin (AM), a potent vasodilatory hypotensive peptide that may play a protective role in the heart [Beltowski and Jamroz, 2004]. Note that cardiac MCs are able to synthesize, store AM and, upon stimulation, to release it near coronary arterioles and venules [Belloni, 2006]. An antifibrotic role has also been suggested for AM, after results showing that AM regulated collagen production from cardiac fibroblasts.

Adenosine, acting via A2a, A2b, and A3 [Cerniway, 2001; Fozard, 1996; Schmidt and Brunner, 1976] receptors, is a potent stimulus for mast cell degranulation [Marquardt, 1978], IL-8 [Feoktistov and Biaggioni, 1995], and VEGF release. Adenosine could play an important role in the pathophysiology of IR injury since A3R knockout mice were tolerant to IR injury [Cerniway, 2001].

OCULAR HYPERSENSITIVITY REACTIONS

The eye and the eyelid are common sites of allergic and other hypersensitivity reactions. Typical ocular allergies develop when pollen, dust, or other irritants come in contact with and activate mast cells on the conjunctiva. Classic ocular allergic reactions include (1) seasonal allergic conjunctivitis, (2) perennial allergic conjunctivitis, (3) vernal keratoconjuctivitis, (4) giant papillary conjunctivitis, and (5) atopic keratoconjuctivitis. Large numbers of mast cells have been detected in the posterior choroid of the eyes of guinea pigs, rabbits, and rats [Steptoe, 1994]. In normal human subjects, few MCs of the MCT type were found in the conjunctival epithelium [Leonardi, 2002]. The substantia propria of the normal conjunctiva contained large numbers of MCs, 95% of which were of the MCTC type. Activated mast cells in the eye secrete histamine, prostanoids, kinins, proteases, and other proinflammatory mediators. Mast cell activation in allergic conjunctivitis is evidenced by an increase in tryptase levels in unstimulated tear fluid of subjects who have symptomatic allergic conjunctivitis [Margrini, 1996]. Activation of appropriate receptors produce the known symptoms of burning, itching and watery discharge, that may be accompanied by nasal discharge and others allergic symptoms.

Mast cells are also involved in chronic inflammatory disorders of the eye, such as vernal conjunctivitis, giant papillary conjunctivitis, and atopic keratoconjunctivitis. This result is suggested by epithelial invasion of the conjunctiva with MCs by an altered MC phenotype and by the release of MC mediators in tear fluid [Anderson, 1997]. Finally, MC ocular involvement has been noted in a number of systematic diseases. Recently, a case of interstitial keratitis has been reported in patients with mastocytosis [Magone, 2006]. Mast cells were abundant in the site of the lesion and the patient complained of allergic symptoms, which responded well to MC stabilizers.

CONCLUSION

Mast cells have emerged as a unique immune cell that can be activated by many inflammatory and neurohormonal triggers, as well as participate in cross-talks with other immune cells in the pathogenesis of inflammatory diseases. Unfortunately, there are few effective treatments for these diseases and great need for *in vitro* models. Matrix CD34-cells may be a unique source of growing mast cells that may have even greater capacity to develop into unique tissue-associated phenotypes that could be used alone or in three-dimensional cultures with other cell types using collagen or other appropriate scaffoldings.

REFERENCES

Aaron LA, Buchwald D. 2001a. A review of the evidence for overlap among unexplained clinical conditions. Ann Intern Med. 134:868–881.

Aaron LA, Buchwald D. 2001b. Fibromyalgia and other unexplained clinical conditions. Curr Rheumatol. 3:116–122.

Aaron LA, Buchwald D. 2003. Chronic diffuse musculoskeletal pain, fibromyalgia and co-morbid unexplained clinical conditions. Best Pract Res Clin Rheumatol. 17:563–574.

Aaron LA et al. 2001. Comorbid clinical conditions in chronic fatigue: A co-twin control study. J Gen Intern Med. 16:24–31.

Abdel-Majid RM, Marshall JS. 2004. Prostaglandin E2 induces degranulation-independent production of vascular endothelial growth factor by human mast cells. J Immunol. 172:1227–1236.

Abeles AM, Pillinger MH, Solitar BM, Abeles M. 2007. Narrative review: The pathophysiology of fibromyalgia. Ann Intern Med. 146:726–734.

Aderem A, Ulevitch RJ. 2000. Toll-like receptors in the induction of the innate immune response. Nature (London). 406:782–787.

Ahn K et al. 2000. Regulation of chymase production in human mast cell progenitors. J Allergy Clin Immunol. 106:321–328.

Akira S, Takeda K, Kaisho T. 2001. Toll-like receptors: Critical proteins linking innate and acquired immunity. Nat Immunol. 2:675–680.

Al'Abadie MS, Senior HJ, Bleehen SS, Gawkrodger DJ. 1995. Neuropeptides and general neuronal marker in psoriasis–an immunohistochemical study. Clin Exp Dermatol. 20:384–389.

Alexandrakis MG et al. 2003a. Inhibitory effect of retinoic acid on proliferation, maturation and tryptase level in human leukemic mast cells (HMC-1). Int J Immunopath Pharmacol. 16:43–47.

Alexandrakis MG et al. 2003b. Flavones inhibit proliferation and increase mediator content in human leukemic mast cells (HMC-1). Eur J Haematol. 71:448–454.

Ali H, Leung KBP, Pearce FL, Hayes NA, Foreman JC. 1986. Comparison of the histamine-releasing action of substance P on mast cells and basophils from different species and tissues. Int Arch Allergy Appl Immunol. 79:413–418.

Allakhverdi Z et al. 2007a. Thymic stromal lymphopoietin is released by human epithelial cell in response to microbes, trauma, or inflammation and potently activates mast cells. J Exp Med. 19:253–258.

Allakhverdi Z, Smith DE, Comeau MR, Delespesse G. 2007b. Cutting edge: The ST2 ligand IL-33 potently activates and drives maturation of human mast cells. J Immunol. 179:2051–2054.

Aloe L, Levi-Montalcini R. 1977. Mast cells increase in tissues of neonatal rats injected with the nerve growth factor. Brain Res. 133:358–366.

Amano H, Kurosawa M, Ishikawa O, Chihara J, Miyachi Y. 2000. Cultured human mast cells derived from umbilical cord blood cells in the presence of stem cell factor and interleukin-6 cannot be a model of human skin mast cells: Fluorescence microscopic analysis of intracellular calcium ion mobilization. J Dermatol Sci. 24:146–152.

Anderson DF et al. 1997. Seasonal allergic conjunctivitis is accompanied by increased mast cell numbers in the absence of leucocyte infiltration. Clin Exp Allergy. 27:1060–1066.

Anthony M, Lance JW. 1971. Whole blood histamine and plasma serotonin in cluster headache. Proc Aust assoc Neurol. 8:43–46.

Arici A, Tazuke SI, Attar E, Kliman HJ, Olive DL. 1996. Interleukin-8 concentration in peritoneal fluid of patients with endometriosis and modulation of interleukin-8 expression in human mesothelial cells. Mol Hum Reprod. 2:40–45.

Ashwood P, Wills S, Van de WJ. 2006. The immune response in autism: A new frontier for autism research. J Leukoc Biol. 80:1–15.

Askenase PW. 2005. Mast cells and the mediation of T-cell recruitment in arthritis.e. N Engl J Med. 349: 1294.

Azzi A et al. 2004. Vitamin E mediates cell signaling and regulation of gene expression. Ann NY Acad Sci. 1031:86–95.

Babina M, Mammeri K, Henz BM. 1999. ICAM-3 (CD50) is expressed by human mast cells: Induction of homotypic mast cell aggregation via ICAM-3. Cell Adhesion Commun. 7:195–209.

Bachelet I, Levi-Schaffer F. 2007. Mast cells as effector cells: A co-stimulating question. Trends Immunol. 28:360–365.

Bani-Sacchi T et al. 1986. The release of histamine by parasympathetic stimulation in guinea pig auricle and rat ileum. J Physiol. 371:29–43.

Barke KE, Hough LB. 1993. Opiates, mast cells and histamine release. Life Sci. 53:1391–1399.

Belloni AS et al. 2006. Identification and localization of adrenomedullin-storing cardiac mast cells. Int J Mol Med. 17:709–713.

Beltowski J, Jamroz A. 2004. Adrenomedullin–what do we know 10 years since its discovery? Pol J Pharmacol. 56:5–27.

Bennett R. 1998. Fibromyalgia, chronic fatigue syndrome, and myofascial pain. Curr Opin Rheumatol. 10:95–103.

Benyon R, Robinson C, Church MK. 1989. Differential release of histamine and eicosanoids from human skin mast cells activated by IgE-dependent and non-immunological stimuli. Br J Pharmacol. 97:898–904.

Bethin KE, Vogt SK, Muglia LJ. 2000. Interleukin-6 is an essential, corticotropin-releasing hormone-independent stimulator of the adrenal axis during immune system activation. Proc Natl Acad Sci USA. 97:9317–9322.

Bhattacharya K et al. 2007. Mast cell deficient W/Wv mice have lower serum IL-6 and less cardiac tissue necrosis than their normal littermates following myocardial ischemia-reperfusion. Int J Immunopath Pharmacol. 20:69–74.

Bidri M, Feger F, Varadaradjalou S, Ben HN, Guillosson JJ, Arock M. 2001. Mast cells as a source and target for nitric oxide. Int Immunopharmacol. 1:1543–1558.

Bienenstock J et al. 1987. The role of mast cells in inflammatory processes: Evidence for nerve mast cell interactions. Int Arch Allergy Appl Immunol. 82:238–243.

Bienenstock J. 2002. Stress and asthma: The plot thickens. Am J Respir Crit Care Med. 165:1034–1035.

Bischoff SC, Sellge G, Lorentz A, Sebald W, Raab R, Manns MP. 1999. IL-4 enhances proliferation and mediator release in mature human mast cells. Proc Natl Acad Sci USA. 96:8080–8085.

Blair RJ et al. 1997. Human mast cells stimulate vascular tube formation. Tryptase is a novel, potent angiogenic factor. J Clin Invest. 99:2691–2700.

Blandina P, Fantozzi R, Mannaioni PF, Masini E. 1980. Characteristics of histamine release evoked by acetylcholine in isolated rat mast cells. J Physiol. 301:281–293.

Blank U, Rivera J. 2004. The ins and outs of IgE-dependent mast-cell exocytosis. Trends Immunol. 25:266–273.

Blennerhassett MG, Tomioka M, Bienenstock J. 1991. Formation of contacts between mast cells and sympathetic neurons *in vitro*. Cell Tissue Res. 265:121–128.

Boe A, Baiocchi M, Carbonatto M, Papoian R, Serlupi-Crescenzi O. 1999. Interleukin 6 knock-out mice are resistant to antigen-induced experimental arthritis. Cytokine. 11:1057–1064.

Boesiger J et al. 1998. Mast cells can secrete vascular permeability factor/vascular endothelial cell growth factor and exhibit enhanced release after immunoglobulin E-dependent upregulation of Fcε receptor I expression. J Exp Med. 188:1135–1145.

Boros M, Kaszaki J, Bako L, Nagy S. 1992. Studies on the relationship between xanthine oxidase and histamine release during intestinal ischemia-reperfusion. Circ Shock. 38:108–114.

Bradding P, Okayama Y, Howarth PH, Church MK, Holgate ST. 1995. Heterogeneity of human mast cells based on cytokine content. J Immunol. 155:297–307.

Bradding P. 2003. The role of the mast cell in asthma: A reassessment. Curr Opin Allergy Clin Immunol. 3:45–50.

Brigelius-Flohe R. 2005. Induction of drug metabolizing enzymes by vitamin E. J Plant Physiol. 162:797–802.

Brightling CE, Bradding P, Pavord ID, Wardlaw AJ. 2003. New insights into the role of the mast cell in asthma. Clin Exp Allergy. 33:550–556.

Brooks AC, Whelan CJ, Purcell WM. 1999. Reactive oxygen species generation and histamine release by activated mast cells: Modulation by nitric oxide synthase inhibition. Br J Pharmacol. 128:585–590.

Brown JM, Swindle EJ, Kushnir-Sukhov NM, Holian A, Metcalfe DD. 2007. Silica-directed mast cell activation is enhanced by scavenger receptors. Am J Respir Cell Mol Biol. 36:43–52.

Brown S, Reynolds NJ. 2006. Atopic and non-atopic eczema. BMJ. 332:584–588.

Buske-Kirschbaum A, Hellhammer DH. 2003. Endocrine and immune responses to stress in chronic inflammatory skin disorders. Ann NY Acad Sci. 992:231–240.

Buske-Kirschbaum A, von Auer K, Krieger S, Weis S, Rauh W, Hellhammer D. 2003. Blunted cortisol responses to psychosocial stress in asthmatic children: A general feature of atopic disease? Psychosom Med. 65:806–810.

Buskila D, Sarzi-Puttini P, Ablin JN. 2007. The genetics of fibromyalgia syndrome. Pharmacogenomics. 8:67–74.

Cairns JA, Walls AF. 1996. Mast cell tryptase is a mitogen for epithelial cells. Stimulation of IL-8 production and intercellular adhesion molecule-1 expression. J Immunol. 156:275–283.

Camarda V, Rizzi A, Galo G, Guerrini R, Salvadori S, Regoli D. 2002. Pharmacological profile of hemokinin 1: A novel member of the tachykinin family. Life Sci. 71:363–370.

Cao J et al. 2005. Human mast cells express corticotropin-releasing hormone (CRH) receptors and CRH leads to selective secretion of vascular endothelial growth factor. J Immunol. 174:7665–7675.

Cao J, Boucher W, Donelan JM, Theoharides TC. 2006a. Acute stress and intravesical corticotropin-releasing hormone induces mast cell-dependent vascular endothelial growth factor release from mouse bladder explants. J Urol. 176:1208–1213.

Cao J, Curtis CL, Theoharides TC. 2006b. Corticotropin-releasing hormone induces vascular endothelial growth factor release from human mast cells via the cAMP/protein kinase A/p38 mitogen-activated protein kinase pathway. Mol Pharmacol. 69:998–1006.

Carraway R. 1982. Neurotensin stimulates exocytotic histamine secretion from rat mast cells and elevates plasma histamine levels. J Physiol. 323:403–414.

Carraway RE, Cochrane DE, Boucher W, Mitra SP. 1989. Structures of histamine-releasing peptides formed by the action of acid proteases on mammalian albumin(s). J Immunol. 143:1680–1684.

Casale TB, Marom Z. 1983. Mast cells and asthma. The role of mast cell mediators in the pathogenesis of allergic asthma. Ann Allergy. 51:2–6.

Centers for Disease Control and Prevention (CDC) 2002. Self-reported increase in asthma severity after the September 11 attacks on the World Trade Center–Manhattaan, New York, 2001. MMWR Morb Mortal Wkly Rep. 51:781–784.

Cerniway RJ, Yang Z, Jacobson MA, Linden J, Matherne GP. 2001. Targeted deletion of A(3) adenosine receptors improves tolerance to ischemia-reperfusion injury in mouse myocardium. Am J Physiol Heart Circ Physiol. 281:H1751–H1758.

Chan J, Smoller BR, Raychauduri SP, Jiang WY, Farber EM. 1997. Intraepidermal nerve fiber expression of calcitonin gene-related peptide, vasoactive intestinal peptide and substance P in psoriasis. Arch Dermatol Res. 289:611–616.

Chen E, Fisher EB, Bacharier LB, Strunk RC. 2003. Socioeconomic status, stress, and immune markers in adolescents with asthma. Psychosom Med. 65:984–992.

Cho CH, Ogle CW. 1977. The effects of zinc sulphate on vagal-induced mast cell changes and ulcers in the rat stomach. Eur J Pharmacol. 43:315–322.

Cho SH, Anderson AJ, Oh CK. 2002. Importance of mast cells in the pathophysiology of asthma. Clin Rev Allergy Immunol. 22:161–174.

Cho SH, Woo CH, Yoon SB, Kim JH. 2004. Protein kinase Cdelta functions downstream of Ca2+ mobilization in FcepsilonRI signaling to degranulation in mast cells. J Allergy Clin Immunol. 114:1085–1092.

Choi IS, Koh YI, Chung SW, Lim H. 2004. Increased releasability of skin mast cells after exercise in patients with exercise-induced asthma. J Korean Med Sci. 19:724–728.

Christian EP, Undem BJ, Weinreich D. 1989. Endogenous histamine excites neurones in the guinea-pig superior cervical ganglion *in vitro*. J Physiol. 409:297–312.

Church MK, Lowman MA, Rees PH, Benyon RC. 1989. Mast cells, neuropeptides and inflammation. Agents Actions. 27:8–16.

Clauw DJ, Schmidt M, Radulovic D, Singer A, Katz P, Bresette J. 1997. The relationship between fibromyalgia and interstitial cystitis. J Psychaitr Res. 31:125–131.

Clejan S, Japa S, Clemetson C, Hasabnis SS, David O, Talano JV. 2002. Blood histamine is associated with coronary artery disease, cardiac events and severity of inflammation and atherosclerosis. J Cell Mol Med. 6:583–592.

Clifton VL, Crompton R, Smith R, Wright IM. 2002. Microvascular effects of CRH in human skin vary in relation to gender. J Clin Endocrinol Metab. 87:267–270.

Cocchiara R, Bongiovanni A, Albeggiani G, Azzolina A, Geraci D. 1998. Evidence that brain mast cells can modulate neuroinflammatory responses by tumor necrosis factor-α production. Neuroreport. 9:95–98.

Cochrane DE, Carraway RE, Feldberg RS, Boucher W, Gelfand JM. 1993. Stimulated rat mast cells generate histamine-releasing peptide from albumin. Peptides. 14:117–123.

Cohly HH, Panja A. 2005. Immunological findings in autism. Int Rev Neurobiol. 71:317–341.

Coleman JW. 2002. Nitric oxide: A regulator of mast cell activation and mast cell-mediated inflammation. Clin Exp Immunol. 129:4–10.

Constantinides P. 1995. Infiltrates of activated mast cells at the site of coronary atheromatous erosion or rupture in myocardial infarction. Circulation. 92:1083–1088.

Conti P et al. 1997. Impact of Rantes and MCP-1 chemokines on *in vivo* basophilic mast cell recruitment in rat skin injection model and their role in modifying the protein and mRNA levels for histidine decarboxylase. Blood. 89:4120–4127.

Conti P, Reale M, Barbacane RC, Letourneau R, Theoharides TC. 1998. Intramuscular injection of hrRANTES causes mast cell recruitment and increased transcription of histidine decarboxylase: Lack of effects in genetically mast cell-deficient W/W^v mice. FASEB J. 12:1693–1700.

Conti P, Reale M, Barbacane RC, Castellani ML, Orso C. 2002. Differential production of RANTES and MCP-1 in synovial fluid from the inflamed human knee. Immunol Lett. 80:105–111.

Conti P et al. 2003. IL-10, an inflammatory/inhibitory cytokine, but not always. Immunol Lett. 86:123–129.

Crisp AJ, Champan CM, Kirkham SE, Schiller AL, Keane SM. 1984. Articular mastocytosis in rheumatoid arthritis. Arthritis Rheum. 27:845–851.

Cristofaro P, Opal SM. 2006. Role of toll-like receptors in infection and immunity: Clinical implications. Drugs. 66:15–29.

Crompton R, Clifton VL, Bisits AT, Read MA, Smith R, Wright IM. 2003. Corticotropin-releasing hormone causes vasodilation in human skin via mast cell-dependent pathways. J Clin Endocrinol Metab. 88:5427–5432.

Cruse G, Cockerill S, Bradding P. 2008. IgE alone promotes human lung mast cell survival through the autocrine production of IL-6. BMC Immunol. 9:2.

Cuttitta F et al. 2002. Adrenomedullin functions as an important tumor survival factor in human carcinogenesis. Microsc Res Tech. 57:110–119.

Czarnetzki BM, Figdor CG, Kolde G, Vroom T, Aalberse R, de Vries JE. 1984. Development of human connective tissue mast cells from purified blood monocytes. Immunology. 51:549–554.

Dakhama A, Lee YM, Gelfand EW. 2005. Virus-induced airway dysfunction: Pathogenesis and biomechanisms. Pediatr Infect Dis J. 24:S159–69, discussion.

de Hooge AS, van De Loo FA, Arntz OJ, van den Berg WB. 2000. Involvement of IL-6, apart from its role in immunity, in mediating a chronic response during experimental arthritis. Am J Pathol. 157:2081–2091.

de Paulis A et al. 1996. Human synovial mast cells. I. Utrastructural in situ and in vitro immunologic characterization. Arthritis Rheum. 39:1222–1233.

De Simone R, Alleva E, Tirassa P, Aloe L. 1990. Nerve growth factor released into the bloodstream following intraspecific fighting induces mast cell degranulation in adult male mice. Brain Behav Immunol. 4:74–81.

De Stefano R et al. 2000. Image analysis quantification of substance P immunoreactivity in the trapezius muscle of patients with fibromyalgia and myofascial pain syndrome. J Rheumatol. 27:2906–2910.

Deanfield JE et al. 1984. Silent myocardial ischaemia due to mental stress. Lancet. 2:1001–1005.

Deedwania PC. 1995. Mental stress, pain perception and risk of silent ischemia. JACC. 25:1504–1506.

Deliargyris EN, Raymond RJ, Theoharides TC, Boucher WS, Tate DA, Dehmer GJ. 2000. Sites of interleukin-6 release in patients with acute coronary syndromes and in patients with congestive heart failure. Am J Cardiol. 86:913–918.

Deliargyris EN et al. 2005. Mast cell tryptase: A new biomarker in patients with stable coronary artery disease. Atherosclerosis. 178:381–386.

Dell'Italia LJ et al. 1997. Compartmentalization of angiotensin II generation in the dog heart. Evidence for independent mechanisms in intravascular and interstitial spaces. J Clin Invest. 100:253–258.

Dendorfer U, Oettgen P, Libermann TA. 1994. Multiple regulatory elements in the interleukin-6 gene mediate induction by prostaglandins, cyclic AMP, and lipopolysaccharide. Mol Cell Biol. 14:4443–4454.

Deschoolmeester ML, Eastmond NC, Dearman RJ, Kimber I, Basketter DA, Coleman JW. 1999. Reciprocal effects of interleukin-4 and interferon-gamma on immunoglobulin E-mediated mast cell degranulation: A role for nitric oxide but not peroxynitrite or cyclic guanosine monophosphate. Immunology. 96:138–144.

Dewald O et al. 2004. Of mice and dogs: Species-specific differences in the inflammatory response following myocardial infarction. Am J Pathol. 164:665–677.

Dhabhar F, McEwen BS. 1996. Stress-induced enhancement of antigen-specific cell-mediated immunity. J Immunol. 156:2608–2615.

Dhabhar FS, McEwen BS. 1999. Enhancing versus suppressive effects of stress hormones on skin immune function. Proc Natl Acad Sci USA. 96:1059–1064.

Dimitriadou V, Lambracht-Hall M, Reichler J, Theoharides TC. 1990. Histochemical and ultrastructural characteristics of rat brain perivascular mast cells stimulated with compound 48/80 and carbachol. Neuroscience. 39:209–224.

Dimitriadou V, Buzzi MG, Moskowitz MA, Theoharides TC. 1991. Trigeminal sensory fiber stimulation induces morphologic changes reflecting secretion in rat dura mast cells. Neuroscience. 44:97–112.

Dimitriadou V et al. 1997. Functional relationships between sensory nerve fibers and mast cells of dura mater in normal and inflammatory conditions. Neuroscience. 77:829–839.

Ditzel HJ. 2005. The K/BxN mouse: A model of human inflammatory arthritis. Trends Mol Med. 10:40–45.

Doggrell SA, Wanstall JC. 2004. Vascular chymase: Pathophysiological role and therapeutic potential of inhibition. Cardiovasc Res. 61:653–662.

Donelan J, Boucher W, Papadopoulou N, Lytinas M, Papaliodis D, Theoharides TC. 2006. Corticotropin-releasing hormone induces skin vascular permeability through a neurotensin-dependent process. Proc Natl Acad Sci USA. 103:7759–7764.

Drexler HG, MacLeod RA. 2003. Malignant hematopoietic cell lines: In vitro models for the study of mast cell leukemia. Leuk Res. 27:671–676.

Dvorak AM et al. 1992a. Ultrastructural evidence for piecemeal and anaphylactic degranulation of human gut mucosal mast cells *in vivo*. Int Arch Allergy Immunol. 99:74–83.

Dvorak AM et al. 1992b. Human gut mucosal mast cells: Ultrastructural observations and anatomic variation in mast cell-nerve associations *in vivo*. Int Arch Allergy Immunol. 98:158–168.

Dvorak AM. 1997. New aspects of mast cell biology. Int Arch Allergy Immunol. 114:1–9.

Eastmond NC, Banks EM, Coleman JW. 1997. Nitric oxide inhibits IgE-mediated degranulation of mast cells and is the principal intermediate in IFN-gamma-induced suppression of exocytosis. J Immunol. 159:1444–1450.

Ehrenreich H et al. 1992. Endothelins belong to the assortment of mast cell-derived and mast cell-bound cytokines. New Biol. 4:147–156.

Ekoff M, Strasser A, Nilsson G. 2007. FcepsilonRI aggregation promotes survival of connective tissue-like mast cells but not mucosal-like mast cells. J Immunol. 178:4177–4183.

Enestrom S, Bengtson A, Lindstrom F, Johan K. 1990. Attachment of IgG to dermal extracellular matrix in patients with fibromyalgia. Clin Exp Rheumatol. 8:127–135.

Enestrom S, Bengtsson A, Frodin T. 1997. Dermal IgG deposits and increase of mast cells in patients with fibromyalgia-relevant findings or epiphenomena? Scand J Rheumat. 26:308–313.

Fagan J, Galea S, Ahern J, Bonner S, Vlahov D. 2003. Relationship of self-reported asthma severity and urgent health care utilization to psychological sequelae of the September 11, 2001 terrorist attacks on the World Trade Center among New York City area residents. Psychosom Med. 65:993–996.

Fantozzi R, Masini E, Blandina P, Mannaioni PF, Bani-Sacchi T. 1978. Release of histamine from rat mast cells by acetylcholine. Nature (London). 273:473–474.

Feoktistov I, Biaggioni I. 1995. Adenosine A_{2b} receptors evoke interleukin-8 secretion in human mast cells—An enprofylline-sensitive mechanism with implications for asthma. J Clin Invest. 96:1979–1986.

Feoktistov I, Ryzhov S, Goldstein AE, Biaggioni I. 2003. Mast cell-mediated stimulation of angiogenesis: Cooperative interaction between A2B and A3 adenosine receptors. Circ Res. 92:485–492.

Fewtrell CMS, Foreman JC, Jordan CC, Oehme P, Renner H, Stewart JM. 1982. The effects of substance P on histamine and 5-hydroxytryptamine release in the rat. J Physiol. 330:393–411.

Foreman JC, Hallett MB, Mongar JL. 1977. The relationship between histamine secretion and 45 calcium uptake by mast cells. J Physiol. 271:193–214.

Foreman JC. 1987a. Neuropeptides and the pathogenesis of allergy. Allergy. 42:1–11.

Foreman JC. 1987b. Peptides and neurogenic inflammation. Brain Res Bull. 43:386–398.

Forman MB, Oates JA, Robertson D, Robertson RM, Roberts LJ, II, Virmani R. 1985. Increased adventitial mast cells in a patient with coronary spasm. N Engl J Med. 313:1138–1141.

Forsythe P, Ebeling C, Gordon JR, Befus AD, Vliagoftis H. 2004. Opposing effects of short- and long-term stress on airway inflammation. Am J Respir Crit Care Med. 169:220–226.

Fortune DG, Richards HL, Griffiths CE. 2005. Psychologic factors in psoriasis: Consequences, mechanisms, and interventions. Dermatol Clin. 23:681–694.

Fozard JR, Pfannkuche HJ, Schuurman HJ. 1996. Mast cell degranulation following adenosine A3 receptor activation in rats. Eur J Pharmacol. 298:293–297.

Frangogiannis NG. 2007. Chemokines in ischemia and reperfusion. Thromb Haemost. 97:738–747.

Freeman LJ, Nixon PGF, Sallabank P, Reaveley D. 1987. Psychological stress and silent myocardial ischemia. Am Heart J. 114:477–482.

Gagari E, Tsai M, Lantz CS, Fox LG, Galli SJ. 1997. Differential release of mast cell interleukin-6 via c-kit. Blood. 89:2654–2663.

Galli SJ, Kalesnikoff J, Grimbaldeston MA, Piliponsky AM, Williams CM, Tsai M. 2005a. Mast cells as "tunable" effector and immunoregulatory cells: Recent advances. Annu Rev Immunol. 23:749–786.

Galli SJ, Nakae S, Tsai M. 2005b. Mast cells in the development of adaptive immune responses. Nat Immunol. 6:135–142.

Gebhardt T et al. 2005. Growth, phenotype, and function of human intestinal mast cells are tightly regulated by transforming growth factor beta1. Gut. 54:928–934.

Genovese A, Spadaro G. 1997. Highlights in cardiovascular effects of histamine and H1-receptor antagonists. Allergy. 52:67–78.

Ghannadan M et al. 1998. Phenotypic characterization of human skin mast cells by combined staining with toluidine blue and CD antibodies. J Invest Dermatol. 111:689–695.

Gilchrist M, McCauley SD, Befus AD. 2004. Expression, localization, and regulation of NOS in human mast cell lines: Effects on leukotriene production. Blood. 104:462–469.

Goetzl EJ, Chernov T, Renold F, Payan DG. 1985. Neuropeptide regulation of the expression of immediate hypersensitivity. J Immunol. 135:802s–805s.

Goetzl EJ, Cheng PPJ, Hassner A, Adelman DC, Frick OL, Speedharan SP. 1990. Neuropeptides, mast cells and allergy: Novel mechanisms and therapeutic possibilities. Clin Exp Allergy. 20:3–7.

Gomez G et al. 2005. TGF-beta1 inhibits mast cell FceRI expression. J Immunol. 174:5987–5993.

Gordon DJ, Rifkind BM. 1989. High-density lipoprotein—the clinical implications of recent studies. N Engl J Med. 321:1311–1316.

Gordon JR, Burd PR, Galli SJ. 1990. Mast cells as a source of multifunctional cytokines. Immunol Today. 11:458–464.

Gordon JR, Galli SJ. 1990. Mast cells as a source of both preformed and immunologically inducible TNF-α/cachectin. Nature (London). 346:274–276.

Gotis-Graham I, Smith MD, Parker A, McNeil HP. 1998. Synovial mast cell responses during clinical improvement in early rheumatoid arthritis. Ann Rheum Dis. 57:664–671.

Grabbe J, Welker P, Möller A, Dippel E, Ashman LK, Czarnetzki BM. 1994. Comparative cytokine release from human monocytes, monocyte-derived immature mast cells and a human mast cell line (HMC-1). J Invest Dermatol. 103:504–508.

Graham DT, Wolf S. 1953. The relation of eczema to attitude and to vascular reactions of the human skin. J Lab Clin Med. 42:238–254.

Grisham MB, Jourd'Heuil D, Wink DA. 1999. Nitric oxide. I. Physiological chemistry of nitric oxide and its metabolites: Implications in inflammation. Am J Physiol. 276:G315–G321.

Gristwood RW, Lincoln JC, Owen DA, Smith IR. 1981. Histamine release from human right atrium. Br J Pharmacol. 74:7–9.

Gruber BL et al. 1997. Human mast cells activate fibroblasts—Tryptase is a fibrogenic factor stimulating collagen messenger ribonucleic acid synthesis and fibroblast chemotaxis. J Immunol. 158:2310–2317.

Grundtman C, Salomonsson S, Dorph C, Bruton J, Andersson U, Lundberg IE. 2007. Immunolocalization of interleukin-1 receptors in the sarcolemma and nuclei of skeletal muscle in patients with idiopathic inflammatory myopathies. Arthritis Rheum. 56:674–687.

Grutzkau A et al. 1998. Synthesis, storage and release of vascular endothelial growth factor/vascular permeability factor (VEGF/VPF) by human mast cells: Implications for the biological significance of $VEGF_{206}$. Mol Biol Cell. 9:875–884.

Gualano RC, Vlahos R, Anderson GP. 2006. What is the contribution of respiratory viruses and lung proteases to airway remodelling in asthma and chronic obstructive pulmonary disease? Pulm Pharmacol Ther. 19:18–23.

Gur A, Karakoc M, Erdogan S, Nas K, Cevik R, Sarac AJ. 2002a. Regional cerebral blood flow and cytokines in young females with fibromyalgia. Clin Exp Rheumatol. 20:753–760.

Gur A et al. 2002b. Cytokines and depression in cases with fibromyalgia. J Rheumatol. 29:358–361.

Harvima IT, Viinamäki H, Naukkarinen A, Paukkonen K, Neittaanmäki H, Horsmanheimo M. 1993. Association of cutaneous mast cells and sensory nerves with psychic stress in psoriasis. Psychother Psychosom. 60:168–176.

Hasler G et al. 2005. Asthma and panic in young adults: A 20-year prospective community study. Am J Respir Crit Care Med. 171:1224–1230.

He S, Gaca MD, Walls AF. 2001. The activation of synovial mast cells: Modulation of histamine release by tryptase and chymase and their inhibitors. Eur J Pharmacol. 412:223–229.

He Y, Ding G, Wang X, Zhu T, Fan S. 2000. Calcitonin gene-related peptide in Langerhans cells in psoriatic plaque lesions. Chin Med J (Engl). 113:747–751.

Heine H, Lien E. 2003. Toll-like receptors and their function in innate and adaptive immunity. Int Arch Allergy Immunol. 130:180–192.

Herrmann M, Scholmerich J, Straub RH. 2000. Stress and rheumatic diseases. Rheum Dis Clin North Am. 26:737–763.

Hirota S, Nomura S, Asada H, Ito A, Morii E, Kitamura Y. 1993. Possible involvement of *c-kit* receptor and its ligand in increase of mast cells in neurofibroma tissues. Arch Pathol Lab Med. 117:996–999.

Ho LH et al. 2007. IL-33 induces IL-13 production by mouse mast cells independently of IgE-FcepsilonRI signals. J Leukocyte Biol. 82:1481–1490.

Hogan AD, Schwartz LB. 1997. Markers of mast cell degranulation. Methods Enzymol. 13:43–52.

Holgate ST. 2007. The epithelium takes centre stage in asthma and atopic dermatitis. Trends Immunol. 28:248–251.

Howarth PH et al. 2005. Tumour necrosis factor (TNFalpha) as a novel therapeutic target in symptomatic corticosteroid dependent asthma. Thorax. 60:1012–1018.

Huang M, Pang X, Letourneau L, Boucher W, Theoharides TC. 2002. Acute stress induces cardiac mast cell activation and histamine release, effects that are increased in apolipoprotein E knockout mice. Cardiovasc Res. 55:150–160.

Huang M, Pang X, Karalis K, Theoharides TC. 2003. Stress-induced interleukin-6 release in mice is mast cell-dependent and more pronounced in Apolipoprotein E knockout mice. Cardiovasc Res. 59:241–249.

Hudson JI, Goldenberg DL, Pope HG Jr, Keck PE, Schlesinger L. 1992. Comorbidity of fibromyalgia with medical and psychiatric disorders. Am J Med. 92:363–367.

Hurst SD et al. 2002. New IL-17 family members promote Th1 or Th2 responses in the lung: In vivo function of the novel cytokine IL-25. J Immunol. 169:443–453.

Iikura M et al. 2007. IL-33 can promote survival, adhesion and cytokine production in human mast cells. Lab Invest. 87:971–978.

Ikeda RK et al. 2003. Accumulation of peribronchial mast cells in a mouse model of ovalbumin allergen induced chronic airway inflammation: Modulation by immunostimulatory DNA sequences. J Immunol. 171:4860–4867.

Inase N, Schreck RE, Lazarus SC. 1993. Heparin inhibits histamine release from canine mast cells. Am J Physiol. 264:L387–L390.

Inoue T, Suzuki Y, Yoshimaru T, Ra C. 2008. Nitric oxide protects mast cells from activation-induced cell death: The role of the phosphatidylinositol-3 kinase-Akt-endothelial nitric oxide synthase pathway. J Leukoc Biol. 83:1218–1229.

Irani AA, Schechter NM, Craig SS, DeBlois G, Schwartz LB. 1986. Two types of human mast cells that have distinct neutral protease compositions. Proc Natl Acad Sci USA. 83:4464–4468.

Ishida S, Kinoshita T, Sugawara N, Yamashita T, Koike K. 2003. Serum inhibitors for human mast cell growth: Possible role of retinol. Allergy. 58:1044–1052.

Janiszewski J, Bienenstock J, Blennerhassett MG. 1994. Picomolar doses of substance P trigger electrical responses in mast cells without degranulation. Am J Physiol. 267:C138–C145.

Jeziorska M, McCollum C, Woolley DE. 1997. Mast cell distribution, activation and phenotype in atherosclerotic lesions of human carotid arteries. J Pathol. 182:115–122.

Jiang WY, Raychaudhuri SP, Farber EM. 1998. Double-labeled immunofluorescence study of cutaneous nerves in psoriasis. Int J Dermatol. 37:572–574.

Joachim RA, Quarcoo D, Arck PC, Herz U, Renz H, Klapp BF. 2003. Stress enhances airway reactivity and airway inflammation in an animal model of allergic bronchial asthma. Psychosom Med. 65:811–815.

Johnson JR et al. 2004. Continuous exposure to house dust mite elicits chronic airway inflammation and structural remodeling. Am J Respir Crit Care Med. 169:378–385.

Jones KD, Deodhar P, Lorentzen A, Bennett RM, Deodhar AA. 2007. Growth hormone perturbations in fibromyalgia: A review. Semin Arthritis Rheum. 36:357–379.

Jorens PG, van Overveld FJ, Bult H, Vermeire PA, Herman AG. 1993. Muramyldipeptide and granulocyte-macrophage colony-stimulating factor enhance interferon-gamma-induced nitric oxide production by rat alveolar macrophages. Agents Actions. 38:100–105.

Kaartinen M, Penttilä A, Kovanen PT. 1994. Accumulation of activated mast cells in the shoulder region of human coronary atheroma, the predilection site of atheromatous rupture. Circulation. 90:1669–1678.

Kanbe N, Kurosawa M, Nagata H, Saitoh H, Miyachi Y. 1999. Cord blood-derived human cultured mast cells produce transforming growth factor beta 1. Clin Exp Allergy. 29:105–113.

Kandere-Grzybowska K et al. 2003a. Stress-induced dura vascular permeability does not develop in mast cell-deficient and neurokinin-1 receptor knockout mice. Brain Res. 980:213–220.

Kandere-Grzybowska K et al. 2003b. IL-1 induces vesicular secretion of IL-6 without degranulation from human mast cells. J Immunol. 171:4830–4836.

Kaneko K, Kawana S, Arai K, Shibasaki T. 2003. Corticotropin-releasing factor receptor type 1 is involved in the stress-induced exacerbation of chronic contact dermatitis in rats. Exp Dermatol. 12:47–52.

Kaneko S, Satoh T, Chiba J, Ju C, Inoue K, Kagawa J. 2000. Interleukin-6 and interleukin-8 levels in serum and synovial fluid of patients with osteoarthritis. Cytokines Cell Mol Ther. 6:71–79.

Kapoor U, Tayal G, Mittal SK, Sharma VK, Tekur U. 2003. Plasma cortisol levels in acute asthma. Indian J Pediatr. 70:965–968.

Karlsson M, Lundin S, Dahlgren U, Kahu H, Pettersson I, Telemo E. 2001. "Tolerosomes" are produced by intestinal epithelial cells. Eur J Immunol. 31:2892–2900.

Kassessinoff TA, Pearce FL. 1988. Histamine secretion from mast cells stimulated with somatostatin. Agents Actions. 23:211–213.

Katsarou-Katsari A, Filippou A, Theoharides TC. 1999. Effect of stress and other psychological factors on the pathophysiology and treatment of dermatoses. Int J Immunopathol Pharmacol. 12:7–11.

Katsarou-Katsari A, Singh LK, Theoharides TC. 2001. Alopecia areata and affected skin CRH receptor upregulation induced by acute emotional stress. Dermatology. 203:157–161.

Kawamura F, Hirashima N, Furuno T, Nakanishi M. 2006. Effects of 2-methyl-1,4-naphtoquinone (menadione) on cellular signaling in RBL-2H3 cells. Biol Pharm Bull. 29:605–607.

Kawana S, Liang Z, Nagano M, Suzuki H. 2006. Role of substance P in stress-derived degranulation of dermal mast cells in mice. J Dermatol Sci. 42:47–54.

Kedzierski RM, Yanagisawa M. 2001. Endothelin system: The double-edged sword in health and disease. Annu Rev Pharmacol Toxicol. 41:851–876.

Kelley J, Hemontolor G, Younis W, Li C, Krishnaswamy G, Chi DS. 2006. Mast cell activation by lipoproteins. Methods Mol Biol. 315:341–348.

Kempuraj D et al. 1999. Characterization of mast cell-committed progenitors present in human umbilical cord blood. Blood. 93:3338–3346.

Kempuraj D et al. 2004. Corticotropin-releasing hormone and its structurally related urocortin are synthesized and secreted by human mast cells. Endocrinology. 145:43–48.

Kempuraj D et al. 2005. Flavonols inhibit proinflammatory mediator release, intracellular calcium ion levels and protein kinase C theta phosphorylation in human mast cells. Br J Pharmacol. 145:934–944.

Kervinen H, Kaartinen M, Makynen H, Palosuo T, Manttari M, Kovanen PT. 2005. Serum tryptase levels in acute coronary syndromes. Int J Cardiol. 104:138–143.

Kiecolt-Glaser JK, Preacher KJ, MacCallum RC, Atkinson C, Malarkey WB, Glaser R. 2003. Chronic stress and age-related increases in the proinflammatory cytokine IL-6. Proc Natl Acad Sci USA. 100:9090–9095.

Kiener HP et al. 1998. Expression of the C5a receptor (CD88) on synovial mast cells in patients with rheumatoid arthritis. Arthritis Rheum. 41:233–245.

Kilpelainen M, Koskenvuo M, Helenius H, Terho EO. 2002. Stressful life events promote the manifestation of asthma and atopic diseases. Clin Exp Allergy. 32:256–263.

Kim JE et al. 2007. Expression of the corticotropin-releasing hormone proopiomelanocortin axis in the various clinical types of psoriasis. Exp Dermatol. 16:104–109.

Kim JY, Ro JY. 2005. Signal pathway of cytokines produced by reactive oxygen species generated from phorbol myristate acetate-stimulated HMC-1 cells. Scand J Immunol. 62:25–35.

Kim KS, Wass CA, Cross AS, Opal SM. 1992. Modulation of blood-brain barrier permeability by tumor necrosis factor and antibody to tumor necrosis factor in the rat. Lymphokine Cytokine Res. 11:293–298.

Kim SH. 2007. Skin biopsy findings: Implications for the pathophysiology of fibromyalgia. Med Hypotheses. 69:141–144.

Kimata H. 2003a. Enhancement of allergic skin wheal responses and in vitro allergen-specific IgE production by computer-induced stress in patients with atopic dermatitis. Brain Behav Immun. 17:134–138.

Kimata H. 2003b. Enhancement of allergic skin wheal responses in patients with atopic eczema/dermatitis syndrome by playing video games or by a frequently ringing mobile phone. Eur J Clin Invest. 33:513–517.

Kimata M, Shichijo M, Miura T, Serizawa I, Inagaki N, Nagai H. 2000. Effects of luteolin, quercetin and baicalein on immunoglobulin E-mediated mediator release from human cultured mast cells. Clin Exp Allergy. 30:501–508.

Kinoshita T, Sawai N, Hidaka E, Yamashita T, Koike K. 1999. Interleukin-6 directly modulates stem cell factor-dependent development of human mast cells derived from $CD34^+$ cord blood cells. Blood. 94:496–508.

Kips JC et al. 2003. Murine models of asthma. Eur Respir J. 22:374–382.

Kirshenbaum AS, Goff JP, Semere T, Foster B, Scott LM, Metcalfe DD. 1999. Demonstration that human mast cells arise from a progenitor cell population that is CD34(+), c-kit(+), and expresses aminopeptidase N (CD13). Blood. 94:2333–2342.

Kirshenbaum AS et al. 2003. Characterization of novel stem cell factor responsive human mast cell lines LAD 1 and 2 established from a patient with mast cell sarcoma/leukemia; activation following aggregation of FcepsilonRI or FcgammaRI. Leuk Res. 27:677–682.

Kirshenbaum AS, Metcalfe DD. 2006. Growth of human mast cells from bone marrow and peripheral blood-derived CD34+ pluripotent progenitor cells. Methods Mol Biol. 315:105–112.

Kjaer A, Larsen PJ, Knigge U, Jorgensen H, Warberg J. 1998. Neuronal histamine and expression of corticotropin-releasing hormone, vasopressin and oxytocin in the hypothalamus: Relative importance of H_1 and H_2 receptors. Eur J Endocrinol. 139:238–243.

Klinkert WEF, Kojima K, Lesslauer W, Rinner W, Lassmann H, Wekerle H. 1997. TNF-α receptor fusion protein prevents experimental auto- immune encephalomyelitis and demyelination in Lewis rats: An overview. J Neuroimmunol. 72:163–168.

Kohno M et al. 2001. Urocortin expression in synovium of patients with rheumatoid arthritis and osteoarthritis: relation to inflammatory activity. J Clin Endocrinol Metab. 86:4344–4352.

Kops SK, Van Loveren H, Rosenstein RW, Ptak W, Askenase PW. 1984. Mast cell activation and vascular alterations in immediate hypersensitivity-like reactions induced by a T cell derived antigen-binding factor. Lab Invest. 50:421–434.

Kops SK, Theoharides TC, Cronin CT, Kashgarian MG, Askenase PW. 1990. Ultrastructural characteristics of rat peritoneal mast cells undergoing differential release of serotonin without histamine and without degranulation. Cell Tissue Res. 262:415–424.

Koranteng RD, Dearman RJ, Kimber I, Coleman JW. 2000. Phenotypic variation in mast cell responsiveness to the inhibitory action of nitric oxide. Inflamm Res. 49:240–246.

Korszun A, Papadopoulos E, Demitrack M, Engleberg C, Crofford L. 1998. The relationship between temporomandibular disorders and stress-associated syndromes. Oral Surg Oral Med Oral Pathol Oral Radiol Endod. 86:416–420.

Kounis NG, Zavras GM. 1996. Allergic angina and allergic myocardial infarction. Circulation. 94:1789.

Kounis NG, Grapsas ND, Goudevenos JA. 1999. Unstable angina, allergic angina, and allergic myocardial infarction. Circulation. 100:e156.

Kounis NG, Hahalis G, Theoharides TC. 2007. Coronary stents, hypersensitivity reactions and the Kounis syndrome. J Invasive Cardiol. 20:314–323.

Kovanen PT. 1996. Mast cells in human fatty streaks and atheromas: Implications for intimal lipid accumulation. Curr Opin Lipidol. 7:281–286.

Kovanen PT. 2007. Mast cells: Multipotent local effector cells in atherothrombosis. Immunol Rev. 217:105–122.

Kraft S, Rana S, Jouvin MH, Kinet JP. 2004. The role of the FcepsilonRI beta-chain in allergic diseases. Int Arch Allergy Immunol. 135:62–72.

Kraneveld AD et al. 2005. Elicitation of allergic asthma by immunoglobulin free light chains. Proc Natl Acad Sci USA. 102:1578–1583.

Kristjansson S et al. 2005. Respiratory syncytial virus and other respiratory viruses during the first 3 months of life promote a local TH2-like response. J Allergy Clin Immunol. 116:805–811.

Krueger J. 2003. Mechanisms underlying the central effects of cytokines. NIH, Bethesda, MD. 04–5497:63–68.

Krüger-Krasagakes S, Möller AM, Kolde G, Lippert U, Weber M, Henz BM. 1996. Production of inteuleukin-6 by human mast cells and basophilic cells. J Invest Dermatol. 106:75–79.

Kulka M, Alexopoulou L, Flavell RA, Metcalfe DD. 2004. Activation of mast cells by double-stranded RNA: Evidence for activation through Toll-like receptor 3. J Allergy Clin Immunol. 114:174–182.

Kulka M, Metcalfe DD. 2005. High-resolution tracking of cell division demonstrates differential effects of TH1 and TH2 cytokines on SCF-dependent human mast cell production in vitro: Correlation with apoptosis and Kit expression. Blood. 105:592–599.

Kulka M, Sheen CH, Tancowny BP, Grammer LC, Schleimer RP. 2007. Neuropeptides activate human mast cell degranulation and chemokine production. Immunology. 123:398–410.

Kurose I, Argenbright LW, Wolf R, Lianxi L, Granger DN. 1997. Ischemia/reperfusion-induced microvascular dysfunction: Role of oxidants and lipid mediators. Am J Physiol. 272: H2976–H2982.

Lagunoff D, Martin TW, Read G. 1983. Agents that release histamine from mast cells. Annu Rev Pharmacol Toxicol. 23:331–351.

Laham RJ, Li J, Tofukuji M, Post M, Simons M, Sellke FW. 2003. Spatial heterogeneity in VEGF-induced vasodilation: VEGF dilates microvessels but not epicardial and systemic arteries and veins. Ann Vasc Surg. 17:245–252.

Laine P, Kaartinen M, Penttilä A, Panula P, Paavonen T, Kovanen PT. 1999. Association between myocardial infarction and the mast cells in the adventitia of the infarct-related coronary artery. Circulation. 99:361–369.

Laube BL, Curbow BA, Costello RW. 2002. A pilot study examining the relationship between stress and serum cortisol concentrations in women with asthma. Respir Med. 96:823–828.

Lawrence DA. 2002. Psychologic stress and asthma: Neuropeptide involvement. Environ Health Perspect. 110:A230–A231.

Leal-Berumen I, Conlon P, Marshall JS. 1994. IL-6 production by rat peritoneal mast cells is not necessarily preceded by histamine release and can be induced by bacterial lipopolysaccharide. J Immunol. 152:5468–5476.

Lee DM, Friend DS, Gurish MF, Benoist C, Mathis D, Brenner MB. 2002a. Mast cells: A cellular link between autoantibodies and inflammatory arthritis. Science. 297:1689–1692.

Lee M, Calabresi L, Chiesa G, Franceschini G, Kovanen PT. 2002b. Mast cell chymase degrades apoE and apoA-II in apoA-I-knockout mouse plasma and reduces its ability to promote cellular cholesterol efflux. Arterioscler Thromb Vasc Biol. 22:1475–1481.

Lee M, Sommerhoff CP, von EA, Zettl F, Fritz H, Kovanen PT. 2002c. Mast cell tryptase degrades HDL and blocks its function as an acceptor of cellular cholesterol. Arterioscler Thromb Vasc Biol. 22:2086–2091.

Lee M, Kovanen PT, Tedeschi G, Oungre E, Franceschini G, Calabresi L. 2003. Apolipoprotein composition and particle size affect HDL degradation by chymase: Effect on cellular cholesterol efflux. J Lipid Res. 44:539–546.

Lee-Rueckert M, Kovanen PT. 2006. Mast cell proteases: Physiological tools to study functional significance of high density lipoproteins in the initiation of reverse cholesterol transport. Atherosclerosis. 189:8–18.

Lemanske RF Jr et al. 2005. Rhinovirus illnesses during infancy predict subsequent childhood wheezing. J Allergy Clin Immunol. 116:571–577.

Leonardi A. 2002. The central role of conjunctival mast cells in the pathogenesis of ocular allergy. Curr Allergy Asthma Rep. 2:325–331.

Letourneau R, Pang X, Sant GR, Theoharides TC. 1996. Intragranular activation of bladder mast cells and their association with nerve processes in interstitial cystitis. Br J Urol. 77:41–54.

Letourneau R, Rozniecki JJ, Dimitriadou V, Theoharides TC. 2003. Ultrastructural evidence of brain mast cell activation without degranulation in monkey experimental allergic encephalomyelitis. J Neuroimmunol. 145:18–26.

Levi-Schaffer F, Austen KF, Gravallese PM, Stevens RL. 1986. Co-culture of interleukin 3-dependent mouse mast cells with fibroblasts results in a phenotypic change of the mast cells. Proc Natl Acad Sci USA. 83:6485–6488.

Levi-Schaffer F, Shalit M. 1989. Differential release of histamine and prostaglandin D_2 in rat peritoneal mast cells activated with peptides. Int Arch Allergy Appl Immunol. 90:352–357.

Levi-Schaffer F, Kelav-Appelbaum R, Rubinchik E. 1995. Human foreskin mast cell viability and functional activity is maintained *ex vivo* by coculture with fibroblasts. Cell Immunol. 162:211–216.

Liao LX, Granger DN. 1996. Role of mast cells in oxidized low-density lipoprotein-induced microvascular dysfunction. Am J Physiol Heart Circ Physiol. 271:H1795–H1800.

Lindstedt L, Lee M, Castro GR, Fruchart JC, Kovanen PT. 1996. Chymase in exocytosed rat mast cell granules effectively proteolyzes apolipoprotein AI-containing lipoproteins, so reducing the cholesterol efflux-inducing ability of serum and aortic intimal fluid. J Clin Invest. 97:2174–2182.

Liu LY, Coe CL, Swenson CA, Kelly EA, Kita H, Busse WW. 2002. School examinations enhance airway inflammation to antigen challenge. Am J Respir Crit Care Med. 165:1062–1067.

Liu Y. 2006. Thymic stromal lymphopoietin: Master switch for allergic inflammation. J Exp Med. 203:269–273.

Lloyd CM, Gutierrez-Ramos JC. 2004. Animal models to study chemokine receptor function: In vivo mouse models of allergic airway inflammation. Methods Mol Biol. 239:199–210.

Lowry PJ, Woods RJ, Baigent S. 1996. Corticotropin releasing factor and its binding protein. Pharmacol Biochem Behav. 54:305–308.

Lucas HJ, Brauch CM, Settas L, Theoharides TC. 2006. Fibromyalgia–new concepts of pathogenesis and treatment. Int J Immunopathol Pharmacol. 19:5–10.

Lund I, Lundeberg T, Carleson J, Sonnerfors H, Uhrlin B, Svensson E. 2006. Corticotropin releasing factor in urine-A possible biochemical marker of fibromyalgia Responses to massage and guided relaxation. Neurosci Lett. 403:166–171.

Lutgendorf SK, Russell D, Ullrich P, Harris TB, Wallace R. 2004. Religious participation, interleukin-6, and mortality in older adults. Health Psychol. 23:465–475.

Lytinas M, Kempuraj D, Huang M, Boucher W, Esposito P, Theoharides TC. 2003. Acute stress results in skin corticotropin-releasing hormone secretion, mast cell activation and vascular permeability, an effect mimicked by intradermal corticotropin-releasing hormone and inhibited by histamine-1 receptor antagonists. Int Arch Allergy Immunol. 130:224–231.

Mackins CJ et al. 2006. Cardiac mast cell-derived renin promotes local angiotensin formation, norepinephrine release, and arrhythmias in ischemia-reperfusion. J Clin Invest. 116:1063–1070.

Magerl M, Lammel V, Siebenhaar F, Zuberbier T, Metz M, Maurer M. 2008. Non-pathogenic commensal Escherichia coli bacteria can inhibit degranulation of mast cells. Exp Dermatol. 17:427–435.

Magone MT, Maric I, Hwang DG. 2006. Peripheral interstitial keratitis: A novel manifestation of ocular mastocytosis. Cornea. 25:364–367.

Malaviya R et al. 1994. Mast cell phagocytosis of FimH-expressing enterobacteria. J Immunol. 152:1907–1914.

Malone ED. 2002. Managing chronic arthritis. Vet Clin North Am Equine Pract. 18:411–437.

Margrini L, Bonini S, Centofanti M, Schiavone M, Bonini S. 1996. Tear tryptase levels and allergic conjunctivitis. Allergy. 51:577–581.

Marone G et al. 2007. Role of superallergens in allergic disorders. Chem Immunol Allergy. 93:195–213.

Marquardt DL, Parker CW, Sullivan TJ. 1978. Potentiation of mast cell mediator release by adenosine. J Immunol. 120:871–878.

Marshall JS. 2004. Mast-cell responses to pathogens. Nat Rev Immunol. 4:787–799.

Mastorakos G, Chrousos GP, Weber JS. 1993. Recombinant interleukin-6 activates the hypothalamic-pituitary-adrenal axis in humans. J Clin Endocrinol Metab. 77:1690–1694.

Masuda A, Yoshikai Y, Aiba K, Matsuguchi T. 2002. Th2 cytokine production from mast cells is directly induced by lipopolysaccharide and distinctly regulated by c-Jun N-terminal kinase and p38 pathways. J Immunol. 169:3801–3810.

Matsuda H, Kawakita K, Kiso Y, Nakano T, Kitamura Y. 1989. Substance P induces granulocyte infiltration through degranulation of mast cells. J Immunol. 142:927–931.

Matsuse H et al. 2005. Naturally occurring parainfluenza virus 3 infection in adults induces mild exacerbation of asthma associated with increased sputum concentrations of cysteinyl leukotrienes. Int Arch Allergy Immunol. 138:267–272.

Matsushima H, Yamada N, Matsue H, Shimada S. 2004. The effects of endothelin-1 on degranulation, cytokine, and growth factor production by skin-derived mast cells. Eur J Immunol. 34:1910–1019.

Matsuzawa S, Sakashita K, Kinoshita T, Ito S, Yamashita T, Koike K. 2003. IL-9 enhances the growth of human mast cell progenitors under stimulation with stem cell factor. J Immunol. 170:3461–3467.

Mattheos S, Christodoulou S, Kempuraj D, Kempuraj B, Karalis K, Theoharides TC. 2003. Mast cells and corticotropin-releasing hormone (CRH) are required for experimental inflammatory arthritis. FASEB J. 17:C44.

Maurer M et al. 2004. Mast cells promote homeostasis by limiting endothelin-1-induced toxicity. Nature (London). 432:512–516.

McCurdy JD, Olynych TJ, Maher LH, Marshall JS. 2003. Cutting edge: Distinct Toll-like receptor 2 activators selectively induce different classes of mediator production from human mast cells. J Immunol. 170:1625–1629.

McEvoy AN, Bresnihan B, FitzGerald O, Murphy EP. 2001. Corticotropin-releasing hormone signaling in synovial tissue from patients with early inflammatory arthritis is mediated by the type 1α corticotropin-releasing hormone receptor. Arthritis Rheum. 44:1761–1767.

Mease P. 2005. Fibromyalgia syndrome: Review of clinical presentation, pathogenesis, outcome measures, and treatment. J Rheumatol Suppl. 75:6–21.

Mease PJ et al. 2005. Fibromyalgia syndrome. J Rheumatol. 32:2270–2277.

Mekori YA, Metcalfe DD. 1999. Mast cell-T cell interactions. J Allergy Clin Immunol. 104:517–523.

Metcalfe DD, Kaliner M, Donlon MA. 1981. The mast cell. CRC Crit Rev Immunol. 3:23–74.

Metcalfe DD, Baram D, Mekori YA. 1997. Mast cells. Physiol Rev. 77:1033–1079.

Metz M, Maurer M. 2007. Mast cells–key effector cells in immune responses. Trends Immunol. 28:234–241.

Middleton E Jr, Kandaswami C, Theoharides TC. 2000. The effects of plant flavonoids on mammalian cells: Implications for inflammation, heart disease and cancer. Pharmacol Rev. 52:673–751.

Mihara M et al. 2001. Humanized antibody to human interleukin-6 receptor inhibits the development of collagen arthritis in cynomolgus monkeys. Clin Immunol. 98:319–326.

Mitchell KE et al. 2003. Matrix cells from Wharton's jelly form neurons and glia. Stem Cells. 21:50–60.

Molino M et al. 1997. Interactions of mast cell tryptase with thrombin receptors and PAR-2. J Biol Chem. 272:4043–4049.

Mousli M, Hugli TE, Landry Y, Bronner C. 1994. Peptidergic pathway in human skin and rat peritoneal mast cell activation. Immunopharmacol. 27:1–11.

Mullington JM, Hinze-Selch D, Pollmacher T. 2001. Mediators of inflammation and their interaction with sleep: Relevance for chronic fatigue syndrome and related conditions. Ann NY Acad Sci. 933:201–210.

Murphy PM. 2001. Chemokines and the molecular basis of cancer metastasis. N Engl J Med. 345:833–835.

Nakae S et al. 2006. Mast cells enhance T cell activation: Importance of mast cell costimulatory molecules and secreted TNF. J Immunol. 176:2238–2248.

Nakae S et al. 2007. Mast cell-derived TNF contributes to airway hyperreactivity, inflammation, and TH2 cytokine production in an asthma model in mice. J Allergy Clin Immunol. 120:48–55.

Naukkarinen A, Jarvikallio A, Lakkakorpi J, Harvima IT, Harvima RJ, Horsmanheimo M. 1996. Quantitative histochemical analysis of mast cells and sensory nerves in psoriatic skin. J Pathol. 180:200–205.

Naureckas ET, Ndukwu IM, Halayko AJ, Maxwell C, Hershenson MB, Solway J. 1999. Bronchoalveolar lavage fluid from asthmatic subjects is mitogenic for human airway smooth muscle. Am J Respir Crit Care Med. 160:2062–2066.

Navarra P, Tsagarakis S, Faria MS, Rees LH, Besser GM, Grossman AB. 1991. Interleukins-1 and -6 stimulate the release of corticotropin-releasing hormone-41 from rat hypothalamus *in vitro* via the eicosanoid cyclooxygenase pathway. Endocrinology. 128:37–44.

Neeck G, Crofford LJ. 2000. Neuroendocrine perturbations in fibromyalgia and chronic fatigue syndrome. Rheum Dis Clin Norht Am. 26:989–1002.

Neeck G. 2002. Pathogenic mechanisms of fibromyalgia. Ageing Res Rev. 1:243–255.

Newson B, Dahlström A, Enerbäck L, Ahlman H. 1983. Suggestive evidence for a direct innervation of mucosal mast cells. Neuroscience. 10:565–570.

Nielsen LA, Henriksson KG. 2007. Pathophysiological mechanisms in chronic musculoskeletal pain (fibromyalgia): The role of central and peripheral sensitization and pain disinhibition. Best Pract Res Clin Rheumatol. 21:465–480.

Nigrovic PA, Lee DM. 2005. Mast cells in inflammatory arthritis. Arthritis Res Ther. 7:1–11.

Niide O, Suzuki Y, Yoshimaru T, Inoue T, Takayama T, Ra C. 2006. Fungal metabolite gliotoxin blocks mast cell activation by a calcium- and superoxide-dependent mechanism: Implications for immunosuppressive activities. Clin Immunol. 118:108–116.

O'sullivan SM. 2005. Asthma death, CD8+ T cells, and viruses. Proc Am Thorac Soc. 2:162–165.

Ochi H, DeJesus NH, Hsieh FH, Austen KF, Boyce JA. 2000. IL-4 and -5 prime human mast cells for different profiles of IgE-dependent cytokine production. Proc Natl Acad Sci USA. 97:10509–10513.

Odum L, Petersen LJ, Skov PS, Ebskov LB. 1998. Pituitary adenylate cyclase activating polypeptide (PACAP) is localized in human dermal neurons and causes histamine release from skin mast cells. Inflamm Res. 47:488–492.

Olivenstein R, Du T, Xu LJ, Martin JG. 1997. Microvascular leakage in the airway wall and lumen during allergen induced early and late responses in rats. Pulm Pharmacol Ther. 10:223–230.

Ott VL, Cambier JC, Kappler J, Marrack P, Swanson BJ. 2003. Mast cell-dependent migration of effector CD8+ T cells through production of leukotriene B4. Nat Immunol. 4:974–981.

Ozgocmen S, Ozyurt H, Sogut S, Akyol O. 2006. Current concepts in the pathophysiology of fibromyalgia: The potential role of oxidative stress and nitric oxide. Rheumatol Int. 26:585–597.

Pang X, Marchand J, Sant GR, Kream RM, Theoharides TC. 1995. Increased number of substance P positive nerve fibers in interstitial cystitis. Br J Urol. 75:744–750.

Pang X et al. 1998. A neurotensin receptor antagonist inhibits acute immobilization stress-induced cardiac mast cell degranulation, a corticotropin-releasing hormone-dependent process. J Pharm Exp Therap. 287:307–314.

Papadopoulou N, Kalogeromitros D, Staurianeas NG, Tiblalexi D, Theoharides TC. 2005. Corticotroopin-releasing hormone receptor-1 and histidine decarboxylase expression in chronic urticaria. J Invest Dermatol. 125:952–955.

Papiris S, Kotanidou A, Malagari K, Roussos C. 2002. Clinical review: Severe asthma. Crit Care. 6:30–44.

Pardo CA, Vargas DL, Zimmerman AW. 2005. Immunity, neuroglia and neuroinflammation in autism. Int Rev Psychiatry. 17:485–495.

Parikh V, Singh M. 2001. Possible role of nitric oxide and mast cells in endotoxin-induced cardioprotection. Pharmacol Res. 43:39–45.

Patella V, de Crescenzo G, Ciccarelli A, Marino I, Adt M, Marone G. 1995. Human heart mast cells: A definitive case of mast cell heterogeneity. Int Arch Allergy Immunol. 106:386–393.

Paus R, Theoharides TC, Arck PC. 2006. Neuroimmunoendocrine circuitry of the 'brain–skin connection'. Trends Immunol. 27:32–39.

Pelaia G et al. 2006. Respiratory infections and asthma. Respir Med. 100:775–784.

Peres MF, Zukerman E, Young WB, Silberstein SD. 2002. Fatigue in chronic migraine patients. Cephalalgia. 22:720–724.

Peters EM, Kuhlmei A, Tobin DJ, Muller-Rover S, Klapp BF, Arck PC. 2005. Stress exposure modulates peptidergic innervation and degranulates mast cells in murine skin. Brain Behav Immun. 19:252–262.

Phillips GD, Pickering EC, Wilkinson K. 1975. Renal responses of the cow to alteration of the dietary intake of nitrogen and sodium chloride. J Physiol. 245:95P–96P.

Pio R et al. 2001. Complement factor H is a serum-binding protein for adrenomedullin, and the resulting complex modulates the bioactivities of both partners. J Biol Chem. 276:12292–12300.

Probert L, Akassoglou K, Kassiotis G, Pasparakis M, Alexopoulou L, Kollias G. 1997. TNF-α transgenic and knockout models of CNS inflammation and degeneration. J Neuroimmunol. 72:137–141.

Puxeddu I, Piliponsky AM, Bachelet I, Levi-Schaffer F. 2003. Mast cells in allergy and beyond. Int J Biochem Cell Biol. 35:1601–1607.

Raphael KG, Janal MN, Nayak S, Schwartz JE, Gallagher RM. 2006. Psychiatric comorbidities in a community sample of women with fibromyalgia. Pain. 124:117–125.

Raposo G, Tenza D, Mecheri S, Peronet R, Bonnerot C, Desaymard C. 1997. Accumulation of major histocompatibility complex class II molecules in mast cell secretory granules and their release upon degranulation. Mol Biol Cell. 8:2631–2645.

Ray A, Chakraborti A, Gulati K. 2007. Current trends in nitric oxide research. Cell Mol Biol (Noisy-le-grand). 53:3–14.

Raymond RJ, Dehmer GJ, Theoharides TC, Deliargyris EN. 2001. Elevated interleukin-6 levels in patients with asymptomatic left ventricular systolic dysfunction. Am Heart J. 141:435–438.

Redegeld FA et al. 2002. Immunoglobulin-free light chains elicit immediate hypersensitivity-like responses. Nat Med. 8:694–701.

Redegeld FA, Nijkamp FP. 2005. Immunoglobulin free light chains and mast cells: Pivotal role in T-cell-mediated immune reactions? Trends Immunol. 24:181–185.

Remrod C, Lonne-Rahm S, Nordliond K. 2007. Study of substance P and its receptor neurokinin-1 in psoriasis and their relation to chronic stress and pruritus. Arch Dermatol Res. 299:85–91.

Ricciarelli R, Tasinato A, Clement S, Ozer NK, Boscoboinik D, Azzi A. 1998. alpha-Tocopherol specifically inactivates cellular protein kinase C alpha by changing its phosphorylation state. Biochem J. 334 (Pt 1):243–249.

Riedel W, Schlapp U, Leck S, Netter P, Neeck G. 2002. Blunted ACTH and cortisol responses to systemic injection of corticotropin-releasing hormone (CRH) in fibromyalgia: Role of somatostatin and CRH-binding protein. Ann NY Acad Sci. 966:483–490.

Rijnierse A, Nijkamp FP, Kraneveld AD. 2007. Mast cells and nerves tickle in the tummy. Implications for inflammatory bowel disease and irritable bowel syndrome. Pharmacol Ther. 116:207–235.

Roberts F, Calcutt CR. 1983. Histamine and the hypothalamus. Neuroscience. 9:721–739.

Robinson-White A, Beaven MA. 1982. Presence of histamine and histamine-metabolizing enzyme in rat and guinea-pig microvascular endothelial cells. J Pharmacol Exp Ther. 223:440–445.

Rock FL, Hardiman G, Timans JC, Kastelein RA, Bazan JF. 1998. A family of human receptors structurally related to Drosophila Toll. Proc Natl Acad Sci USA. 95:588–593.

Romanov YA, Svintsitskaya VA, Smirnov VN. 2003. Searching for alternative sources of postnatal human mesenchymal stem cells: Candidate MSC-like cells from umbilical cord. Stem Cells. 21:105–110.

Rothrock NE, Lutgendorf SK, Kreder KJ, Ratliff T, Zimmerman B. 2001. Stress and symptoms in patients with interstitial cystitis: A life stress model. Urology. 57:422–427.

Rottem M, Mekori YA. 2005. Mast cells and autoimmunity. Autoimmun Rev. 4:21–27.

Roupe van der Voort C, Heijnen CJ, Wulffraat N, Kuis W, Kavelaars A. 2000. Stress induces increases in IL-6 production by leucocytes of patients with the chronic inflammatory disease juvenile rheumatoid arthritis: A putative role for alpha(1)-adrenergic receptors. J Neuroimmunol. 110:223–229.

Rozanski A et al. 1988. Mental stress and the induction of silent myocardial ischemia in patients with coronary artery disease. N Engl J Med. 318:1005–1012.

Rozniecki JJ, Dimitriadou V, Lambracht-Hall M, Pang X, Theoharides TC. 1999. Morphological and functional demonstration of rat dura mast cell-neuron interactions *in vitro* and *in vivo*. Brain Res. 849:1–15.

Russell IJ et al. 1994. Elevated cerebrospinal fluid levels of substance P in patients with the fibromyalgia syndrome. Arthritis Rheum. 37:1593–1601.

Saavedra Y, Vergara P. 2003. Somatostatin inhibits intestinal mucosal mast cell degranulation in normal conditions and during mast cell hyperplasia. Regul Pept. 111:67–75.

Sackesen C, Pinar A, Sekerel BE, Akyon Y, Saraclar Y. 2005. Use of polymerase chain reaction for detection of adenovirus in children with or without wheezing. Turk J Pediatr. 47:227–231.

Sadakane K et al. 2002. Murine strain differences in airway inflammation induced by diesel exhaust particles and house dust mite allergen. Int Arch Allergy Immunol. 128:220–228.

Sakata Y et al. 1996. Elevation of the plasma histamine concentration in the coronary circulation in patients with variant angina. Am J Cardiol. 77:1121–1126.

Salamon P, Shoham NG, Gavrieli R, Wolach B, Mekori YA. 2005. Human mast cells release Interleukin-8 and induce neutrophil chemotaxis on contact with activated T cells. Allergy. 60:1316–1319.

Salemi S et al. 2003. Detection of interleukin 1beta (IL-1beta), IL-6, and tumor necrosis factor-alpha in skin of patients with fibromyalgia. J Rheumatol. 30:146–150.

Saraceno R, Kleyn CE, Terenghi G, Griffiths CE. 2006. The role of neuropeptides in psoriasis. Br J Dermatol. 155:876–882.

Sarugaser R, Lickorish D, Baksh D, Hosseini MM, Davies JE. 2005. Human umbilical cord perivascular (HUCPV) cells: A source of mesenchymal progenitors. Stem Cells. 23:220–229.

Sawai N et al. 1999. Thrombopoietin augments stem cell factor-dependent growth of human mast cells from bone marrow multipotential hematopoietic progenitors. Blood. 93:3703–3712.

Schmaling KB, McKnight PE, Afari N. 2002. A prospective study of the relationship of mood and stress to pulmonary function among patients with asthma. J Asthma. 39:501–510.

Schmidlin F, Bunnett NW. 2001. Protease-activated receptors: How proteases signal to cells. Curr Opin Pharmacol. 1:575–582.

Schmidt HD, Brunner H. 1976. [Ascending choledochal papillomatosis (author's transl)]. MMW Munch Med Wochenschr. 118:163–166.

Scully MF, Ellis V, Kakkar VV. 1986. Localisation of heparin in mast cells. Lancet. 2:718–719.

Seebeck J, Kruse ML, Schmidt-Choudhury A, Schmidt WE. 1998. Pituitary adenylate cyclase activating polypeptide induces degranulation of rat peritoneal mast cells via high-affinity PACAP receptor-independent activation of G proteins. Ann NY Acad Sci. 865:141–146.

Sekar Y, Moon TC, Munoz S, Befus AD. 2005. Role of nitric oxide in mast cells: Controversies, current knowledge, and future applications. Immunol Res. 33:223–239.

Serafin WE, Katz HR, Austen KF, Stevens RL. 1986. Complexes of heparin proteoglycans, chondroitin sulfate E proteoglycans, and [3H]diisopropyl fluorophosphate-binding proteins are exocytosed from activated mouse bone marrow-derived mast cells. J Biol Chem. 261:15017–15021.

Shanahan F, Denburg JA, Fox J, Bienenstock J, Befus D. 1985. Mast cell heterogeneity: Effects of neuroenteric peptides on histamine release. J Immunol. 135:1331–1337.

Shanas U, Bhasin R, Sutherland AK, Silverman A-J, Silver R. 1998. Brain mast cells lack the c-kit receptor: Immunocytochemical evidence. J Neuroimmunol. 90:207–211.

Sillaber C, Sperr WR, Agis H, Spanblöchl E, Lechner K, Valent P. 1994. Inhibition of stem cell factor dependent formation of human mast cells by interleukin-3 and interleukin-4. Int Arch Allergy Immunol. 105:264–268.

Silver R, Silverman A-J, Vitkovic L, Lederhendler II. 1996. Mast cells in the brain: Evidence and functional significance. Trends Neurosci. 19:25–31.

Silver RB, Reid AC, Mackins CJ, Askwith T, Schaefer U, Herzlinger D et al. 2004. Mast cells: A unique source of renin. Proc Natl Acad Sci USA. 101:13607–13612.

Singh LK et al. 1999a. Potent mast cell degranulation and vascular permeability triggered by urocortin through activation of CRH receptors. J Pharmacol Exp Ther. 288:1349–1356.

Singh LK, Pang X, Alexacos N, Letourneau R, Theoharides TC. 1999b. Acute immobilization stress triggers skin mast cell degranulation via corticotropin-releasing hormone, neurotensin and substance P: A link to neurogenic skin disorders. Brain Behav Immunity. 13:225–239.

Skokos D et al. 2003. Mast cell-derived exosomes induce phenotypic and functional maturation of dendritic cells and elicit specific immune responses in vivo. J Immunol. 170:3037–3045.

Slominski A, Wortsman J. 2000. Neuroendocrinology of the skin. Endocr Rev. 21:457–487.

Slominski A, Wortsman J, Luger T, Paus R, Solomon S. 2000. Corticotropin releasing hormone and proopiomelanocortin involvement in the cutaneous response to stress. Physiol Rev. 80:979–1020.

Slominski A et al. 2001. Cutaneous expression of corticotropin-releasing hormone (CRH), urocortin, and CRH receptors. FASEB J. 15:1678–1693.

Slominski A et al. 2006. Corticotropin releasing hormone and the skin. Front Biosci. 11:2230–2248.

Smith A, Nicholson K. 2001. Psychosocial factors, respiratory viruses and exacerbation of asthma. Psychoneuroendocrinology. 26:411–420.

Smith MS, Martin-Herz SP, Womack WM, Marsigan JL. 2003. Comparative study of anxiety, depression, somatization, functional disability, and illness attribution in adolescents with chronic fatigue or migraine. Pediatrics. 111:e376–e381.

Solinas G et al. 2006. Corticotropin-releasing hormone directly stimulates thermogenesis in skeletal muscle possibly through substrate cycling between de novo lipogenesis and lipid oxidation. Endocrinology. 147:31–38.

Spinedi E, Hadid R, Daneva T, Gaillard RC. 1992. Cytokines stimulate the CRH but not the vasopressin neuronal system: Evidence for a median eminence site of interleukin-6 action. Neuroendocrinology. 56:46–53.

Stauder A, Kovacs M. 2003. Anxiety symptoms in allergic patients: Identification and risk factors. Psychosom Med. 65:816–823.

Stead RH, Tomioka M, Quinonez G, Simon GT, Felten SY, Bienenstock J. 1987. Intestinal mucosal mast cells in normal and nematode-infected rat intestines are in intimate contact with peptidergic nerves. Proc Natl Acad Sci USA. 84:2975–2979.

Stead RH, Dixon MF, Bramwell NH, Riddell RH, Bienenstock J. 1989. Mast cells are closely apposed to nerves in the human gastrointestinal mucosa. Gastroenterology. 97:575–585.

Stead RH et al. 2006. Vagal influences over mast cells. Auton Neurosci. 125:53–61.

Steffel J, Akhmedov A, Greutert H, Luscher TF, Tanner FC. 2005. Histamine induces tissue factor expression: Implications for acute coronary syndromes. Circulation. 112:341–349.

Steptoe RJ, McMenamin PG, McMenamin C. 1994. Distribution and characterisation of rat choroidal mast cells. Br J Ophthalmol. 78:211–218.

Supajatura V et al. 2002. Differential responses of mast cell Toll-like receptors 2 and 4 in allergy and innate immunity. J Clin Invest. 109:1351–1359.

Suzuki M, Inaba S, Nagai T, Tatsuno H, Kazatani Y. 2003. Relation of C-reactive protein and interleukin-6 to culprit coronary artery plaque size in patients with acute myocardial infarction. Am J Cardiol. 91:331–333.

Suzuki R et al. 1999. Direct neurite-mast cell communication *in vitro* occurs via the neuropeptide substance P. J Immunol. 163:2410–2415.

Swieter M, Hamawy MM, Siraganian RP, Mergenhagen SE. 1993. Mast cells and their microenvironment: The influence of fibronectin and fibroblasts on the functional repertoire of rat basophilic leukemia cells. J Periodontol. 64:492–496.

Swindle EJ, Metcalfe DD, Coleman JW. 2004. Rodent and human mast cells produce functionally significant intracellular reactive oxygen species but not nitric oxide. J Biol Chem. 279:48751–48759.

Swindle EJ, Coleman JW, DeLeo FR, Metcalfe DD. 2007. FcepsilonRI- and Fcgamma receptor-mediated production of reactive oxygen species by mast cells is lipoxygenase- and cyclooxygenase-dependent and NADPH oxidase-independent. J Immunol. 179:7059–7071.

Szebeni J. 2005. Complement activation-related pseudoallergy: A new class of drug-induced acute immune toxicity. Toxicology. 216:106–121.

Tagen M et al. 2007. Skin corticotropin-releasing hormone receptor expression in psoriasis. J Invest Dermatol. 127:1789–1791.

Tainsh KR, Pearce FL. 1992. Mast cell heterogeneity: Evidence that mast cells isolated from various connective tissue locations in the rat display markedly graded phenotypes. Int Arch Allergy Immunol. 98:26–34.

Tal M, Liberman R. 1997. Local injection of nerve growth factor (NGF) triggers degranulation of mast cells in rat paw. Neurosci Lett. 221:129–132.

Tang C, Lan C, Wang C, Liu R. 2005. Amelioration of the development of multiple organ dysfunction syndrome by somatostatin via suppression of intestinal mucosal mast cells. Shock. 23:470–475.

Taube C et al. 2004. Mast cells, Fc epsilon RI, and IL-13 are required for development of airway hyperresponsiveness after aerosolized allergen exposure in the absence of adjuvant. J Immunol. 172:6398–6406.

Taylor DD, Akyol S, Gercel-Taylor C. 2006. Pregnancy-associated exosomes and their modulation of T cell signaling. J Immunol. 176:1534–1542.

Taylor PC. 2003. Anti-TNFalpha therapy for rheumatoid arthritis: An update. Intern Med. 42:15–20.

Tetlow LC, Woolley DE. 1995a. Distribution, activation and tryptase/chymase phenotype of mast cells in the rheumatoid lesion. Ann Rheum Dis. 54:549–555.

Tetlow LC, Woolley DE. 1995b. Mast cells, cytokines, and metalloproteinases at the rheumatoid lesion: Dual immunolocalisation studies. Ann Rheum Dis. 54:896–903.

Theoharides TC. 1980. Polyamines spermidine and spermine as modulators of calcium dependent immune processes. Life Sci. 27:703–713.

Theoharides TC, Sieghart W, Greengard P, Douglas WW. 1980. Antiallergic drug cromolyn may inhibit histamine secretion by regulating phosphorylation of a mast cell protein. Science. 207:80–82.

Theoharides TC, Bondy PK, Tsakalos ND, Askenase PW. 1982. Differential release of serotonin and histamine from mast cells. Nature (London). 297:229–231.

Theoharides TC. 1987. Substance P-induced release of brain mast cell nociceptive mediators. Pain. 40, 262. Ref Type: Abstract.

Theoharides TC, Sant GR, El-Mansoury M, Letourneau RJ, Ucci AA Jr, Meares EM Jr. 1995a. Activation of bladder mast cells in interstitial cystitis: A light and electron microscopic study. J Urology. 153:629–636.

Theoharides TC et al. 1995b. Stress-induced intracranial mast cell degranulation. A corticotropin releasing hormone-mediated effect. Endocrinology. 136:5745–5750.

Theoharides TC. 1996. Mast cell: A neuroimmunoendocrine master player. Int J Tissue React. 18:1–21.

Theoharides TC et al. 1998. Corticotropin-releasing hormone induces skin mast cell degranulation and increased vascular permeability, a possible explanation for its pro-inflammatory effects. Endocrinology. 139:403–413.

Theoharides TC et al. 2000. Chondroitin sulfate inhibits connective tissue mast cells. Br J Pharmacol. 131:1039–1049.

Theoharides TC, Kempuraj D, Sant GR. 2001. Mast cell involvement in interstitial cystitis: A review of human and experimental evidence. Urology. 57:47–55.

Theoharides TC. 2004. Panic disorder, interstitial cystitis and mast cells. J Clin Psychopharmacol. 24:361–364.

Theoharides TC, Cochrane DE. 2004. Critical role of mast cells in inflammatory diseases and the effect of acute stress. J Neuroimmunol. 146:1–12.

Theoharides TC, Donelan JM, Papadopoulou N, Cao J, Kempuraj D, Conti P. 2004a. Mast cells as targets of corticotropin-releasing factor and related peptides. Trends Pharmacol Sci. 25:563–568.

Theoharides TC, Weinkauf C, Conti P. 2004b. Brain cytokines and neuropsychiatric disorders. J Clin Psychopharmacol. 24:577–581.

Theoharides TC, Donelan J, Kandere-Grzybowska K, Konstantinidou A. 2005. The role of mast cells in migraine pathophysiology. Brain Res Rev. 49:65–76.

Theoharides TC, Kalogeromitros D. 2006. The critical role of mast cell in allergy and inflammation. Ann NY Acad Sci. 1088:78–99.

Theoharides TC. 2007. Treatment approaches for painful bladder syndrome/interstitial cystitis. Drugs. 67:215–235.

Theoharides TC, Kempuraj D, Iliopoulou BP. 2007a. Mast cells, T cells, and inhibition by luteolin: Implications for the pathogenesis and treatment of multiple sclerosis. Adv Exp Med Biol. 601:423–430.

Theoharides TC, Kempuraj D, Tagen M, Conti P, Kalogeromitros D. 2007b. Differential release of mast cell mediators and the pathogenesis of inflammation. Immunol Rev. 217:65–78.

Thomason BT, Brantley PJ, Jones GN, Dyer HR, Morris JL. 1992. The relation between stress and disease activity in rheumatoid arthritis. J Behav Med. 15:215–220.

Torpy DJ, Papanicolaou DA, Lotsikas AJ, Wilder RL, Chrousos GP, Pillemer SR. 2000. Responses of the sympathetic nervous system and the hypothalamic-pituitary-adrenal axis to interleukin-6: A pilot study in fibromyalgia. Arthritis Rheum. 43:872–880.

Toru H, Eguchi M, Matsumoto R, Yanagida M, Yata J, Nakahata T. 1998. Interleukin-4 promotes the development of tryptase and chymase double-positive human mast cells accompanied by cell maturation. Blood. 91:187–195.

Toyoda M, Makino T, Kagoura M, Morohashi M. 2000. Immunolocalization of substance P in human skin mast cells. Arch Dermatol Res. 292:418–421.

Tsinkalovsky OR, Laerum OD. 1994. Flow cytometric measurement of the production of reactive oxygen intermediate in activated rat mast cells. APMIS. 102:474–480.

Tsuji K et al. 1990. Effects of interleukin-3 and interleukin-4 on the development of "connective tissue-type" mast cells: Interleukin-3 supports their survival and interleukin-4 triggers and supports their proliferation synergistically with interleukin-3. Blood. 75:421–427.

Uzuki M et al. 2001. Urocortin in the synovial tissue of patients with rheumatoid arthritis. Clin Sci. 100:577–589.

Valadi H, Ekstrom K, Bossios A, Sjostrand M, Lee JJ, Lotvall JO. 2007. Exosome-mediated transfer of mRNAs and microRNAs is a novel mechanism of genetic exchange between cells. Nat Cell Biol. 9:654–659.

Valent P et al. 1992. Induction of differentiation of human mast cells from bone marrow and peripheral blood mononuclear cells by recombinant human stem cell factor/*kit*-ligand in long-term culture. Blood. 80:2237–2245.

Valent P. 1994. The phenotype of human eosinophils, basophils, and mast cells. J Allergy Clin Immunol. 94:1177–1183.

Valent P et al. 2001. Variable expression of activation-linked surface antigens on human mast cells in health and disease. Immunol Rev. 179:74–81.

van der Kleij HP, Ma D, Redegeld FA, Kraneveld AD, Nijkamp FP, Bienenstock J. 2003. Functional expression of neurokinin 1 receptors on mast cells induced by IL-4 and stem cell factor. J Immunol. 171:2074–2079.

van Haaster CM, Engels W, Lemmens PJMR, Hornstra G, van der Vusse GJ, Heemskerk JWM. 1995. Differential release of histamine and prostaglandin D_2 in rat peritoneal mast cells; roles of cytosolic calcium and protein tyrosine kinases. Biochim Biophys Acta. 1265:79–88.

Van Loveren H, Kops SK, Askenase PW. 1984. Different mechanisms of release of vasoactive amines by mast cells occur in T cell-dependent compared to IgE-dependent cutaneous hypersensitivity responses. Eur J Immunol. 14:40–47.

Van Rensen EL, Hiemstra PS, Rabe KF, Sterk PJ. 2002. Assessment of microvascular leakage via sputum induction: The role of substance P and neurokinin A in patients with asthma. Am J Respir Crit Care Med. 165:1275–1279.

Van Rijt LS, van Kessel CH, Boogaard I, Lambrecht BN. 2005. Respiratory viral infections and asthma pathogenesis: A critical role for dendritic cells? J Clin Virol. 34:161–169.

Varadaradjalou S et al. 2003. Toll-like receptor 2 (TLR2) and TLR4 differentially activate human mast cells. Eur J Immunol. 33:899–906.

von Eckardstein A, Nofer JR, Assmann G. 2001. High density lipoproteins and arteriosclerosis. Role of cholesterol efflux and reverse cholesterol transport. Arterioscler Thromb Vasc Biol. 21:13–27.

von Hertzen LC. 2002. Maternal stress and T-cell differentiation of the developing immune system: Possible implications for the development of asthma and atopy. J Allergy Clin Immunol. 109:923–928.

von Kockritz-Blickwede M et al. 2008. Phagocytosis-independent antimicrobial activity of mast cells by means of extracellular trap formation. Blood. 111:3070–3080.

Wallace DJ, Linker-Israeli M, Hallegua D, Silverman S, Silver D, Weisman MH. 2001. Cytokines play an aetiopathogenetic role in fibromyalgia: A hypothesis and pilot study. Rheumatology (Oxford). 40:743–749.

Wallace DJ. 2006. Is there a role for cytokine based therapies in fibromyalgia? Curr Pharm Des. 12:17–22.

Wang HS et al. 2004. Mesenchymal stem cells in the Wharton's jelly of the human umbilical cord. Stem Cells. 22:1330–1337.

Wasserman SI. 1984. The mast cell and synovial inflammation. Arthritis Rheum. 27:841–844.

Watt FM. 1991. Cell culture models of differentiation. FASEB J. 5:287–294.

Weber S, Babina M, Feller G, Henz BM. 1997. Human leukaemic (HMC-1) and normal skin mast cells express beta 2-integrins: Characterization of beta 2-integrins and ICAM-1 on HMC-1 cells. Scand J Immunol. 45:471–481.

Wedemeyer J, Tsai M, Galli SJ. 2000. Roles of mast cells and basophils in innate and acquired immunity. Curr Opin Immunol. 12:624–631.

Weiss ST. 2001. Epidemiology and heterogeneity of asthma. Ann Allergy Asthma Immunol. 87:5–8.

Weissman MM et al. 2004. Interstitial cystitis and panic disorder: A potential genetic syndrome. Arch gen Psychiatry. 61:273–279.

Weller K, Foitzik K, Paus R, Syska W, Maurer M. 2006. Mast cells are required for normal healing of skin wounds in mice. FASEB J. 20:2366–2368.

Wiesner-Menzel L, Schulz B, Vakilzadeh F, Czarnetzki BM. 1981. Electron microscopical evidence for a direct contact between nerve fibers and mast cells. Acta Derm Venereol (Stockh). 61:465–469.

Wilhelm M, Silver R, Silverman AJ. 2005. Central nervous system neurons acquire mast cell products via transgranulation. Eur J Neurosci. 22:2238–2248.

Williams DA, Gracely RH. 2007. Biology and therapy of fibromyalgia. Functional magnetic resonance imaging findings in fibromyalgia. Arthritis Res Ther. 8:224.

Williams JV et al. 2005. Human metapneumovirus infection plays an etiologic role in acute asthma exacerbations requiring hospitalization in adults. J Infect Dis. 192:1149–1153.

Wills FL, Gilchrist M, Befus AD. 1999. Interferon-gamma regulates the interaction of RBL-2H3 cells with fibronectin through production of nitric oxide. Immunology. 97:481–489.

Wilson J. 2000. The bronchial microcirculation in asthma. Clin Exp Allergy. 30:51–53.

Wipke BT, Wang Z, Nagengast W, Reichert DE, Allen PM. 2005. Staging the initiation of autoantibody-induced arthritis: A critical role for immune complexes. J Immunol. 172:7694–7702.

Wolfreys K, Oliveira DB. 1997. Alterations in intracellular reactive oxygen species generation and redox potential modulate mast cell function. Eur J Immunol. 27:297–306.

Wong LY, Cheung BM, Li YY, Tang F. 2005. Adrenomedullin is both proinflammatory and antiinflammatory: Its effects on gene expression and secretion of cytokines and macrophage migration inhibitory factor in NR8383 macrophage cell line. Endocrinology. 146:1321–1327.

Woolley DE, Tetlow LC. 2000. Mast cell activation and its relation to proinflammatory cytokine production in the rheumatoid lesion. Arthritis Res. 2:65–74.

Woolley DE. 2003. The mast cell in inflammatory arthritis. N Engl J Med. 348:1709–1711.

Wright RJ, Cohen S, Carey V, Weiss ST, Gold DR. 2002. Parental stress as a predictor of wheezing in infancy: A prospective birth-cohort study. Am J Respir Crit Care Med. 165:358–365.

Wright RJ et al. 2004. Community violence and asthma morbidity: The Inner-City Asthma Study. Am J Public Health. 94:625–632.

Xiang Z, Nilsson G. 2000. IgE receptor-mediated release of nerve growth factor by mast cells. Clin Exp Allergy. 30:1379–1386.

Yoshida M et al. 2001. Adrenomedullin and proadrenomedullin N-terminal 20 peptide induce histamine release from rat peritoneal mast cells. Regul Pept. 101:163–168.

Yoshimaru T, Suzuki Y, Inoue T, Niide O, Ra C. 2006. Silver activates mast cells through reactive oxygen species production and a thiol-sensitive store-independent Ca2+ influx. Free Radic Biol Med. 40:1949–1959.

Young JL, Redmond JC. 2007. Fibromylagia, Chronic fatigue, and adult attention deficit hyperactivity disorder in the adult: A case study. Psychopharmacol Bull. 40:118–126.

Yu XJ, Li CY, Wang KY, Dai HY. 2006. Calcitonin gene-related peptide regulates the expression of vascular endothelial growth factor in human HaCaT keratinocytes by activation of ERK1/2 MAPK. Regul Pept. 137:134–139.

Zudaire E et al. 2006. Adrenomedullin is a cross-talk molecule that regulates tumor and mast cell function during human carcinogenesis. Am J Pathol. 168:280–291.

9

AMNIOTIC FLUID DERIVED STEM CELLS

Ming-Song Tsai

Cathay General Hospital, Fu-Jen Catherine University, Taipei, Taiwan

INTRODUCTION

Despite the well-established culturing of amniotic fluid cells (AFCs) in routine second trimester amniocentesis, many questions concerning the nature and *in vivo* origin of these cells have not been entirely resolved [Gospodarowicz, 1977; Hoehn, 1975; Medina-Gomez and Bard, 1983]. The discovery of pluripotnet-like stem cells in amniotic fluid has postulated that the amniotic fluid may represent an attractive alternative to embryonic and adult stem cells for cellular therapy in regenerative medicine and in other potential applications, such as tissue engineering and gene therapy. This chapter elaborates on the characteristics and the current therapeutic capacity of amniotic fluid derived stem cells.

AMNIOTIC FLUID CELLS

Amniotic fluid cells, usually obtained at the sixteenth to eighteenth week of gestation through amniocentesis, were first used for determination of fetal sex by X-chromatin analysis in the mid-1950s and for prenatal karyotyping in 1966. Importantly, it was found that the cellular composition of amniotic fluid varies with gestational age and if there are fetal abnormalities, such as open neural tube defect and abdominal wall abnormality. The average amniotic fluid volume has been found to be ~207 mL at 16 weeks and close to 400 mL at 20 weeks. It has long been proposed that cells of all three germ layers (ectoderm, mesoderm, and endoderm) can be detected in human amniotic fluid, while very little is

Perinatal Stem Cells. Edited by C. L. Cetrulo, K. J. Cetrulo, and C. L. Cetrulo, Jr.

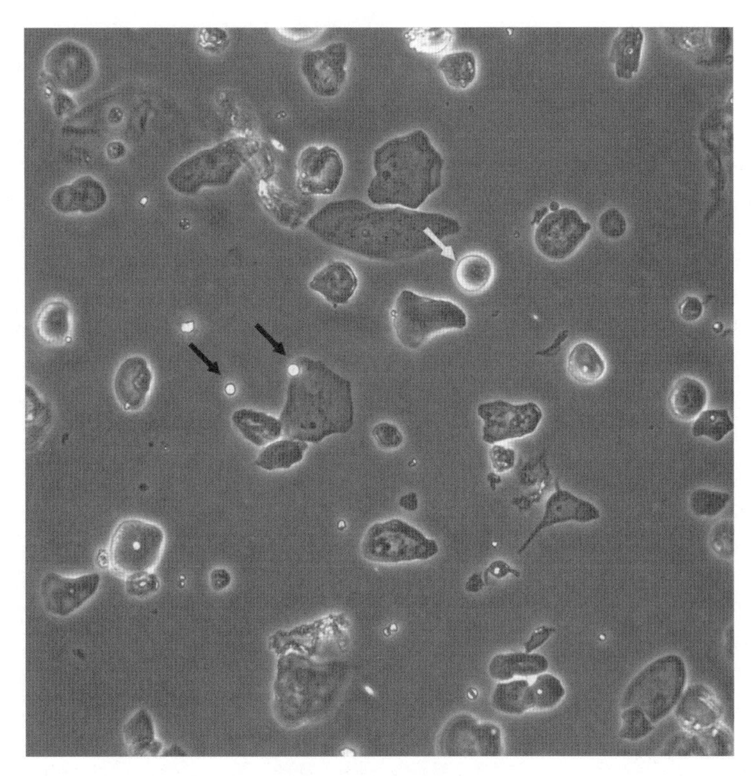

Figure IX.1. Amniotic fluid cells obtained from second trimester amniocentesis, two subpopulations, large cell (black arrow) and small cell (red arrow), of colony-formation cells are noted in fresh amniotic fluid. (See color insert.)

known about their lifespan throughout pregnancy. Amniotic fluid contains heterogeneous population of cells, which are contributed mainly from the fetal skin; the digestive, respiratory, and urinary tract; and possibly from placental tissue as well. Although the total number of nucleated cells per milliliter of amniotic fluid varies between 10^3 and $>10^6$, the number of colony-formation cells is quiet low. An average of 3.5 ± 1.8 clones/mL amniotic fluid, which was obtained between 16 and 18 weeks of gestation, are typically scored at day 12 after the initial plating. However, only 1.5 ± 1.9 colonies/mL amniotic

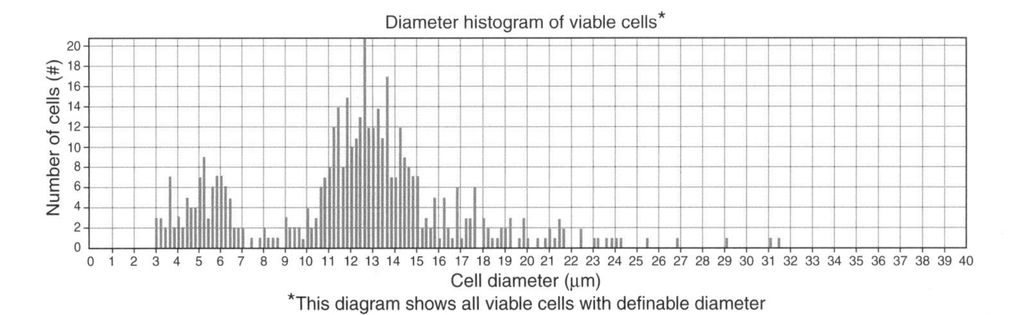

Figure IX.2. Automatic cell counting system showed two the subpopulations of viable cells in fresh amniotic fluid. Small cell (size: 4–8 μm, average: 5); big cells (size: 10–30 μm, average: 13).

fluid reach a clone size of at least 10^6 cells, which is $\sim$20 cumulative population doublings [Milunsky, 1998a]. It is clear that the vast majority of AFCs within native amniotic fluid are incapable of attachment and proliferation that can be divided into adhering and nonadhering cells (NA–AFCs) two major categories. Even though, many of these cells may exclude trypan blue. The number of NA–AFCs is between 5 and 8×10^5 cells/mL and 5% of it is morphologically small, rounded (Figs. IX.1 and IX.2), and alive, which can adhere to a culture flask under specific conditions. Furthermore, AFCs can also be classified according to their morphological aspects and growth characteristics into three groups: epitheloid E-type cells; amniotic fluid specific AF type cells; and fibroblastic F type cells [Hoehn, 1974]. E type cells have been thought to be derived from fetal skin and urine, AF type cells from fetal membranes and trophoblasts, and F type cells from fibrous connective tissue and dermal fibroblasts [Milunsky, 1998a].

THE DISCOVERY OF STEM CELLS IN AMNIOTIC FLUID

The history of stem cell discovery in amniotic fluid dates back to 1993, when Torricelli et al. reported that hematopoietic progenitor cells were found in the amniotic fluid before 12 weeks of gestation [Torricelli, 1993]. Three years later, multilineage potential of nonhematopoietic cells were presented in the amniotic fluid, by showing myogenic conversion of amniocytes [Streubel, 1996]. In 2001, Fauza's group introduced the notion that amniotic fluid cells could be used in autologous fetal tissue engineering for the surgical repair of congenital diaphragmatic hernia and abdominal wall defects [Kaviani, 2001]. However, those studies mentioned above did not specify the identity of amniotic fluid-derived stem cells. The first evidence of human amniotic fluid containing stem cells was reported by Prusa et al. in 2003. They identified Oct-4 expression cells in human amniotic fluid and proposed that human amniotic fluid might be a new source for pluripotent-like stem cells without raising any ethical debates associated with human embryonic stem cells [Prusa, 2003]. In the same year, In 't Anker et al. demonstrated that human amniotic fluid was an abundant source of fetal mesenchymal stem cells (MSCs), and these cells exhibited a phenotype and multilineage differentiation potential that similar to that of bone marrow derived MSCs. They suggested that these amniotic fluid derived MSCs could be used for cotransplantation in conjunction with umbilical cord blood derived hematopoietic stem cells [In 't Anker, 2003]. In 2004, Tsai et al. successfully obtained amniotic fluid derived mesenchymal stem cells (AFMSCs) using a novel two-stage culture protocol without interfering with the process of fetal karyotyping, and his study showed that AFMSCs could be expanded rapidly and maintained the capacity to differentiate into multiple cell types, such as adipocytes, osteoblasts, and neurons *in vitro* [Tsai, 2004]. Beyond being a breakthrough in obtaining human fetal stem cells from nonadhering cells of backup culture, this major discovery also offers a solution to the thorny ethical problems that surround obtaining stem cells from human embryos and fetuses. In 2007, Atala and co-workers isolated the c-kit (CD117) positive population by immunoselection from second trimester amniotic fluid samples and demonstrated that amniotic fluid derived stem cells have the pluripotent potential like embryonic stem cells that can differentiate into adipogenic, osteogenic, myogenic, endothelial, neurogenic, and hepatic lineages, inclusive of all three germ layers [De Coppi, 2007a].

Base on these findings, amniotic fluid will become an important and promising source of isolating pluripotent stem cells both for cellular therapy and for stem cell engineering in the near future.

CULTIVATION OF AMNIOTIC FLUID DERIVED STEM CELLS

The common culture protocol for amniotic fluid stem cells (AFSCs) is similar to that for postnatal mensenchymal stem cell: AFCs are centrifuged and plated in 5 mL of alpha-modified Minimum Essential Medium (α-MEM, Gibco BRL) supplemented with 20% fetal bovine serum (FBS, Hyclone, Logan, UT) 4 ng/mL basic fibroblast growth factor (b-FGF; R&D systems, Minneapolis, MN), 1% L-glutamine, and 1% penicillin/streptomycin (Gibco) in a 25-cm^2 flask and incubated at 37°C with 5% humidified CO_2 for the AFSCs culture [Tsai, 2004]. Currently, there are three major ways for isolation of AFSCs from human amniotic fluid. The first one is direct culture from native amniotic fluid obtained from second-trimester amniocentesis described by In 't Anker et al. [In 't Anker, 2003]. The second one is immunoselection with magnetic microspheres. Atala et al. isolated CD117 positive cells from amniocentesis specimens using c-kit (a rabbit polyclonal antibody to CD117, Santa Cruz Biotechnology) and found that these cells can be easily expanded in culture as stable lines [De Coppi, 2007a]. While Schmidt et al. sorted human AFSCs by CD133 magnetic beads. Their study showed that a subpopulation of CD133+ and CD133− cells exhibited similar characteristics of mesenchymal progenitors and functional endothelial cells, respectively [Schmidt, 2007]. The third one is the two-stage culture protocol established by Tsai et al., using nonadhering amniotic fluid cells of the primary AFCs cultures to isolate AFSCs without interfering with the routine process of fetal karyotyping [Tsai, 2004]. In the two-stage culture protocol, the NA–AFCs are collected from supernatant of primary amniocytes culture (first stage) that using serum-free Chang's medium [Chang and Jones, 1985]. Then these NA–AFCs are plated for AFSCs culturing after the completion of fetal chromosome analysis (second stage). The principal advantage of this two-stage culture protocol comparing to the other two is that instead of the adhering cells of the primary AFCs culture it isolates the AFSCs from the nonadhering cells, which is being left in the incubator without any added nutrition for 7–10 days, a condition similar to serum deprivation as reported by Pochampally RR et al. that may enhance some embryonic gene expressions of the stem cells [Pochampally, 2004].

CHARACTERISTICS OF AMNIOTIC FLUID DERIVED STEM CELLS

The characteristics of AFSCs obtained by different methods are very similar, suggesting a close relationship between these fetal stem cell populations. Reverse-transcriptase polymerase chain reaction (RT–PCR), flowcytometry, and immunocytochemical staining are used to assess the cellular markers expressed by human AFSCs. The AFSCs are positive for Class I major histocompatibility (MHC) antigens (HLA–ABC), and negative for MHC Class II (HLA-DR, DP, DQ). The AFSCs are negative for markers of the hematopoietic lineage (CD45) and of hematopoietic stem cells (CD34, CD133). However, the AFSCs also have phenotypic characteristics similar to those of mesenchymal stem cells derived from other sources, such as term umbilical cord blood, placenta, and first-trimester fetal tissues (blood, liver, and bone marrow), which are positive for SH2, SH3, SH4, CD29, CD44, and CD166, and negative for CD10, CD11b, CD14, CD34, CD117, and EMA [Colter, 2001; Pittenger, 1999; Young, 2001]. Most importantly, human AFSCs express Oct-4 mRNA and Oct-4 protein, a transcription factor expressed in embryonic carcinoma cells, embryonic stem cells, and embryonic germ cells, reflecting a key role in the maintenance of pluripotency of mammalian stem cells both *in vivo* and *in vitro* [Cogle, 2003;

Henderson, 2002; Jiang, 2002]. In addition, AFSCs do not express other surface markers characteristic of embryonic stem cells and embryonic germ cells, SSEA-3, and Tra-1-81, but some cell lines are weakly positive for Tra-1-60. It is clear that amniotic fluid contains a subpopulation of high potential stem cells, which are shed from embryonic and extra-embryonic tissues during the process of fetal development and growth. The characteristics of AFSCs are summarized in Table IX.1.

TABLE IX.1. Characteristics of Amniotic Fluid Derived Stem Cells from Different Study Groups

	Ref.[a]	Ref.[b]	Ref.[c]	Ref.[d]
	Mesenchymal Antigen[e]			
CD73 (SH3, SH4)	+	+	+	+
CD90 (Thy-1)	+	+	+	ND
CD105 (SH2)	+	+	+	±
CD166	+	+*	ND	+
	Endothelial Markers			
CD31	−	ND	ND	−
CD144	ND	ND	ND	−
	Hematopoietic Markers[e]			
CD34	−	−	−	−
CD45	−	ND	−	−
CD133	ND	ND	−	ND
	HLA Class[e]			
HLA-I (A, B, C)	+	+	+	+
HLA-II (DR, DP, DQ)	−	−	−	−
	Embryonic Stem Cells Markers[e]			
SSEA-3	ND	−*	−	ND
SSEA-4	ND	+*	+	ND
Tra-1-60	ND	±*	±	ND
Tra-1-81	ND	−*	−	ND
Oct4	ND	+	+	ND
NANOG	ND	+	ND	ND
	Others[e]			
CD29	ND	+	+	+
CD44	+	+	+	±
CD117	ND	−	+	ND

[a]In 't Anker, 2003.
[b]Tsai, 2006, 2007.
[c]De Coppi, 2007a.
[d]Sessarego, 2008.
[e]+ = Positive expression; ± = Dim expression; − = Negative expression; ND = No detection; * = our unpublication data.

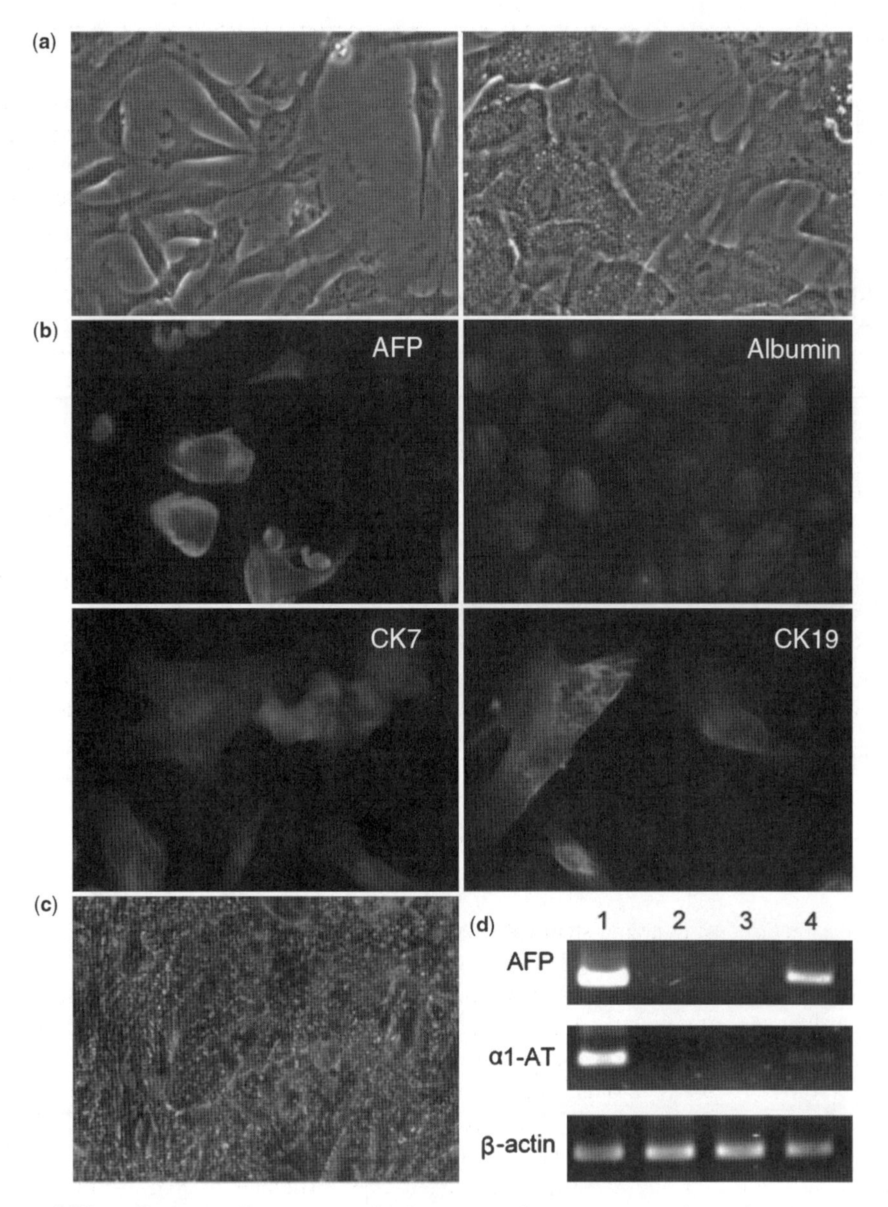

Figure IX.3. Differentiated amniotic fluid stem cells express markers for hepatocytes. Morphological of AFSCs before hepatic induction (a left) and at 4-weeks postinduction (a right). Differentiated cells at 6-weeks postinduction express immunofluorescence stains of AFP, Albumin, CK7 and CK19, markers for hepatic differentiation (b). Positive glycogen storage was noted at 4-weeks postinduction (c). The RT–PCR analyses are shown in (d). Lane 1: Hepa G2 for positive control, Lane 2: noninduction, Lane 3: at 2 weeks posthepatocyte-induction, Lane 4: at 3-weeks posthepatocyte induction. The AFP (α-fetoprotein): early hepatocyte marker, α1-AT(α1-antitrypsin): hepatocyte expression marker, a 52-kDa glycoprotein produced by hepatocytes and mononuclear phagocytes. β-Actin as a internal control. (See color insert.)

DIFFERENTIATION POTENTIAL OF AMNIOTIC FLUID STEM CELLS

The AFSCs can be expanded rapidly and they maintain the capacity to differentiate into multiple cell types *in vitro*. Aside from the common mesenchymal lineages (adipocytes, chondrocytes, osteocytes, and myocytes), they can also be differentiated into neural lineages. These phenotypes are quite similar to bone marrow derived MSCs that can be differentiated into multiple mesodermal cell and neural cell types. Several reports show that adult bone marrow contains a subpopulation of pluripotent-like mesenchymal stem cells that can be isolated at the single-cell level without losing their capacity for differentiation into cells of all three germ layers [Jiang, 2002; Lee, 2004; Smith, 2004; Yoon, 2005]. Do AFSCs have a similar pluripotent capability to marrow stromal cells? Whether they contain a heterogeneous population of stem cells from different germ layers, or a subpopulation of primitive precursors that are capable to overcome germ layers commitment was unknown till 2006. Tsai et al. demonstrated that a population of single-cell derived clonal AFSCs also exhibited a capacity to differentiate into multiple mesodermal cell types (adipocytes, oesteocytes, and chondrocytes) and ectodermal cells types (neuronal cells and glial cells) [Tsai, 2006]. In 2007, Atala et al. successfully isolated a population of CD117 positive cells from amniotic fluid. Besides the mesodermal and ectodermal lineages, the cells also represented the ability to differentiate into endodermal hepatocytes *in vitro*. In the same year, our single-cell derived AFSC clones, which were negative for CD117 expression, were successfully differentiated into hepatocytes as shown in Fig. IX.3. Based on these findings, amniotic fluid may provide a promising alternative source for investigation of human pluripotent stem cells without raising any ethical debates that associated with human embryonic research.

AMNIOTIC FLUID AS A PROMISING ALTERNATIVE SOURCE FOR NEURAL STEM CELLS

Cellular therapy using neural stem cells (NSCs) is a potential strategy for the treatment of neurodegenerative disorders and central nervous system injuries [Baksh, 2004; Dezawa, 2004; Korbling and Estrov, 2003]. Commonly, NSCs can be derived from embryonic stem cells or two principal adult neurogenic regions of the brain, the hippocampus and the subventricular zone [Johansson, 1999; Steindler and Pincus, 2002; Temple, 2001]. Each of these sources, though successful, has their limitations, whether they involve inadequate tissue supply or are immersed in ethical controversy. Therefore, looking for alternative sources of NSCs has become a meaningful research [Goodwin, 2001; Hofstetter, 2002; Hung, 2002; Kakishita, 2003; Mitchell, 2003; Safford, 2004; Woodbury, 2000]. Prusa et al. first reported that native amniotic fluid contains neurogenic cells that differentiation occurs sporadically in standard culture conditions, but is strongly increased in neurogenic induction medium [Prusa, 2004]. In the past, unexpected presence of neural cells in AFC cultures would indicate that the fetus had a neural tube defect [Baksh, 2004; Medina-Gomez and Bard, 1983; Milunsky, 1998a; Milunsky, 1998b]. In 2006, both Tsai et al. and Mclaughlin et al. confirmed the Prusa's report and demonstrated that during routine cultivation a small portion of AFSCs may receive a specific commitment into the direction of neural cells, as identified by both morphological transformation and immunochemical expression of markers (NES, TUBB3, NEFH, NEUNA60, GALC, and GFAP) for neural progenitor and mature neural cells [McLaughlin, 2006; Tsai, 2006]. However, when these cells are treated with neural differentiation medium, more than three-fourths of them show a neuronal-like appearance and the percentage of mature neural marker

TABLE IX.2. Cases of Application of AFSCs on Cellular Therapy

Organism	Citation	Sources of Stem Cells	Target Organ	Results
In vitro study	[Schmidt, 2007]	Fetal human amniotic progenitors	Heart valve leaflet scaffolds	Engineered heart valve leaflets demonstrated endothelialized tissue formation with production of extracellular matrix elements.
Sprague–Dawley rats	[Cipriani, 2007]	Fetal human amniotic fluid mesenchymal cells	Injured brain	Grafted mesenchymal cells could survive and migrate toward multiple brain regions in the normal animals, while they moved toward the injured region in the ischemic rat.
Sprague–Dawley rats with or without cyclosporine treatment	[Chiavegato, 2007]	Human amniotic fluid derived stem cells	Myocardial infarction	Despite amniotic fluid stem cells *in vitro* can differentiate to some extent to cells of cardiovascular lineages, their *in vivo* use in xenotransplantation for cell therapy of myocardial infarction is hampered by their peculiar immunogenic properties and phenotypic instability.
Nude athymic rats	[De Coppi, 2007b]	Rat mesenchymal stem cells	Cryo-injured rat bladder	Stem cell transplantation has a limited effect on smooth muscle cell regeneration. Instead it can regulate postinjury bladder remodeling, possibly via a paracrine mechanism.
Newborn lambs	[Kunisaki, 2006]	Amniotic fluid cells from ewes	Experimental diaphragmatic defect	Diaphragmatic repair with a mesenchymal amniocyte-based engineered tendon leads to improved structural outcomes when compared with equivalent fetal myoblast-based and acellular grafts.
Sprague–Dawley rats	[Pan, 2006, 2007]	Rat amniotic fluid mesenchymal stem cells	Injured sciatic nerve	1. Amniotic fluid MSC can augment growth of injured nerve across a nerve gap. 2. Motor function recovery, the compound muscle action potential and nerve conduction latency showed significant improvement in rats treated with amniotic fluid mesenchymal stem cells.
Pigs	[Sartore, 2005]	Porcine amniotic fluid cells	Acute myocardial infaction	Porcine amniotic fluid cells were able to transdifferentiate to cells of vascular cell lineages but not to cardiomyocytes in the ischemic area of porcine hearts.

expressions are increased. Furthermore, high-performance liquid chromatography (HPLC) analysis showed evidence of dopamine release in the extract of dopaminogenic-induced AFSCs when they are depolarized by high K^+. These findings indicate that AFSCs may be a favorable candidate for celluar therapy of neurodegenerative disorders and central nervous system injuries. As an alternative source of NSCs, amniotic fluid provides several advantages, compared to those derived from other sources, such as adult bone marrow, human amniotic epithelial cells, umbilical cord blood, Wharton's Jelly, and adipose tissue. First, amniotic fluid contains a subpopulation of primitive stem cells in early fetal life that have a higher proliferation capacity than that of postnatal adult stem cells. The AFSCs reveal relatively longer telomere length compared to that of umbilical cord blood derived MSCs and adult bone marrow derived MSCs. Second, amniotic fluid contains a pool of various stem cells and progenitors, which may coordinate their capacities and improve the efficiency of tissue repair during cellular transplantation. Third, amniotic fluid is easily obtained and has a higher success rate in isolating stem cells. Finally, amniotic fluid-derived stem cells open an unprecedented approach for autologous intrauterine fetal gene and cellular therapy.

CURRENT APPLICATIONS OF AMNIOTIC FLUID STEM CELLS IN THERAPY

The therapeutic application of amniotic fluid stem cells began with Fauza et al. He first used the mesenchymal anmiocytes as a preferred cell source to improved mechanical and functional outcome on diaphragmatic reconstruction [Kaviani, 2001; Kunisaki, 2006]. Table IX.2 summarizes the application of AFSCs on different fields, such as myocardial infarction, brain injury, heart valve leaflet, sciatic nerve injury, and bladder injury. These cases shed the light on replacing current source of mesenchymal stem cell obtained from postnatal bone marrow and umbilical cord blood on the cellular therapy.

AMNIOTIC FLUID STEM CELLS BANKING

With all the advantage of AFSCs, the human amniotic fluid may provide an excellent source for stem cell research and cellular therapy. There are couples of institutions that have already commenced AFSCs cell banking combining with the cord blood and placenta tissue for their individual identity and capacity. Definitely, cotransplantation with these cells on therapies is going to be a very attractive field to be explored.

REFERENCES

Baksh D, Song L, Tuan RS. 2004. Adult mesenchymal stem cells: Characterization, differentiation, and application in cell and gene therapy. J Cell Mol Med. 8(3):301–316.

Chang HC, Jones OW. 1985. Reduction of sera requirements in amniotic fluid cell culture. Prenat Diagn. 5(5):305–312.

Chiavegato A et al. 2007. Human amniotic fluid-derived stem cells are rejected after transplantation in the myocardium of normal, ischemic, immuno-suppressed or immuno-deficient rat. J Mol Cell Cardiol. 42(4):746–759.

Cipriani S, Bonini D, Marchina E, Balgkouranidou I, Caimi L, Grassi Zucconi G, Barlati S. 2007. Mesenchymal cells from human amniotic fluid survive and migrate after transplantation into adult rat brain. Cell Biol Int. 31(8):845–850.

Cogle CR, Guthrie SM, Sanders RC, Allen WL, Scott EW, Petersen BE. 2003. An overview of stem cell research and regulatory issues. Mayo Clin Proc. 78(8):993–1003.

Colter DC, Sekiya I, Prockop DJ. 2001. Identification of a subpopulation of rapidly self-renewing and multipotential adult stem cells in colonies of human marrow stromal cells. Proc Natl Acad Sci. 141221698.

De Coppi P et al. 2007a. Isolation of amniotic stem cell lines with potential for therapy. Nat Biotechnol. 25(1):100–106.

De Coppi P et al. 2007b. Amniotic fluid and bone marrow derived mesenchymal stem cells can be converted to smooth muscle cells in the cryo-injured rat bladder and prevent compensatory hypertrophy of surviving smooth muscle cells. J Urol. 177(1):369–376.

Dezawa M et al. 2004. Specific induction of neuronal cells from bone marrow stromal cells and application for autologous transplantation. J Clin Invest. 113(12):1701–1710.

Goodwin HS, Bicknese AR, Chien SN, Bogucki BD, Quinn CO, Wall DA. 2001. Multilineage differentiation activity by cells isolated from umbilical cord blood: Expression of bone, fat, and neural markers. Biol Blood Marrow Transplant. 7(11):581–588.

Gospodarowicz D, Moran JS, Owashi ND. 1977. Effects of fibroblast growth factor and epidermal growth factor on the rate of growth of amniotic fluid-derived cells. J Clin Endocrinol Metab. 44(4):651–659.

Henderson JK, Draper JS, Baillie HS, Fishel S, Thomson JA, Moore H, Andrews PW. 2002. Preimplantation Human Embryos and Embryonic Stem Cells Show Comparable Expression of Stage-Specific Embryonic Antigens. Stem Cells. 20(4):329.

Hoehn H, Bryant EM, Karp LE, Martin GM. 1974. Cultivated cells from diagnostic amniocentesis in second trimester pregnancies. I. Clonal morphology and growth potential. Pediatr Res. 8(8):746–754.

Hoehn H, Bryant EM, Fantel AG, Martin GM. 1975. Cultivated cells from diagnostic amniocentesis in second trimester pregnancies. Human Genet. 29(4):285–290.

Hofstetter CP, Schwarz EJ, Hess D, Widenfalk J, El Manira A, Prockop DJ, Olson L. 2002. Marrow stromal cells form guiding strands in the injured spinal cord and promote recovery. Proc Natl Acad Sci USA. 99(4):2199–2204.

Hung SC, Cheng H, Pan CY, Tsai MJ, Kao LS, Ma HL. 2002. In vitro differentiation of size-sieved stem cells into electrically active neural cells. Stem Cells. 20(6):522–529.

In 't Anker PS, Scherjon SA, Kleijburg-van der Keur C, Noort WA, Claas FH, Willemze R, Fibbe WE, Kanhai HH. 2003. Amniotic fluid as a novel source of mesenchymal stem cells for therapeutic transplantation. Blood. 102(4):1548–1549.

Jiang Y et al. 2002. Pluripotency of mesenchymal stem cells derived from adult marrow. Nature (London). 418(6893):41–49.

Johansson CB, Momma S, Clarke DL, Risling M, Lendahl U, Frisen J. 1999. Identification of a neural stem cell in the adult mammalian central nervous system. Cell. 96(1):25–34.

Kakishita K, Nakao N, Sakuragawa N, Itakura T. 2003. Implantation of human amniotic epithelial cells prevents the degeneration of nigral dopamine neurons in rats with 6-hydroxydopamine lesions. Brain Res. 980(1):48–56.

Kaviani A, Perry TE, Dzakovic A, Jennings RW, Ziegler MM, Fauza DO. 2001. The amniotic fluid as a source of cells for fetal tissue engineering. J Pediatr Surg. 36(11):1662–1665.

Korbling M, Estrov Z. 2003. Adult stem cells for tissue repair—a new therapeutic concept? N Engl J Med. 349(6):570–582.

Kunisaki SM, Fuchs JR, Kaviani A, Oh JT, LaVan DA, Vacanti JP, Wilson JM, Fauza DO. 2006. Diaphragmatic repair through fetal tissue engineering: A comparison between mesenchymal amniocyte- and myoblast-based constructs. J Pediatr Surg. 41(1):34–39; discussion 34–39.

Lee OK, Kuo TK, Chen WM, Lee KD, Hsieh SL, Chen TH. 2004. Isolation of multipotent mesenchymal stem cells from umbilical cord blood. Blood. 103(5):1669–1675.

McLaughlin D, Tsirimonaki E, Vallianatos G, Sakellaridis N, Chatzistamatiou T, Stavropoulos-Gioka C, Tsezou A, Messinis I, Mangoura D. 2006. Stable expression of a neuronal dopaminergic progenitor phenotype in cell lines derived from human amniotic fluid cells. J Neurosci Res. 83(7):1190–1200.

Medina-Gomez P, Bard JB. 1983. Analysis of normal and abnormal amniotic fluid cells in vitro by cinemicrography. Prenat Diagn. 3(4):311–326.

Milunsky A. 1998a. Fluid cell culture, Genetic disorders and the fetus: Diagnosis, prevention, and treatment: Baltimore: Johns Hopkins University Press. pp. 128–149.

Milunsky A. 1998b. Maternal serum screening for neural tube and other defects; Genetic disorders and the fetus: Diagnosis, prevention, and treatment: Johns Hopkins University Press. pp. 635–701.

Mitchell KE et al. 2003. Matrix cells from Wharton's jelly form neurons and glia. Stem Cells. 21(1):50–60.

Pan HC, Yang DY, Chiu YT, Lai SZ, Wang YC, Chang MH, Cheng FC. 2006. Enhanced regeneration in injured sciatic nerve by human amniotic mesenchymal stem cell. J Clin Neurosci. 13(5): 570–575.

Pan HC, Cheng FC, Chen CJ, Lai SZ, Lee CW, Yang DY, Chang MH, Ho SP. 2007. Post-injury regeneration in rat sciatic nerve facilitated by neurotrophic factors secreted by amniotic fluid mesenchymal stem cells. J Clin Neurosci. 14(11):1089–1098.

Pittenger MF, Mackay AM, Beck SC, Jaiswal RK, Douglas R, Mosca JD, Moorman MA, Simonetti DW, Craig S, Marshak DR. 1999. Multilineage potential of adult human mesenchymal stem cells. Science. 284(5411):143–147.

Pochampally RR, Smith JR, Ylostalo J, Prockop DJ. 2004. Serum deprivation of human marrow stromal cells (hMSCs) selects for a subpopulation of early progenitor cells with enhanced expression of OCT-4 and other embryonic genes. Blood. 103(5):1647–1652.

Prusa AR, Marton E, Rosner M, Bernaschek G, Hengstschlager M. 2003. Oct-4-expressing cells in human amniotic fluid: A new source for stem cell research? Hum Reprod. 18(7):1489–1493.

Prusa AR, Marton E, Rosner M, Bettelheim D, Lubec G, Pollack A, Bernaschek G, Hengstschlager M. 2004. Neurogenic cells in human amniotic fluid. Am J Obstet Gynecol. 191(1):309–314.

Safford KM, Safford SD, Gimble JM, Shetty AK, Rice HE. 2004. Characterization of neuronal/glial differentiation of murine adipose-derived adult stromal cells. Exp Neurol. 187(2):319–328.

Sartore S, Lenzi M, Angelini A, Chiavegato A, Gasparotto L, De Coppi P, Bianco R, Gerosa G. 2005. Amniotic mesenchymal cells autotransplanted in a porcine model of cardiac ischemia do not differentiate to cardiogenic phenotypes. Eur J Cardiothorac Surg. 28(5):677–684.

Schmidt D, Achermann J, Odermatt B, Breymann C, Mol A, Genoni M, Zund G, Hoerstrup SP. 2007. Prenatally fabricated autologous human living heart valves based on amniotic fluid derived progenitor cells as single cell source. Circulation. 116(11 Suppl):164–170.

Sessarego N et al. 2008. Multipotent mesenchymal stromal cells from amniotic fluid: Solid perspectives for clinical application. Haematologica. 93(3):339–346.

Smith JR, Pochampally R, Perry A, Hsu SC, Prockop DJ. 2004. Isolation of a highly clonogenic and multipotential subfraction of adult stem cells from bone marrow stroma. Stem Cells. 22(5): 823–831.

Steindler DA, Pincus DW. 2002. Stem cells and neuropoiesis in the adult human brain. Lancet. 359(9311):1047–1054.

Streubel B, Martucci-Ivessa G, Fleck T, Bittner RE. 1996. In vitro transformation of amniotic cells to muscle cells—background and outlook. Wien Med Wochenschr. 146(9–10):216–217.

Temple S. 2001. The development of neural stem cells. Nature (London). 414(6859):112–117.

Torricelli F, Brizzi L, Bernabei PA, Gheri G, Di Lollo S, Nutini L, Lisi E, Di Tommaso M, Cariati E. 1993. Identification of hematopoietic progenitor cells in human amniotic fluid before the 12th week of gestation. Ital J Anat Embryol. 98(2):119–126.

Tsai MS, Hwang SM, Tsai YL, Cheng FC, Lee JL, Chang YJ. 2006. Clonal amniotic fluid-derived stem cells express characteristics of both mesenchymal and neural stem cells. Biol Reprod. 74(3):545–551.

Tsai MS, Lee JL, Chang YJ, Hwang SM. 2004. Isolation of human multipotent mesenchymal stem cells from second-trimester amniotic fluid using a novel two-stage culture protocol. Hum Reprod. 19(6):1450–1456.

Tsai MS et al. 2007. Functional network analysis of the transcriptomes of mesenchymal stem cells derived from amniotic fluid, amniotic membrane, cord blood, and bone marrow. Stem Cells. 25(10):2511–2523.

Woodbury D, Schwarz EJ, Prockop DJ, Black IB. 2000. Adult rat and human bone marrow stromal cells differentiate into neurons. J Neurosci Res. 61(4):364–370.

Yoon YS et al. 2005. Clonally expanded novel multipotent stem cells from human bone marrow regenerate myocardium after myocardial infarction. J Clin Invest. 115(2):326–338.

Young HE, Steele TA, Bray RA, Hudson J, Floyd JA, Hawkins K, Thomas K, Austin T, Edwards C, Cuzzourt J. 2001. Human reserve pluripotent mesenchymal stem cells are present in the connective tissues of skeletal muscle and dermis derived from fetal, adult, and geriatric donors. The Anatomical Record. 264(1):51–62.

10

AMNIOTIC EPITHELIAL STEM CELLS IN REGENERATIVE MEDICINE

Fabio Marongiu, Roberto Gramignoli, Toshio Miki, Aarati Ranade, Ewa C.S. Ellis, Kenneth Dorko and Stephen C. Strom

University of Pittsburgh, Department of Pathology, Pittsburgh, PA 15261

Julio C. Davila

Pfizer Inc. St. Louis, Missouri 63198, USA

INTRODUCTION

As recently described by Parolini et al., placenta represents a reserve of different types of progenitor/stem cells, which can be easily isolated from different placental compartments [Parolini, 2008]. Furthermore, placenta is involved in maintaining fetal tolerance, and it has been reported that several placenta cell types have immunomodulatory properties.

The term placenta has a discoid shape and comprises the endometrial decidua and the fetal membranes chorion and amnion, which extend from the margin of the chorionic plate to form the amniotic cavity. The amnion is a thin, avascular membrane that surrounds the fetus during its development. Despite its thin size, the amnion is surprisingly strong and elastic, working as a shock absorber while protecting the fetus against infections. The amnion is also metabolically active and is involved in maintaining amniotic fluid homeostasis [Bryant-Greenwood, 1998]. The amnion is comprised of an epithelial and outer layer of connective tissue [Bourne, 1966; van Herendael, 1978]. The human amniotic epithelium (hAE) is a single layer of cuboidal and columnar cells that is in contact with amniotic fluid on the internal side. It rests on the connective tissue of the amniotic mesoderm (hAM) on the inner side. The hAE cells secrete adhesive molecules, such as fibronectin and collagen I and III, that constitute an acellular compact basal layer that separates hAE from a deeper layer of mesenchymal and monocyte-like cells [Magatti, 2008].

Perinatal Stem Cells. Edited by C. L. Cetrulo, K. J. Cetrulo, and C. L. Cetrulo, Jr.

Although the decidua is maternally derived, amnion and chorion membranes are genetically identical to the embryo, and will eventually form a highly specialized interface between mother and fetus. While the chorion is derived from the trophoblast layer (the outer layer of the blastocyst, which mediates the implantation of the embryo into the endometrium and provide nutrients to the embryo), the amnion is derived from the epiblast on or about the eight day following fertilization. The epiblast gives rise to the amnion as well as all of the germ layers of the embryo. These lineage relationships open the possibility that cells derived from the amnion may retain some or all of the multipotent phenotype of the epiblast.

The ease of isolation of the cells and the availability of placenta as a tissue that is otherwise discarded make the amnion a potentially useful and noncontroversial source of cells for transplantation and regenerative medicine.

AMNION AND ITS PROPERTIES

The amnion membrane has several clinical applications and has been utilized for $\sim$100 years for the treatment of skin lesions. When applied onto an open wound, the amnion membrane demonstrated to have protective capacity against microbial infections, also reducing the loss of fluids and nutrients, and therefore shortening the healing time [Gruss and Jirsch, 1978; Robson, 1973; Robson and Krizek, 1973]. Allograft of amniotic membranes is still a widely used technique for the treatment of skin burns and ulcers [Ravishanker, 2003].

The amnion is also commonly utilized for the treatment of ophthalmic defects (either inborn or traumatic), such as limbal deficiency, acute chemical or thermal burns, or conjunctival surface reconstruction after surgical intervention [Meller, 2000; Nubile, 2007, 2008]. Use of dried or preserved amnion membrane also has been proven to be effective in the same treatments [Ravishanker, 2003].

Besides its great accessibility, one of the most important advantages of the amniotic tissue is its lack of immunogenicity. Several reports indicate that the amniotic membrane and amniotic epithelial cells do not induce immune reaction when transplanted in diverse settings [Akle, 1981; Dua, 2004]. In addition, the amnion has been reported to have anti-inflammatory, bacteriostatic, antiangiogenic, and analgesic properties, although the mechanisms behind this processes are not yet clear [Hao, 2000; Robson and Krizek, 1973].

ISOLATION AND CULTURE OF hAE CELLS

The detailed protocol for the isolation of hAE cells recently has been published providing a step by step procedure along with photos of the process [Miki, 2007].

For the isolation of hAE cells, the amnion membrane is stripped from the underlying chorion. After a brief wash to remove blood clots, the amnion is digested in trypsin. Following two 40 min digestions each, hAE cells are selectively released from underlying mesenchyme.

After isolation, hAE cells readily attach to the culture dish without the need for a feeder layer or any particular pretreatment of the culture substrate with adhesive molecules. Although hAE cells easily replicate in Dulbecco's modified eagle's medium (DMEM) supplemented with 5–10% serum and epidermal growth factor (EGF), the cells do not replicate indefinitely. The possibility of culturing these cells in a serum-free environment is currently under investigation. The presence of EGF in the medium seems to be required for a longer culture of hAE cells. When EGF is removed from the culture, cell proliferation slows until the cells stop replicating, even in the presence of serum. Normally, two to six passages are possible before proliferation ceases. It has also been observed that hAE cells do not proliferate well at low densities. However, more recent studies with different culture conditions revealed that hAE cells replicate faster and for a longer period of time, when cultured in the presence of low calcium media. In the same culture condition, hAE cells also show clonogenic potential when plated at low densities (unpublished data).

STEM CELL CHARACTERISTICS OF AMNIOTIC EPITHELIUM-DERIVED CELLS

Recently, Miki et al. reported that hAE cells express markers typical of hES cells and have the ability to differentiate in cell types from all three germ layers [Miki, 2005]. Stem cell surface antigens expressed on hAE cells include stage specific embryonic antigens (SSEA) three and four and tumor rejection antigens (TRA) 1–60 and 1–81 [Miki, 2005, 2007; Miki and Strom, 2006]. These surface molecules are selectively express on undifferentiated hES cells, and they are downregulated when these cells differentiate. Other surface antigens are summarized in Table X.1. Furthermore, hAE cells express molecular markers of pluripotency, such as octamer-binding-protein-4 (OCT-4), SRY-related HMG-box gene 2 (SOX-2), and Nanog [Miki, 2005]. These transcription factors are known to be required for self-renewal and pluripotency [Boiani and Scholer, 2005; Chambers, 2003; Nichols, 1998; Niwa, 2007; Pan, 2006].

Furthermore, hAE cells showed the ability to differentiate into cell types from all three germ layers including cardiomyocytes, neurons, pancreatic alpha and beta cells, and hepatocytes [Davila, 2004; Miki, 2005; Miki and Strom, 2006]. This potential also has been reported from other research groups [Elwan and Sakuragawa, 1997]. Sakuragua et al. induced hAE cells to differentiate into neural-like cells that expressed genes and proteins typical of neural cells along with the production of neurotransmitters [Ilancheran, 2007; Kakishita, 2000, 2003; Tamagawa, 2004]. Evidence of hAE differentiation into pancreatic cells capable of producing and releasing insulin has been reported. Transplantation of these cells also decreased blood glucose levels in diabetic mice [Wei, 2003].

Human AE cells do not express telomerase, an enzyme usually highly expressed in immortal and tumor cells, but also in hES and germ cells. Freshly isolated cells were examined for tumorigenicity by injection into Severe Combined Immunodeficiency (SCID) and Rag-2 knockout mice; no evidence of tumor formation was found up to 7 months after transplant [Ilancheran, 2007; Miki, 2005]. Furthermore, proliferating hAE cells display a normal kariotype [Miki, 2005].

TABLE X.1. The hAE Cell Surface Markers

	Fresh Cells (%)	Adherent Cells (%)
SSEA3	12.3	11.0
TRA1-60	14.8	12.1
TRA1-81	14.5	10.7
CD49f	93.1	90.6
HLA-ABC	75.4	83.3
EpCAM	87.0	93.4
CD90	11.4	8.9
ABCG2	0.5	0.4
AC133/2	0.1	0

DIFFERENTIATION OF hAE CELLS

Differentiation of hAE toward a hepatic lineage also has been reported. Sakuragawa et al. [Sakuragawa, 2000], reported that cultured hAE cells produce albumin (Alb) and α-fetoprotein (AFP). Following transplantation, Alb and AFP positive cells were identified in the liver chords of SCID mice. In later studies the hepatic potential of hAE cells was confirmed [Davila, 2004; Miki, 2005; Takashima, 2004]. These investigators also found Alb and AFP production, but in addition other hepatic functions (glycogen storage and the expression of liver-enriched transcriptions factors), such as hepatocyte nuclear factor (HNF) 3γ and HNF4α, CCAAT/enhancer-binding protein (CEBP α and β) and several of the drug metabolizing genes, cytochrome P450, were reported.

Another endodermal tissue, pancreas, was also reported to be produced from hAE. Wei et al. [Wei, 2003], cultured hAE with media supplemented with nicotinamide that resulted in cells that followed a pancreatic lineage. When the insulin producing cells were transplanted they corrected the hyperglycemia of diabetic mice. Interestingly, hAM cells were ineffective in correcting the hyperglycemia, suggesting that hAM cells did not differentiate to β-cells when exposed to the same culture conditions.

Previous work from our group indicates that the differentiation of hAE cells to different cell types is modulated by the culture substrate and the types and concentration of growth factors included in the media. A typical hepatic differentiation protocol would include plating the cells on type 1 collagen coated culture dishes in DMEM media with standard supplements [Miki, 2005, 2008; Miki and Strom, 2006]. Slowly over the next week or so some cells begin to stain positive for albumin, as well as the endodermal–hepatic marker genes hepatocyte nuclear factor-4 (HNF4-α) and Alpha 1-antitrypsin (A1AT) and CAAT enhancer binding protein-alpha (C/EBP-α). The presence of steroid hormones in the media enhances hepatic differentiation of hAE cells.

Exposure to Activin-A, a member of the TGF-β superfamily is known to induce endodermal differentiation of human ES cells. Activin-A effects its biological activity via binding to the TGF-β superfamily Type II receptor ActRII and the coreceptor Type I ALK4 [de Caestecker, 2004; Rodgarkia-Dara, 2006; Valdimarsdottir and Mummery, 2005]; this initiates a cascade of intracellular signaling events that enhance the differentiation of hES cells to endoderm lineages (i.e., hepatic or pancreatic) [D'Amour, 2005, 2006; McLean, 2007; Soto-Gutierrez, 2006]. Other investigators report that Activin-A helps maintaining self-renewal of ES cells [Xiao, 2006] so the state of differentiation of the cells at the time of exposure and also the timing of the Activin-A exposure may be critical to the overall biological effect.

Two types of endoderm are generated during embryonic development. Visceral endoderm contributes to the extraembryonic placental structures, while definitive endoderm gives rise to liver pancreas and other internal organs. Although some markers of early hepatic differentiation, such as albumin and A1AT, are expressed by both visceral and definitive endoderm, CYP7A1 is a gene whose expression is thought to be specific to definitive endoderm [Asahina, 2004]. This gene is located on the endoplasmic reticulum and encodes cholesterol 7-α-hydroxylase, an enzyme involved in the conversion of cholesterol to bile acids in hepatocytes [Ellis, 1998]. The expression of CYP7A1 is readily detected in cultured human amnion cells exposed to a hepatic differentiation protocol. In liver, the transcriptional regulation of the CYP7A1 gene is critical to the regulation of cholesterol metabolism to bile acids. Just as with authentic human hepatocytes in culture, exposure of hAE culture to bile acids results in a feedback inhibition of CYP7A1 expression [Ellis, 2003].

There is a long list of mature hepatic genes whose expression can be detected in cultured hAE cells. They include transcription factors, such as HNF4, C/EBP-alpha and beta, pregnane x receptor (PXR) and constitutive androstane receptor (CAR), members of the CYP1A, 2B, 2C, and the 3A family, as well as a number of transport proteins. Metabolic activities associated with these CYPs are also reported, but less frequently. Other laboratories have reported similar observations of hepatic differentiation of hAE cells [Takashima, 2004]. Remember that the expression of the hepatic genes in cultured hAE cells induced to a hepatic lineage is normally in the range of 1% or less of that observed in a normal human liver. So while the initial studies are quite promising, a great deal of work may be needed before efficient and effective differentiation protocols are developed that would transdifferentiate amnion into cells with a mature liver phenotype.

FUTURE DIRECTIONS

There is solid evidence that cells isolated from term human amnion express stem cell surface markers (see Table X.1) and also express genes normally associated with human embryonic stem cells. However, there are still no reports of efficient and effective protocols to induce the differentiation of hAE to mature cell types, such as liver, pancreas, cardiac, or neural cells. In most reports, only a small fraction of the cells differentiate into the desired cell type. Thus, the reports that cultured hAE cells have characteristics of these differentiated tissues is encouraging enough to promote additional research, however, it remains that no one has reported full differentiation of hAE cells to another mature cell type. The problem may be resolved with additional effort and improved differentiation protocols. However, another possibility exists. The cells isolated from a typical human amnion may be a heterogeneous population of cells. Some cells within this population may have a more primitive phenotype, while others may express more differentiated functions. In an effort to address the possible problem of heterogeneity, we analyzed the population of hAE cells isolated from typical membranes by flow analysis for a number of stem cell surface markers. As reported previously, the results presented in Table X.1 clearly show that there are different populations of cells in the initial cell isolates. While most or all of the cells react with antibodies to stage specific embryonic antigen 4 (SSEA-4) [Miki, 2007], HLA-ABC, integrin alpha 6 or EpCam, only 10–15% of the cells express the other hES markers like SSEA-3 or the TRA antigens. These observations suggest that perhaps within the population of isolated hAE cells there are subpopulations of cells, some as high as 10–15%, that express markers of a more primitive phenotype. It may be that these subpopulations of hAE are the more primitive stem cells in the membrane and when hAE cells are put into unselected-mass culture, it is actually only these more primitive stem cells that differentiate in response to the growth factor and substrate signals received. It is even conceivable that the cells that do not respond to the differentiation signals might inhibit the efficient differentiation of the more primitive stem cells. If this hypothesis is correct, then a useful strategy to improve differentiation of hAE to specific cell types would be to start the differentiation protocols with cells preselected and enriched for those that express the most primitive phenotype. By this reasoning, those cells that express SSEA-3, the TRA antigens or the transport protein ABCG2 might be the most primitive stem cells within the membrane and might be those most likely to differentiate into other cell types. If the differentiation protocols were started with cells enriched for these stem cell markers they might result in cultures that respond to the differentiation signals in a more uniform manner and result in more effective differentiation to the desired cell type, such as liver or pancreas.

A different hypothesis for more efficient differentiation to specific cells types could be based on enrichment for the cell type in the membrane that already expresses the mature gene of interest. As shown in Fig. X.1 freshly isolated hAE cells express a number of genes

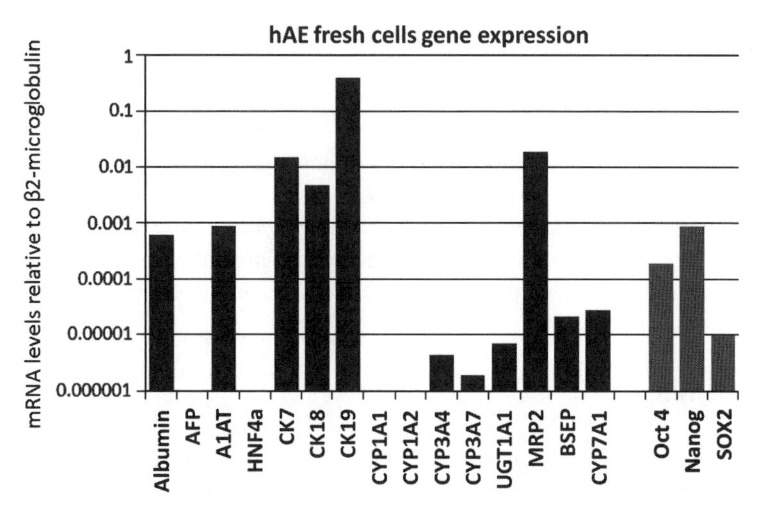

Figure X.1. Gene expression levels of hepatic markers in freshly isolated hAE cells assessed by quantitative RT–PCR. Absolute values were normalized to the internal control β2-microglobulin and expressed as this ratio.

normally found in normal liver such as albumin, CYP 3A4 or UGT1A1. While the level of expression of these genes may be 10,000 times lower than that observed in mature liver, one could hypothesize that at least a subpopulation of hAE cells already express the genes of interest and these cells may be "primed" to hepatic differentiation if additional appropriate signals were received. It would be quite easy to devise a selection strategy based on the expression of these liver genes in a reporter-type assay. For example, if a reporter gene, such as green or yellow fluorescent protein (GFP, YFP), were fused to the promoter sequences for albumin or CYP3A4, and transfected into hAE, only those cells that expressed the authentic albumin or CYP3A4 gene would also express the GFP or YFP protein. These cells could be easily identified, quantified, and even isolated and recovered with FACS sorting procedures. In this manner, cells that respond to the differentiation signals could be identified and separated from those that do not. Such a strategy would likely enrich for cells with a more mature phenotype. The mature cells could be analyzed separately or even placed back into culture for additional functional or transplanted into animals to assess regenerative potential.

CONCLUSIONS

The human amnion contains cells with stem cell characteristics. There are now a number of reports that the hAE cells can differentiate to cells with characteristics of liver, pancreas, cardiac muscle, and neural cells, as well as other tissues. However, to date no efficient methods have been reported to produce highly enriched cultures of a desired cell type. Additionally, when mature gene expression was quantified for a specific cell type, the cultured hAE cells frequently express the mature tissue genes at levels hundreds to thousands of times lower that the corresponding mature tissue. Thus efficient and effective methods to induce differentiation of these cells to the desired cell types must be developed. A number of potentially useful strategies have been proposed to accomplish these goals.

When differentiation protocols are optimized, hAE cells may become an extremely plentiful, useful, and noncontroversial source of stem cells for regenerative medicine.

ACKNOWLEDGMENT

This work was supported in part by a grant from Pfizer Inc.

REFERENCES

Akle CA, Adinolfi M, Welsh KI, Leibowitz S, McColl I. 1981. Immunogenicity of human amniotic epithelial cells after transplantation into volunteers. Lancet. 28254:1003–1005.

Asahina K, Fujimori H, Shimizu-Saito K, Kumashiro Y, Okamura K, Tanaka Y, Teramoto K, Arii S, Teraoka H. 2004. Expression of the liver-specific gene Cyp7a1 reveals hepatic differentiation in embryoid bodies derived from mouse embryonic stem cells. Genes Cells. 912:1297–1308.

Boiani M, Scholer HR. 2005. Regulatory networks in embryo-derived pluripotent stem cells. Nat Rev Mol Cell Biol. 611:872–884.

Bourne GL. 1966. The anatomy of the human amnion and chorion. Proc R Soc Med. 5911 Part 1:1127–1128.

Bryant-Greenwood GD. 1998. The extracellular matrix of the human fetal membranes: Structure and function. Placenta. 191:1–11.

Chambers I, Colby D, Robertson M, Nichols J, Lee S, Tweedie S, Smith A. 2003. Functional expression cloning of Nanog, a pluripotency sustaining factor in embryonic stem cells. Cell. 1135:643–655.

D'Amour KA, Agulnick AD, Eliazer S, Kelly OG, Kroon E, Baetge EE. 2005. Efficient differentiation of human embryonic stem cells to definitive endoderm. Nat Biotechnol. 2312:1534–1541.

D'Amour KA, Bang AG, Eliazer S, Kelly OG, Agulnick AD, Smart NG, Moorman MA, Kroon E, Carpenter MK, Baetge EE. 2006. Production of pancreatic hormone-expressing endocrine cells from human embryonic stem cells. Nat Biotechnol. 2411:1392–1401.

Davila JC, Cezar GG, Thiede M, Strom S, Miki T, Trosko J. 2004. Use and application of stem cells in toxicology. Toxicol Sci. 792:214–223.

de Caestecker M. 2004. The transforming growth factor-beta superfamily of receptors. Cytokine Growth Factor Rev. 151:1–11.

Dua HS, Gomes JA, King AJ, Maharajan VS. 2004. The amniotic membrane in ophthalmology. Surv Ophthalmol. 491:51–77.

Ellis E, Goodwin B, Abrahamsson A, Liddle C, Mode A, Rudling M, Bjorkhem I, Einarsson C. 1998. Bile acid synthesis in primary cultures of rat and human hepatocytes. Hepatology. 272:615–620.

Ellis ECS. 2003. Use of primary human hepatocytes for the elucidation of bile acid synthesis. Department of Medicine. Stockholm, Karolinska Institute.

Elwan MA, Sakuragawa N. 1997. Evidence for synthesis and release of catecholamines by human amniotic epithelial cells. Neuroreport. 816:3435–3438.

Gruss JS, Jirsch DW. 1978. Human amniotic membrane: A versatile wound dressing. Can Med Assoc J. 11810:1237–1246.

Hao Y, Ma DH, Hwang DG, Kim WS, Zhang F. 2000. Identification of antiangiogenic and anti-inflammatory proteins in human amniotic membrane. Cornea. 193:348–352.

Ilancheran S, Michalska A, Peh G, Wallace EM, Pera M, Manuelpillai U. 2007. Stem Cells Derived from Human Fetal Membranes Display Multi-Lineage Differentiation Potential. Biol Reprod. 77:577–588.

Kakishita K, Elwan MA, Nakao N, Itakura T, Sakuragawa N. 2000. Human amniotic epithelial cells produce dopamine and survive after implantation into the striatum of a rat model of Parkinson's disease: A potential source of donor for transplantation therapy. Exp Neurol. 1651:27–34.

Kakishita K, Nakao N, Sakuragawa N, Itakura T. 2003. Implantation of human amniotic epithelial cells prevents the degeneration of nigral dopamine neurons in rats with 6-hydroxydopamine lesions. Brain Res. 9801:48–56.

Magatti M, De Munari S, Vertua E, Gibelli L, Wengler GS, Parolini O. 2008. Human amnion mesenchyme harbors cells with allogeneic T-cell suppression and stimulation capabilities. Stem Cells. 261:182–192.

McLean AB, D'Amour KA, Jones KL, Krishnamoorthy M, Kulik MJ, Reynolds DM, Sheppard AM, Liu H, Xu Y, Baetge EE, Dalton S. 2007. Activin a efficiently specifies definitive endoderm from human embryonic stem cells only when phosphatidylinositol 3-kinase signaling is suppressed. Stem Cells. 251:29–38.

Meller D, Maskin SL, Pires RT, Tseng SC. 2000. Amniotic membrane transplantation for symptomatic conjunctivochalasis refractory to medical treatments. Cornea. 196:796–803.

Meller D, Pires RT, Mack RJ, Figueiredo F, Heiligenhaus A, Park WC, Prabhasawat P, John T, McLeod SD, Steuhl KP, Tseng SC. 2000. Amniotic membrane transplantation for acute chemical or thermal burns. Ophthalmology 1075:980–990.

Miki T, Lehmann T, Cai H, Stolz DB, Strom SC. 2005. Stem cell characteristics of amniotic epithelial cells. Stem Cells. 2310:1549–1559.

Miki T, Marongiu F, Ellis E, Strom SC. 2007. Isolation of Amniotic Epithelial Cells. Current Protocols in Stem Cell Biol. 1E.3:1E.3.1–1E.3.9.

Miki T, Marongiu F, Ellis ECS, Dorko K, Mitamura K, Ranade A, Gramignoli R, Davila J, Strom SC. 2008. Production of Hepatocyte-Like Cells from Human Amnion. In: Dhawan A and Hughes RD. Hepatocyte Transplantation. Dhawan A and Hughes RD Humana Press.

Miki T, Mitamura K, Ross MA, Stolz DB, Strom SC. 2007. Identification of stem cell marker-positive cells by immunofluorescence in term human amnion. J Reprod Immunol. 752:91–96.

Miki T, Strom SC. 2006. Amnion-derived pluripotent/multipotent stem cells. Stem Cell Rev. 22:133–142.

Nichols J, Zevnik B, Anastassiadis K, Niwa H, Klewe-Nebenius D, Chambers I, Scholer H, Smith A. 1998. Formation of pluripotent stem cells in the mammalian embryo depends on the POU transcription factor Oct4. Cell. 953:379–391.

Niwa H. 2007. How is pluripotency determined and maintained? Development. 1344:635–646.

Nubile M, Carpineto P, Lanzini M, Ciancaglini M, Zuppardi E, Mastropasqua L. 2007. Multilayer amniotic membrane transplantation for bacterial keratitis with corneal perforation after hyperopic photorefractive keratectomy: Case report and literature review. J Cataract Refract Surg. 339:1636–1640.

Nubile M, Dua HS, Lanzini TE, Carpineto P, Ciancaglini M, Toto L, Mastropasqua L. 2008. Amniotic membrane transplantation for the management of corneal epithelial defects: An in vivo confocal microscopic study. Br J Ophthalmol. 921:54–60.

Pan G, Li J, Zhou Y, Zheng H, Pei D. 2006. A negative feedback loop of transcription factors that controls stem cell pluripotency and self-renewal. Faseb J. 2010:1730–1732.

Parolini O, Alviano F, Bagnara GP, Bilic G, Buhring HJ, Evangelista M, Hennerbichler S, Liu B, Magatti M, Mao N, Miki T, Marongiu F, Nakajima H, Nikaido T, Portmann-Lanz CB, Sankar V, Soncini M, Stadler G, Surbek D, Takahashi TA, Redl H, Sakuragawa N, Wolbank S, Zeisberger S, Zisch A, Strom SC. 2008. Concise review: Isolation and characterization of cells from human term placenta: Outcome of the first international Workshop on Placenta Derived Stem Cells. Stem Cells. 262:300–311.

Ravishanker R, Bath AS, Roy R. 2003. "Amnion Bank"—the use of long term glycerol preserved amniotic membranes in the management of superficial and superficial partial thickness burns. Burns. 294:369–374.

Robson MC, Krizek TJ. 1973. The effect of human amniotic membranes on the bacteria population of infected rat burns. Ann Surg. 1772:144–149.

Robson MC, Krizek TJ, Koss N, Samburg JL. 1973. Amniotic membranes as a temporary wound dressing. Surg Gynecol Obstet. 1366:904–906.

Rodgarkia-Dara C, Vejda S, Erlach N, Losert A, Bursch W, Berger W, Schulte-Hermann R, Grusch M. 2006. The activin axis in liver biology and disease. Mutat Res. 6132-3:123–137.

Sakuragawa N, Enosawa S, Ishii T, Thangavel R, Tashiro T, Okuyama T, Suzuki S. 2000. Human amniotic epithelial cells are promising transgene carriers for allogeneic cell transplantation into liver. J Hum Genet. 453:171–176.

Soto-Gutierrez A, Kobayashi N, Rivas-Carrillo JD, Navarro-Alvarez N, Zhao D, Okitsu T, Noguchi H, Basma H, Tabata Y, Chen Y, Tanaka K, Narushima M, Miki A, Ueda T, Jun HS, Yoon JW, Lebkowski J, Tanaka N, Fox IJ. 2006. Reversal of mouse hepatic failure using an implanted liver-assist device containing ES cell-derived hepatocytes. Nat Biotechnol. 2411:1412–1419.

Takashima S, Ise H, Zhao P, Akaike T, Nikaido T. 2004. Human amniotic epithelial cells possess hepatocyte-like characteristics and functions. Cell Struct Funct. 293:73–84.

Tamagawa T, Ishiwata I, Saito S. 2004. Establishment and characterization of a pluripotent stem cell line derived from human amniotic membranes and initiation of germ layers in vitro. Hum Cell. 173:125–130.

Valdimarsdottir G, Mummery C. 2005. Functions of the TGFbeta superfamily in human embryonic stem cells. Apmis. 11311-12:773–789.

van Herendael BJ, Oberti C, Brosens I. 1978. Microanatomy of the human amniotic membranes. A light microscopic, transmission, and scanning electron microscopic study. Am J Obstet Gynecol. 1318:872–880.

Wei JP, Zhang TS, Kawa S, Aizawa T, Ota M, Akaike T, Kato K, Konishi I, Nikaido T. 2003. Human amnion-isolated cells normalize blood glucose in streptozotocin-induced diabetic mice. Cell Transplant 125:545–552.

Xiao L, Yuan X, Sharkis SJ. 2006. Activin A maintains self-renewal and regulates fibroblast growth factor, Wnt, and bone morphogenic protein pathways in human embryonic stem cells. Stem Cells. 246:1476–1486.

11

HUMAN UMBILICAL VEIN ENDOTHELIAL CELLS AND HUMAN DERMAL MICROVASCULAR ENDOTHELIAL CELLS OFFER NEW INSIGHTS INTO THE RELATIONSHIP BETWEEN LIPID METABOLISM, ANGIOGENESIS, AND ABDOMINAL AORTIC ANEURYSM

Ho-Jin Park, Yali Zhang, Jack Naggar, Serban P. Georgescu, Dequen Kong, and Jonas B. Galper

Molecular Cardiology Research Institute, Department of Medicine, Tufts New England Medical Center, Boston, MA 02111

HUMAN UMBILICAL VEIN ENDOTHELIAL CELLS

The human umbilical cord has become one of the most important sources of vascular endothelial cells. The availability of these cells has played a major role in the development of the field of vascular biology [Bevilacqua and Gimbrone, 1987; Davies, 1997; Libby, 2000; Yamada, 1992]. Perfusion of the human umbilical cord vein with collagenase results in a pure preparation of the single layer of endothelial cells that line this vessel [Gimbrone, 1976]. Initial passages of these cells, which are grown in the presence of heparin and pituitary extract, maintain nearly all of the features of native vascular endothelial cells including the expression of endothelial cell specific markers, such as von Willebrand factor and an endothelial specific adhesion molecule, CD31, expression of receptors for growth factors, cytokines, vasoactive ligands, and specific signaling pathways for vascular endothelial growth factor (VEGF), fibroblast growth factor (FGF), transforming growth factor β (TGFβ), tumor necrosis factor α (TNFα), and angiotensin II [Goldberger, 1994;

Perinatal Stem Cells. Edited by C. L. Cetrulo, K. J. Cetrulo, and C. L. Cetrulo, Jr.

Muscella, 1997; Namiki, 1995; Nozawa, 2001]. Human umbilical vein endothelial cells (HUVECs) have provided a critical *in vitro* model for major breakthroughs in molecular medicine including seminal insights into cellular and molecular events in the pathophysiology of atherosclerosis and plaque formation, and mechanisms for the control of angiogenesis or neovascularization in response to hypoxia and inflammation in tumors, ischemic tissue and in embryogenesis [Burns, 2005; Kokura, 1999; Zhang, 2003].

Monolayers of HUVECs have been used for the study of the interaction of leukocytes and macrophages with the endothelial cell layer in vascular tissues resulting in the discovery of adhesion molecules, chemokines, and kinases that mediate the interaction of inflammatory cells with the endothelial surface and their migration into the media [Bevilacqua, 1989]. Monolayers of HUVECs have been generated on deformable surfaces or in chambers that allow the study of the effects of shear stress and pulsitile flow on cell signaling in order to reproduce the effects of blood flow on endothelial cell function *in vivo*. These monolayers have been used to identify transcription factors, such as KLF2, that regulate the expression of adhesion molecules, such as vascular cell adhesion molecule-1 (VCAM-1) and endothelial adhesion molecule E-selectin in response to stress and proinflammatory cytokines, such as TNFα, that mediate changes in cell adhesion and migration that play a role in the early changes of atherosclerosis [Dai, 2004; Parmar, 2005, 2006].

ROLE OF ANGIOGENESIS IN THE PATHOPHYSIOLOGY OF DISEASE

Angiogenesis plays a role in the pathophysiology of atherosclerosis, rheumatoid arthritis, diabetic retinopathy, psoriasis, and tumor growth [Kumar, 2001] might also play a role in the development of aortic aneurysms [Daugherty, 2004]. Angiogenesis may also play a therapeutic role in the response to hypoxia in occlusive vascular disease [Yoon, 2004]. Angiogenesis involves the proliferation, migration, and remodeling of endothelial cells in the process of tube formation. Remodeling and migration involves in part the secretion of metalloproteinases by endothelial cells [Arenas, 2004; Herron, 1986]. Studies of the pathophysiology of AAA have demonstrated that human aneurysmal tissues are characterized by (1) chronic inflammation of the aortic wall with the accumulation of macrophages [Freestone, 1995]; (2) progressive degradation of extracellular matrix including elastin and collagen [Dobrin, 1984]; (3) increased activity of matrix metalloproteinases (MMPs), particularly MMP-2 and MMP-9 [Freestone, 1995]; (4) reendothelialization of the dilated luminal surface of the vessel wall and pronounced neovascularization of the media and adventitia [Choke, 2006; Herron, 1991].

The HUVECs have offered an important *in vitro* model for the study of the role of angiogenesis in these processes. Culture of HUVECs on Matrigel, an extract of endothelial basement membrane, results in the formation of honeycomb-like structures that simulate tube formation by endothelial cells *in vivo* [Nagata, 2003]. Study of the effects of hypoxia, VEGF, FGF, and TGFβ and inhibitors of angiogenesis on the formation and organization of honeycombs by HUVECs have helped elucidate the mechanisms by which these factors regulate both pathologic and therapeutic angiogenesis and the role of small GTP binding proteins in this process [Nagata, 2003; Namiki, 1995]. An important example has been the demonstration that 3-hydroxy-3-methylglutaryl (HMG)–CoA (cofactor A) reductase inhibitors, cholesterol lowering drugs, exert a dose-dependent effect on honeycomb formation and signaling pathways in HUVECs suggesting that under appropriate conditions and at different concentrations, statins might both decrease the progression of atherosclerosis and stimulate the revascularization of ischemic tissues via an effect on angiogenesis [Park, 2002; Weis, 1990]. A second model system for the study of angiogenesis has been the growth of human dermal microvascular endothelial cells, HDMECs, derived from the foreskins of newborns. When cultured on a thick collagen gel these cells form tube-like structures that penetrate the collagen surface resembling capillary-like structures.

HMG–CoA REDUCTASE INHIBITORS HAVE THERAPEUTIC EFFECTS THAT ARE BOTH DEPENDENT AND INDEPENDENT OF CHOLESTEROL LOWERING

3-Hydroxy-3-methylglutaryl–coenzyme A reductase inhibitors, commonly referred to as "statins," are in wide use for the treatment of hypercholesterolemia [Goldstein, 1990; Grundy, 1998], which often coexists with vascular disease and other conditions, such as diabetic retinopathy, whose pathogenesis may be dependent on angiogenesis. For this reason, intense interest has developed in studying the effect of HMG–CoA reductase inhibitors on angiogenesis. Here we review some of our recent studies of the effect of statins on angiogenesis and their role in the treatment and prevention in a mouse model for abdominal aortic aneurysm.

The HMG–CoA reductase inhibitors decrease low-density lipoproteins (LDL) cholesterol by inhibiting the rate-limiting enzyme in cholesterol biosynthesis [Goldstein, 1990; Grundy, 1998]. Increasing evidence suggests that cholesterol-lowering alone does not account for the therapeutic effects of statins in the prevention of atherosclerosis [Massy, 1996; Sacks, 1998]. Thus HMG–CoA reductase inhibitors have been shown to increase the production of NO and inhibit vascular smooth muscle cell proliferation both of which might interfere with atherogenesis [Laufs, 1998]. It has been suggested that these effects involve the inhibition of the post-translational lipidation of small guanosine triphosphate (GTP) binding proteins, such as Ras and RhoA. The HMG–CoA reductase inhibitors interfere with the biosynthesis of farnesylpyrophosphate (FPP), which is not only a precursor to cholesterol, but is also required for the post-translational lipidation of Ras. The FPP condenses with isopentenylpyrophosphate, whose synthesis is also blocked by HMG–CoA reductase inhibitors, to form geranylgeranyl-pyrophosphate (GGPP), which is required for the post-translational lipidation, membrane localization, and function of RhoA.

STUDIES OF THE EFFECTS OF STATINS ON ANGIOGENESIS USING BOTH *IN VIVO* MODELS AND *IN VITRO* HUVEC AND HDMEC MODELS FOR ANGIOGENESIS

Simvastatin, an HMG–CoA Reductase Inhibitor, Interferes with Growth Factor Stimulated Angiogenesis *In Vivo*

In a mouse corneal pocket assay for angiogenesis, in which a pellet is inserted into a small pocket in the avascular cornea, inclusion of FGF-2 in the pellet stimulated capillary growth into the cornea from the surrounding limbic vessels. Microscopic examination demonstrated that FGF-2 resulted in the marked proliferation of tiny capillaries. Addition of Simvastatin to the pellet markedly suppressed new blood vessel growth, as demonstrated by a dose-dependent disappearance of the new blood vessels in corneas treated with FGF-2 plus various concentrations of simvastatin.

In a chick chorioallantoic membrane (CAM) assay for angiogenesis, in which a matrix-polymer containing VEGF is placed on the CAM, capillary growth was stimulated markedly over control. While Simvastatin alone had no effect on the basal level of capillary vessel formation, the incorporation of VEGF plus Simvastatin into the matrix-polymer suppressed VEGF stimulated capillary growth in a dose-dependent manner. Hence, Simvastatin inhibited capillary blood vessel formation in response to two different growth factors in two *in vivo* angiogenesis assays.

Effect of HMG–CoA Inhibitors on Angiogenesis in *In Vitro* Models

In order to study the mechanism of the effect of HMG–CoA reductase inhibitors on angiogenesis, two *in vitro* assays were used: a three-dimensional (3D)-collagen gel assay in which human dermal microvascular endothelial cells (HDMECs) are cultured on a thick collagen matrix and a Matrigel assay in which HUVECs are cultures on Matrigel-coated cultures dishes. In the presence of FGF, PMA, and VEGF; HDMECs became elongated, invaded the collagen gel and differentiated to form tubes that could be visualized by focusing below the surface of the gel (Fig. XI.1a, cf. panels 1 and 2). Addition of various doses of Simvastatin to the growth medium resulted in a dose-dependent inhibition of tube formation with a decrease of $30 \pm 7\%$ ($N = 24$, $p < 0.05$) at 0.05 μM, a decrease of $62 \pm 5\%$ ($N = 24$, $p < 0.01$) at 0.1 μM, and complete inhibition of tube formation at 0.5-μM Simvastatin (Fig. XI.1a, panels 2–4). As discussed above, plating of HUVECs on Matrigel resulted in the formation of honeycomb-like structures (Fig. XI.2, panel 1). Incubation of these cultures with increasing concentrations of Simvastatin resulted in the disruption of the honeycombs and the accumulation of single cells within the honeycomb structure (Fig. XI.2, panel 2). One explanation for this effect of Simvastatin on endothelial cell function might be related to a pro-apoptotic effect of Simvastatin on endothelial cells. Although high doses of Simvastatin resulted in apoptosis of both HDMECs and HUVECs (see Fig. XI.1c), comparison of the dose response curves for Simvastatin inhibition of tube formation and honeycomb formation and the pro-apoptotic effect of Simvastatin on endothelial cells demonstrate that inhibition of angiogenesis occurred at a significantly lower concentration of Simvastatin than that at which apoptosis was observed (Fig. XI.1, cf. panels b and c). Thus 0.5-μM Simvastatin, a concentration that inhibited tube formation completely, increased apoptosis only slightly above basal level.

Mechanism of the Antiangiogenic Effect of Simvastatin. Since HMG–CoA reductase inhibitors interfere with the biosynthesis of farnesylpyrophosphate (FPP) and geranylgeranyl-pyrophosphate (GGPP), and small GTP binding proteins, such as Ras and Rho depend on farnesylation and geranylgeranylation in order to localize to the membrane, one mechanism by which HMG–CoA reductase inhibitors might inhibit angiogenesis would be via the inhibition of post-translational protein lipidation. To test this hypothesis, the effect of exogenously added FPP and GGPP on Simvastatin inhibition of tube formation by HDMECs and honeycomb formation by HUVECs were studied. Incubation of cells with Simvastatin plus FPP demonstrated no effect on Simvastatin inhibition of tube formation by HDMECs (Fig. XI.1a, cf. panels 4 and 5; Fig. XI.1d, lanes 2 and 3). However, incubation of cells with Simvastatin plus GGPP reversed Simvastatin inhibition of tube formation to $80 \pm 9\%$ ($N = 24$, $p < 0.01$) of control (Fig. XI.2a, cf. panels 2, 4, and 6; Fig. XI.1d, lanes 1, 2, and 4).

Similarly, in the honeycomb assay FPP treatment of HUVECs had no effect on the disruption of honeycomb formation by HUVECs seen in cells cultured in Simvastatin

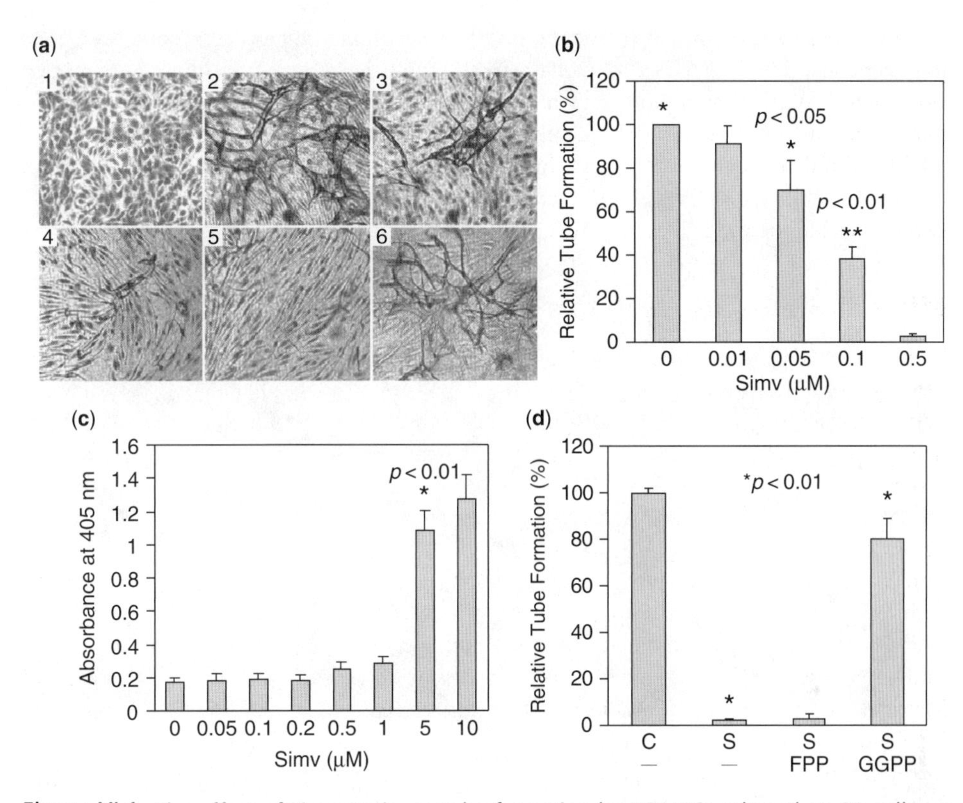

Figure XI.1. The effect of Simvastatin on tube formation by HDMECs cultured on 3D collagen gels. (a) The HDMECs cultured with (1) FGF (40 ng/mL) alone; (2) FGF, PMA (50 ng/mL), and VEGF (40 ng/mL); FGF, PMA, and VEGF, plus (3) 0.1-μM Simvastatin; (4) 0.5-μM Simvastatin; (5) 0.5-μM Simvastatin plus 10-μM FPP; (6) 0.5-μM Simvastatin plus 10-μM GGPP added at the time of plating and incubation continued for 4–5 days. Photomicrographs are typical of four independent experiments each carried out in triplicate. Note the presence of elongated endothelial cells in the plane above the tubes. (b) Dose response curve for Simvastatin inhibition of tube formation. Cells were incubated as in Fig. XI.1a with growth factors and the indicated concentrations of Simvastatin. Data are the mean of the total number of branch points in eight low power fields (100×) per well from three replicate wells. Data are presented as a percent of control and are typical of four similar experiments. (c) The HDMECs were cultured on gelatin-coated dishes treated with various concentrations of Simvastatin for 5 days and cell death was quantitated by measuring cytoplasmic histone- associated DNA fragments. (d) Quantitation of the effect of FPP and GGPP on Simvastatin inhibition of tube formation. C = control; S = 0.5-μM Simvastatin; S + FPP (10 μM) and S + GGPP (10 μM). Data are the mean of tube branch points in eight fields from each of three replicate wells similar to those in Fig. XI.1a panels 2, 4, 5, and 6, respectively, plotted as percentage of control. Data are typical of four independent studies. (Figure reprinted with the permission of Circulation Research.)

(Fig. XI.2, panel 3). However, culture of HUVECs in Simvastatin plus GGPP resulted in complete reversal of the effects of Simvastatin on honeycomb formation (Fig. XI.2, panel 4). These data supported the conclusion that Simvastatin might interfere with angiogenesis via the inhibition of the protein geranylgeranylation while an effect on protein farnesylation appeared not to be involved.

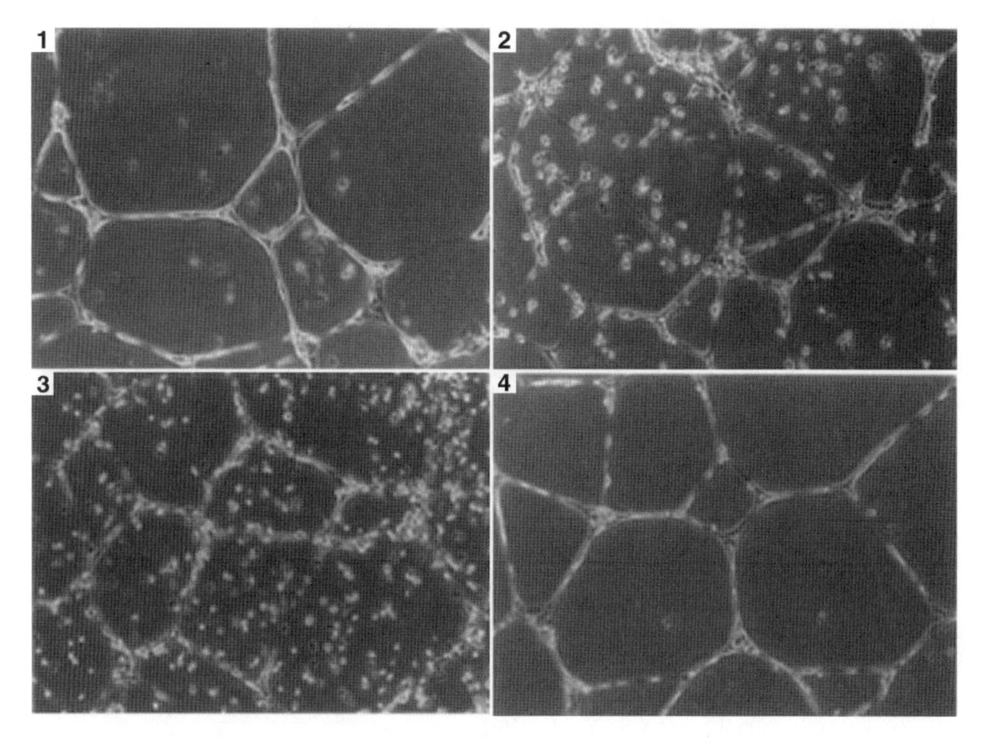

Figure XI.2. Simvastatin inhibits the tube formation of HUVECs on Matrigel and Simvastatin effect is reversed by GGPP, not by FPP. The HUVECs were cultured on Matrigel with (1) Control media; (2) Simvastatin (1 μM); (3) Simvastatin plus FPP (10 μM); (4) Simvastatin plus GGPP (10 μM).

Effect of Inhibitors of Protein Lipidation on Tube Formation by HDMECs and Honeycomb Formation by HUVECs

The conclusion that Simvastatin interfered with tube formation via the inhibition of protein geranylgeranylation was further supported by the finding that FTI-277, a specific inhibitor of protein farnesylation by FPP [Lerner, 1995] had no effect on tube formation by HDMECs (Fig. XI.3a, cf. panels 2 and 4; Fig. XI.3b, cf. lanes 1 and 2). However, GGTI-298, a specific inhibitor of protein geranylgeranylation by GGPP [Vogt, 1996] inhibited tube formation by $70 \pm 5\%$ ($N = 24$, $p < 0.01$) compared to cells cultured under control conditions (Fig. XI.3a, cf. panels 2 and 5; Fig. XI.3b, cf. lanes 1 and 3). Finally, C3 exotoxin, which interferes with the function of Rho GTPases by catalyzing their ADP ribosylation [Ernst, 2000] markedly inhibited tube formation (Fig. XI.3a, panel 6; Fig. XI.3b, lane 4). Similarly in the Matrigel assay for honeycomb formation by HUVECs FTI had no significant effect on honeycomb structure (Fig. XI.4, cf. panels 1 and 3) while GGTI caused disruption of the continuity of the honeycomb wall and an accumulation of individual cells within the honeycombs similar to the effect of Simvastatin (Fig. XI.4, panel 4). Treatment of HUVECs with C3 exotoxin also interfered with the formation of individual honeycombs (cf. Fig. XI.4, panels 1 and 5). These data support the conclusion that tube formation by HDMECs was dependent on Rho and that HMG–CoA reductase inhibitors interfered with angiogenesis via the inhibition of the geranylgeranylation of Rho.

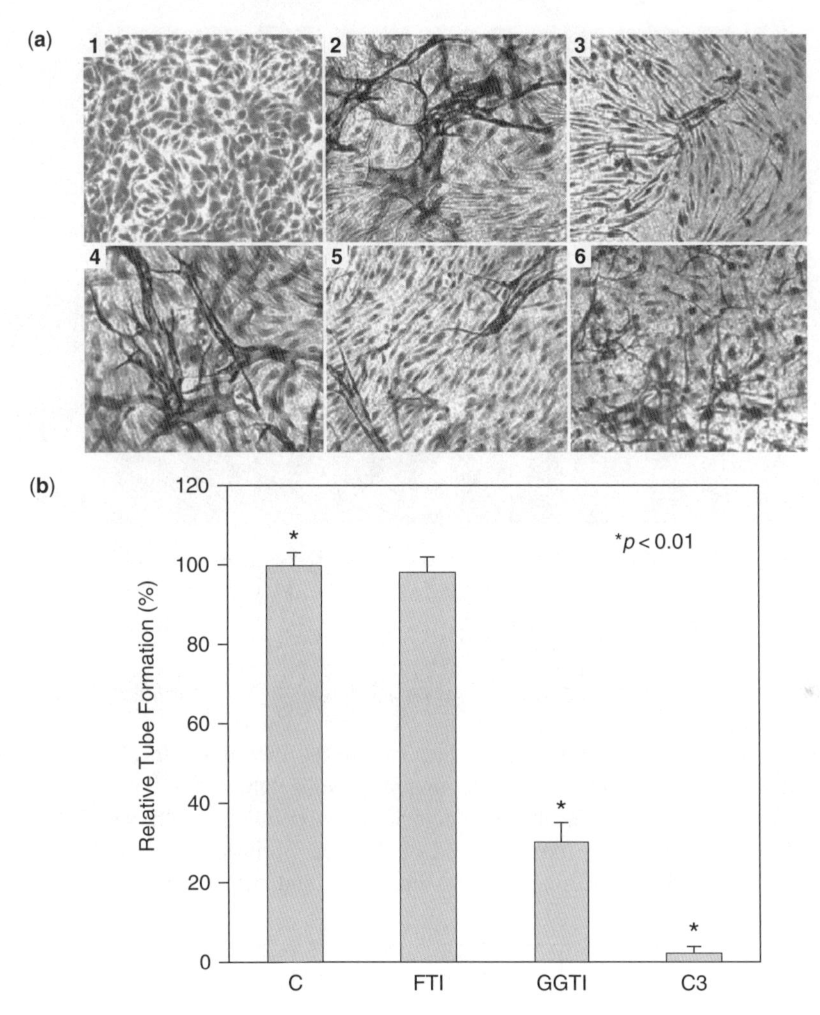

Figure XI.3. Effect of FTI, GGTI, and C3 exotoxin on tube formation. (a) The HDMECs cultured: (1) with FGF alone; (2) with growth factors as in Fig. XI.1a; with growth factors plus (3) 0.5-μM Simvastatin; (4) 10-μM FTI; (5) 10-μM GGTI; (6) 0.5-μg/mL C3 exotoxin. Photomicrographs are typical of three independent experiments each carried out in triplicate. (b) Quantitation of the effect of FTI and GGTI on tube formation. Data are the mean number of branch points in eight low power fields in each of three replicate wells similar to those in Fig. XI.3a panels 2, 4, 5, and 6, respectively, plotted as percentage of control. Data are typical of three similar studies. (Figure reprinted with the permission of Circulation Research.)

Effect of Simvastatin on the Membrane Localization of RhoA

As discussed previously, small GTP binding proteins must be associated with the membrane in order to function. The post-translational lipidation of small GTP binding proteins is critical for their localization to the membrane. In order to determine whether there was a correlation between Simvastatin inhibition of tube formation by HDMECs and honeycomb formation by HUVECs and the inhibition of the membrane localization of Rho, HDMECs were incubated under the same culture conditions and over the same time course as in the

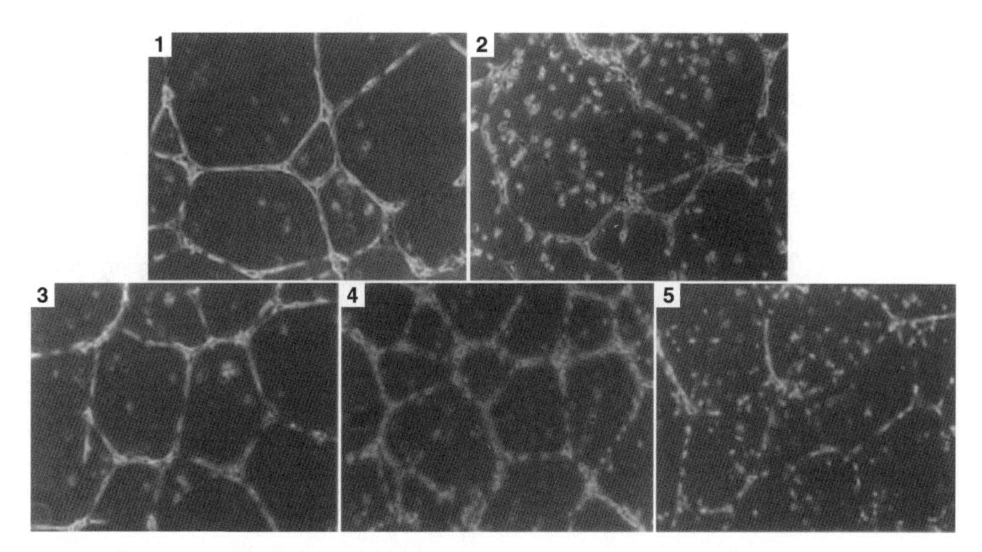

Figure XI.4. GGTI and C3 toxin mimics the effect of Simvastatin on the tube formation of HUVECs on Matrigel. (1). Control; (2). Simvastatin (1 μM); (3). FTI (1 μM, $IC_{50} = 0.2$ μM); (4). GGTI (10 μM, $IC_{50} = 3$ μM); (5). C3 exotoxin (0.5 μg/mL).

tube-formation assays in Fig. XI.1a. The effect of increasing concentrations of Simvastatin on the distribution of RhoA between the membrane and cytoplasm were determined. Simvastatin mediated a dose-dependent decrease in the membrane localization of RhoA and a parallel dose-dependent increase in cytoplasmic RhoA (Fig. XI.5). This effect was significant at 0.1-μM Simvastatin, a concentration at which tube formation was inhibited by 62 ± 5%, and even more marked at 0.5-μM Simvastatin, a concentration that completely inhibited tube formation (see Fig. XI.1) and the formation of honeycombs

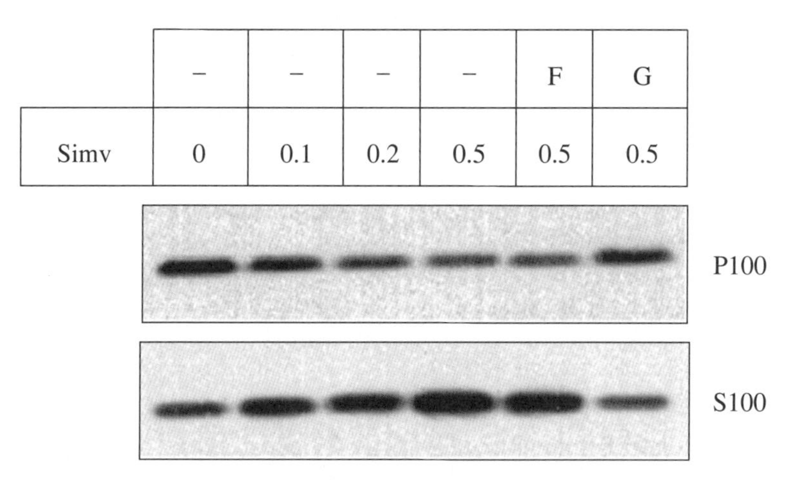

Figure XI.5. Effect of Simvastatin on the membrane localization of RhoA. HDMECs were cultured in medium 2% in serum with growth factors and treated with the indicated concentrations of Simvastatin for 5 days. FPP or GGPP (10 μM) were added 2 h prior to harvest and RhoA determined in S100 and P100 fractions by SDS-PAGE followed by immunoblot analysis using an anti-RhoA antibody. F, FPP; G, GGPP. These data are typical of three independent experiments. (Figure reprinted with the permission of Circulation Research.)

(see Fig. XI.2). Furthermore, GGPP, which reversed the effect of Simvastatin on tube formation and honeycomb formation, completely reversed the effect of Simvastatin on the membrane localization of RhoA while FPP, which had no effect on Simvastatin disruption of tube formation and the formation of honeycombs, had no effect on the localization of RhoA (Fig. XI.5).

Effect of a Dominant Negative (DN)–RhoA Mutant on Angiogenesis In Vitro. To establish the relationship between the inhibition of RhoA function by Simvastatin and Simvastatin inhibition of honeycomb formation, HUVECs were coinfected with an adenovirus expressing a DN–RhoA mutant under the control of a tetracycline-regulated transactivator (tTA) and an adenovirus constitutively expressing the transactivator tTA. The effect on honeycomb formation was determined. To demonstrate the specificity of the effect of tetracycline on gene expression using this system, HUVECs were plated on collagen-coated culture dishes and coinfected with an adenovirus in which GFP expression was under the control of a tetracycline-regulated promoter and an adenovirus expressing the transactivator tTA. Fluorescence microscopy demonstrated GFP expression in these cells (Fig. XI.6a, panels 1 and 3), which was reversed by the addition of doxycycline (Fig. XI.6a, panel 2 and 4). By using this system, HUVECs cultured on Matrigel were infected with an adenovirus expressing a DN–RhoA mutant plus the transactivator tTA. In the presence of doxycycline (expression turned off) the DN–RhoA had no effect on honeycomb formation (Fig. XI.6b, panel 2) while in the absence of doxycycline honeycomb formation was disrupted (Fig. XI.6b, panel 3). Similarly, infection of HDMECs with the adenovirus constitutively expressing the transactivator alone had no effect on tube formation (Fig. XI.7a, panels 1 and 2; Fig. XI.7c, columns 1 and 2). Infection with the adenovirus expressing the DN–RhoA mutant alone decreased tube formation by 25 $\pm$ 4% ($N = 15$, $p < 0.01$) compared to control (Fig. XI.7a, panels 2 and 3; Fig. XI.7c, columns 2 and 3). However, coinfection of cells with both the adenoviruses expressing the DN–RhoA and the transactivator decreased tube formation by 80 $\pm$ 6% ($N = 15$, $p < 0.01$) compared to control cells (Fig. XI.7a, panels 2 and 4; Fig. XI.7c, columns 2 and 4). Under these conditions the DN–RhoA was expressed at high levels (Fig. XI.7b). The small decrease in tube formation in cells infected with the adenovirus expressing the DN–RhoA mutant alone should be due to basal expression of the DN–RhoA mutant [Gossen, 1992].

Biphasic Effect of HMG–CoA Reductase Inhibitors on Angiogenesis

Recently, several studies have demonstrated that HMG–CoA reductase inhibitors have a biphasic effect on angiogenesis. Using both an *in vivo* wound healing model and honeycomb formation by adult human microvascular endothelial cells, low doses of the HMG–CoA reductase inhibitor atorvastatin, 0.005–0.01 μM, stimulated angiogenesis and pro-angiogenic signaling pathways, while higher doses >0.01-μM Atorvastatin interfered with angiogenesis [Urbich, 2002; Weis, 2002]. The antiangiogenic effects were observed at concentrations of Atorvastatin, which decreased lipid levels similar to those used in our studies presented above. The relationship between the antiangiogenic effects of statins and possible therapeutic pro-angiogenic effects of statins in revascularization in peripheral vascular disease and coronary disease remains the subject of continuing investigation.

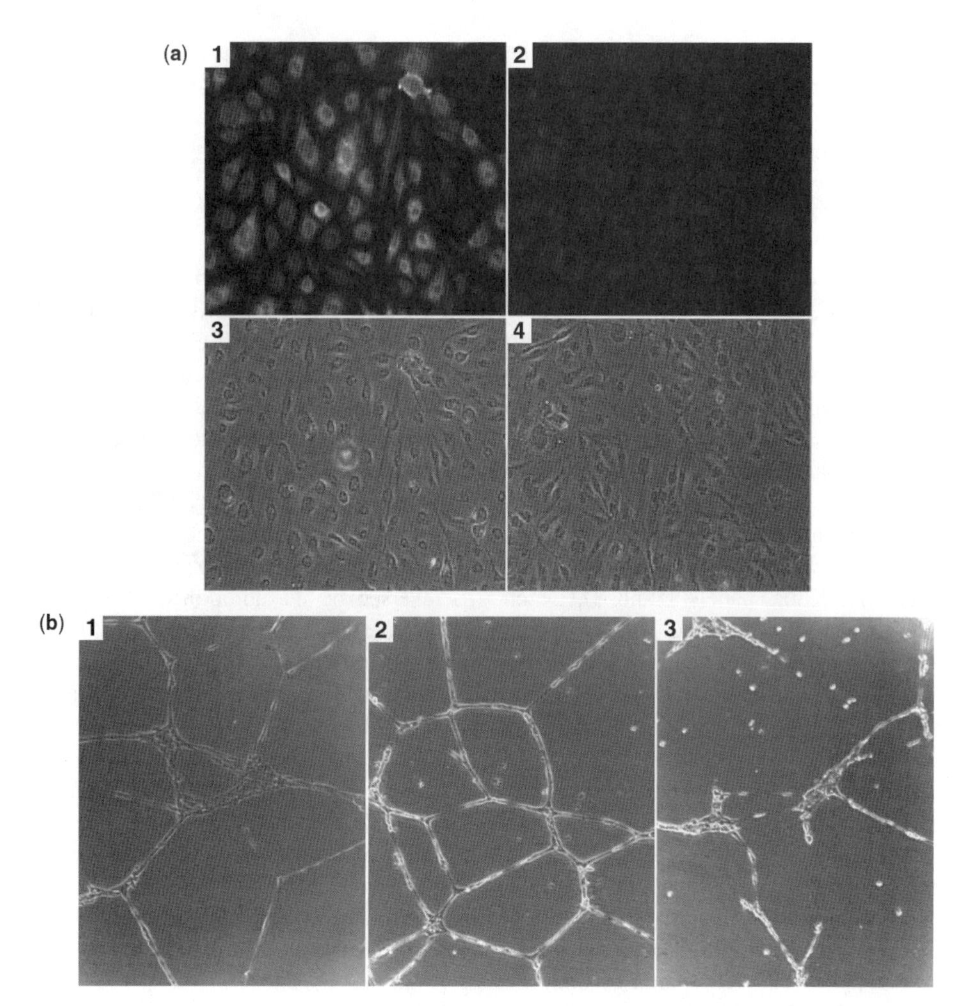

Figure XI.6. Adenoviral expression of DN-RhoA mimics the effect of Simvastatin on honeycomb formation of HUVECs on Matrigel. (a) The HUVECs infected with Ad-tetO-GFP and Ad-tTA in the absence (panels 1 and 3) or presence (panels 2 and 4) of doxycycline (2 ng/mL); Panels 1 and 2, fluorescence photomicrographs; Panels 3 and 4, light photomicrographs. (b) Tube formation of HUVECs on Matrigel infected with Ad-tetO-DN-RhoA and Ad-tTA. (1). Ad-Control; (2). Ad-tetO-DN-RhoA with doxycycline (expression off); (3). Ad-tetO-DN-RhoA without doxycycline (expression on).

Role of Angiotensin II in Atherosclerosis and Aneurysm Formation

Hypertension is a significant risk factor for the development of abdominal aortic aneurysm (AAA) formation. However, the rennin angiotensin system not only plays a role in the development of hypertension, but also stimulates molecular and cellular processes that may play a role in the inflammation and neovascularization associated with atherosclerosis and aneurysm formation independent of its effect on blood pressure [Mehta, 2007]. AngII has been shown to activate macrophages in the kidney [Ozawa, 2007] and stimulate the secretion of MMPs in AAA [Daugherty, 2004]. AngII also has been shown to stimulate neovascularization both in *in vitro* and *in vivo* models for angiogenesis [Richard, 2001]. Statins have

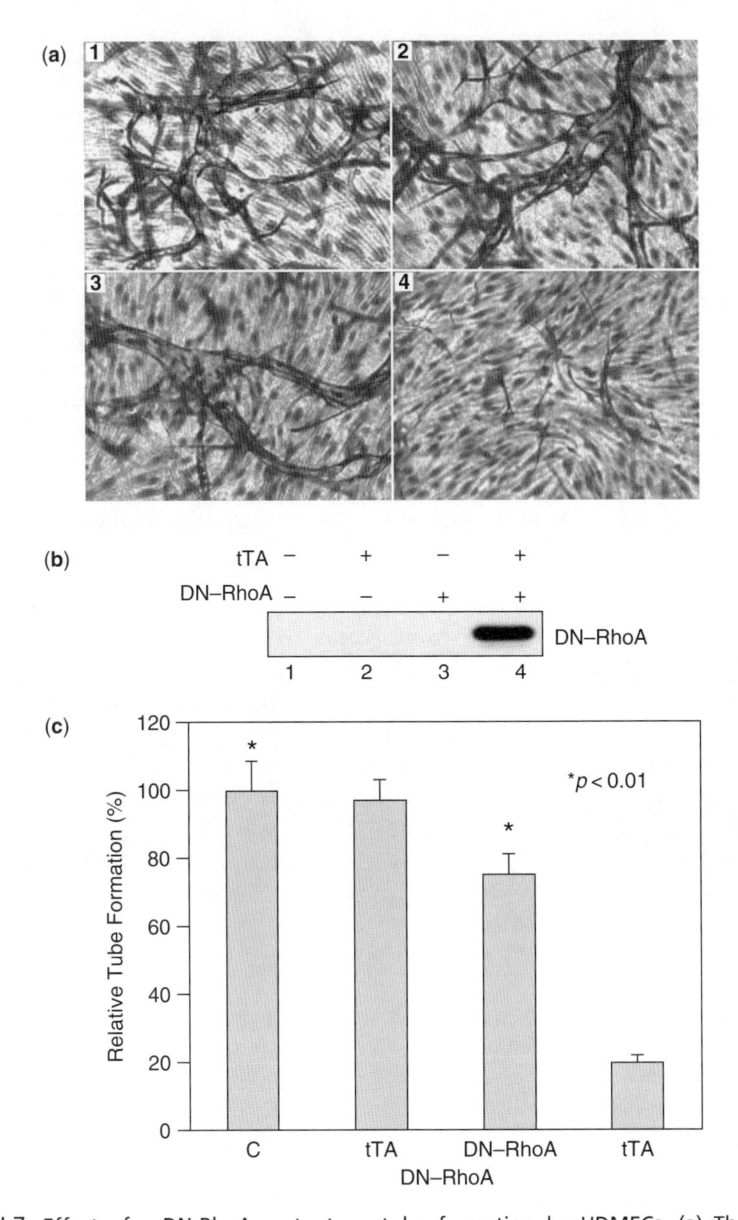

Figure XI.7. Effect of a DN-RhoA mutant on tube formation by HDMECs. (a) The HDMECs plated on 3D-collagen gels (1) control; infected with an adenovirus expressing (2) a tetracycline-controlled transactivator, (3) a DN–RhoA, (4) coinfected with adenoviruses expressing a transactivator and a DN–RhoA. Cells were incubated, fixed, and stained and tube formation was determined as described above. Photomicrographs are typical of four similar experiments. (b) Immunoblot analysis of DN–RhoA expression. Cell extracts from infected cells were analyzed by using an anti-RhoA antibody. The tTA denotes a tetracycline transactivator. These data are typical of three independent experiments. (c) Quantitation of the effects of a DN–RhoA mutant on tube formation by HDMECs. Data are presented as the mean number of branch points per five low-power fields from each of three replicate determinations plotted as the % of total tubes in a control well. Data are typical of four similar experiments. (Figure reprinted with the permission of Circulation Research.)

been shown to interfere with the secretion of MMPs in atherosclerotic plaques [Bellosta, 1998; Koh, 2007] and the migration of macrophages [Tsiara, 2003]. Our data demonstrate that statins inhibit neovascularization in response to FGF and VEGF in the mouse corneal pocket and the chick chorioallontoic membrane assays [Park, 2002]. Infusion of AngII via subcutaneous osmotic minipumps in ApoE−/− mice also has been shown to result in reproducible formation of suprarenal AAAs that exhibit many characteristics of the human disease including infiltration of macrophages, secretion of MMPs, disruption of the media, rupture of the elastic layer and neovascularization [Daugherty, 2004]. This result is consistent with the finding that AngII not only plays a role in controlling cardiovascular and renal homeostasis, but also affects vascular endothelial cell function, macrophage activation [Ozawa, 2007] and the contraction, migration, and proliferation of vascular smooth muscle cells [Mehta, 2007]. Hence, AngII infusion is an important model for the study of mechanisms of AAA formation.

Statins Interfere with Angiotensin II Mediated Angiogenesis

In order to determine the role of statins in AngII mediated angiogenesis, we used both the HDMEC and the HUVEC Matrigel models for angiogenesis. To demonstrate AngII stimulation of tube formation by HDMECs is was necessary to treat the cells with a submaximal dose of VEGF. In the presence of 20-μg/mL VEGF, AngII stimulated tube formation by HDMECs by greater than twofold ($N = 4$) in a dose-dependent manner (Fig. XI.8a). This effect was specific as demonstrated by the finding that it was blocked by 10-μM AT1 receptor blocker losartan while 10-μM AT2 receptor blocker PD123319 had no effect. As shown in Fig. XI.2, culture of HUVECs on Matrigel led to the formation of capillary-like structures. Pretreatment of HUVECs with 100-nM AngII enhanced tube formation by 1.55 ± 0.06-fold ($n = 6$, $p < 0.05$) compared to control (Fig. XI.8b and c). The AngII stimulated angiogenesis was completely reversed by the addition of 1-μM simvastatin to the medium (Fig. XI.8c).

Mechanism of AngII Stimulated Angiogenesis

Vascular endothelial cell growth factor (VEGF) stimulates tube formation by endothelial cells. In order to determine the mechanism of AngII stimulation of tube formation, we treated monolayer cultures of HUVECs with AngII and determined the effect on VEGF expression. AngII stimulated VEGF expression in a dose-dependent manner. This effect was inhibited by the AT-1 antagonist losartan, but was unaffected by the AT2 receptor antagonist PD123319 (Fig. XI.9a). Furthermore, Simvastatin inhibited the AngII stimulated VEGF expression in a dose-dependent manner with a maximal effect at 0.2 μM (Fig. XI.9b). To further determine the effect of AngII on VEGF signaling, the effect of AngII on the expression of the VEGF receptor KDR was determined. AngII stimulated KDR expression in a dose-dependent manner with a maximum effect at 100 nM. Simvastatin reversed the effect of AngII on KDR expression (Fig. XI.9c). These data demonstrate that AngII stimulated angiogenesis is mediated via an increase in VEGF signaling. Finally, in order to determine whether the effect of Simvastatin on VEGF signaling is dependent on the inhibition of Rho, we determined the effect of C3 exotoxin, which interferes with Rho GTPases by catalyzing their ADP ribosylation, on AngII stimulation of VEGF expression. Both Simvastatin and C3 exotoxin reversed the effect of AngII on VEGF expression consistent

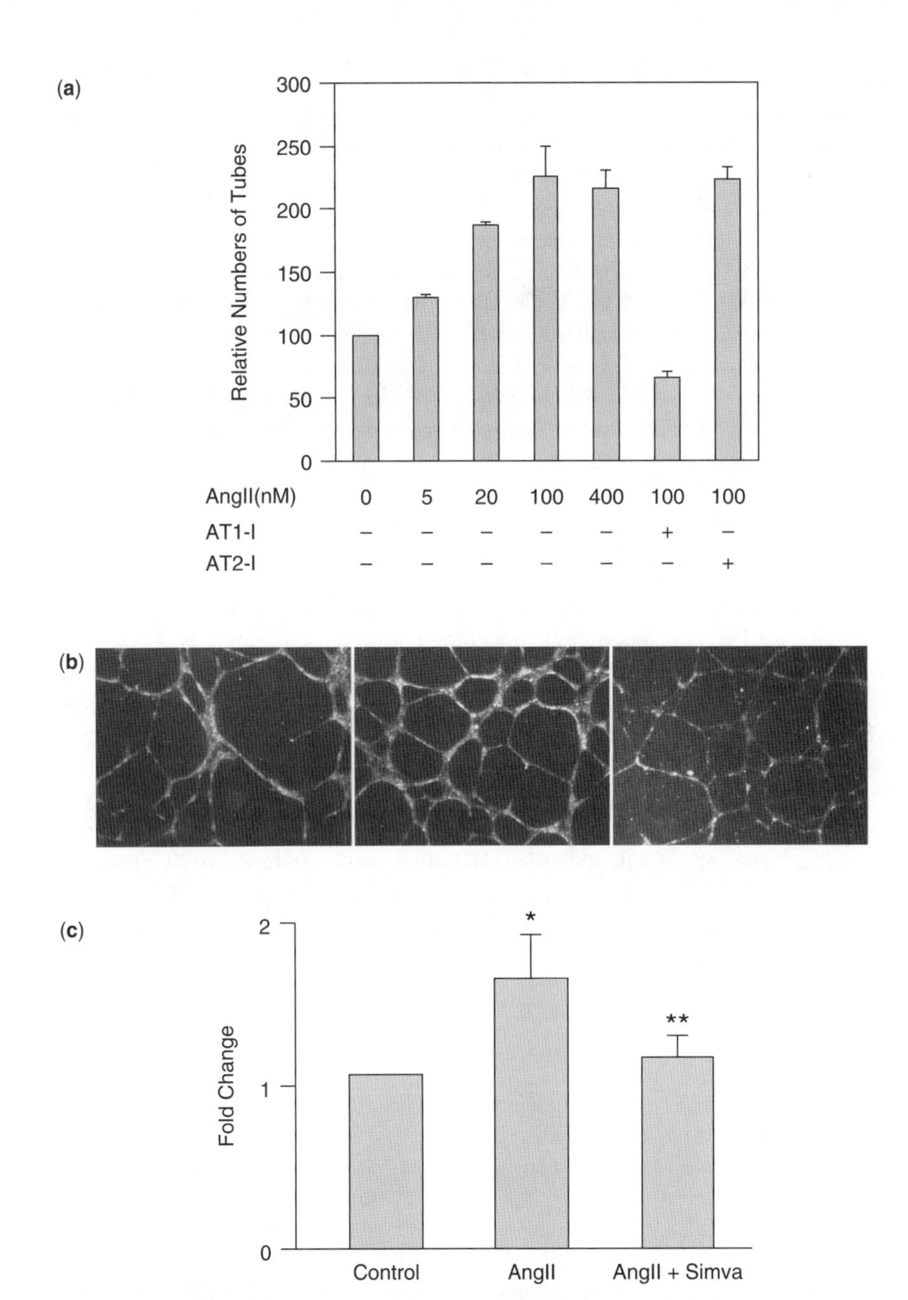

Figure XI.8. Effect of AngII on *in vitro* angiogenesis. (a) Quantitation of the effect of AngII on tube formation by HDMECs. Cells were plated as in Fig. XI.1 except that VEGF was decreased to 20 ng/mL. Cells were incubated with the indicated concentrations of AngII with and without AT1 and AT2 receptor inhibitors. Tube formation was quantified as in Fig. XI.1 ($n = 4$). (b) Simvastatin attenuates AngII induced tube formation by HUVECs. Representative photomicrographs of HUVECs cultured on Matrigel: left panel, control; middle panel, 200-nM AngII; right panel, AngII plus 0.1-μM simvastatin. (c) Quantitation of the area of honeycomb formation compared to control, determined using Scion Image for determination of total area of the honeycomb. Data are expressed as (±SE). $^*p < 0.05$ compared with control, $^{**}p < 0.05$ compared with AngII.

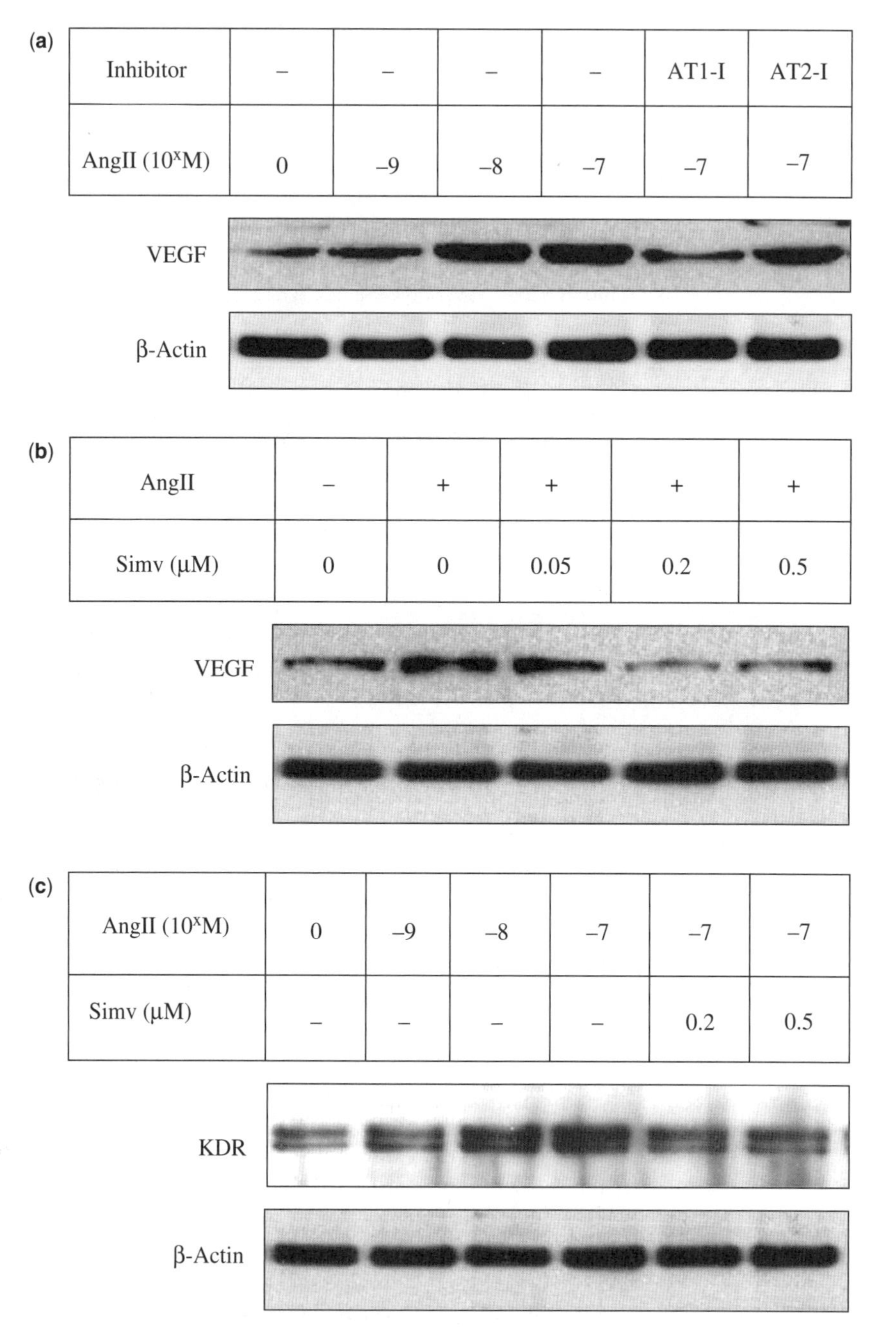

Figure XI.9. Effect of AngII on the VEGF signaling pathway. (a) Effect of AngII on VEGF expression. HUVECs were plated on collagen coated dishes and incubated with the indicated concentrations of AngII with or without 10-μM AT1 or AT2 receptor inhibitor. Cells were harvested and VEGF determined by Western Blot analysis. (b) Effect of Simvastatin on AngII stimulated VEGF expression. As in (a) except that cells were incubated with AngII plus the indicated concentrations of Simvastatin. (c) Effect of Ang II and Simvastatin on the expression of KDR. Cells were incubated with the indicated concentrations of Ang II and Simvastatin.

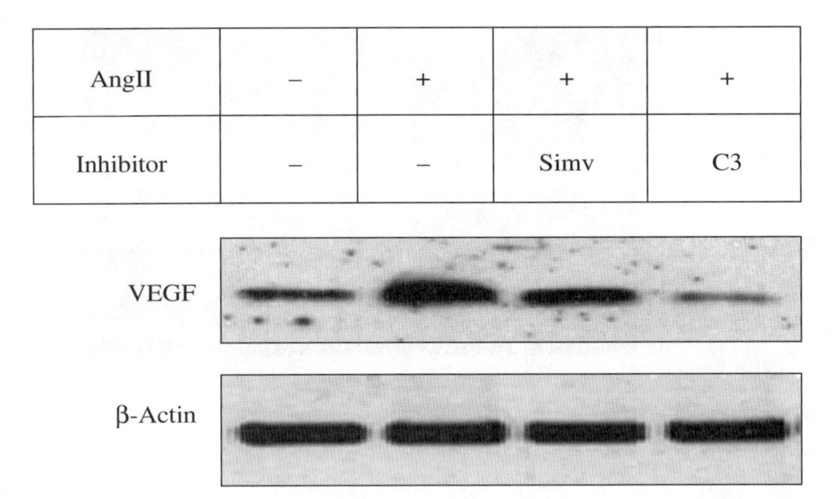

Figure XI.10. Role of Rho in AngII stimulation of VEGF expression. The HUVECs were incubated with either vehicle or 200-nM AngII with or without either 0.2-μM Simvastatin or 0.5-μg/mL C3 exotoxin and VEGF determined by Western Blot analysis.

with the conclusion that AngII stimulates VEGF signaling and angiogenesis via the activation of Rho (Fig. XI.10).

Simvastatin Decreases the Incidence and Severity of Aneurysms in AngII-Infused ApoE−/− Mice

The AngII treatment of ApoE−/− mice has been shown to result in the development of abdominal aortic aneurysms with marked similarities to those found in humans. These aneurysms have been shown to be associated with the development of extensive neovascularization involving the surface of the aneurysm and the media and adventitia of the aneurysmal tissues. New capillary formation associated with AAA has been found to be more concentrated at the site of the AAA rupture suggesting that neovascularization might play a role in the pathogenesis of AAA [Choke, 2006; Herron, 1991]. In order to determine the effect of Simvastatin on AngII stimulated AAA formation, AngII infused mice were treated subcutaneously with Simvastatin (10 mg/kg/day) or vehicle. All AngII treated mice developed suprarenal aortic dilatation and/or thrombus formation (Fig. XI.11a) while animals infused with saline (control) did not develop aneurysms. This effect was significantly reversed in mice treated with Simvastatin. Forty percent of Simvastatin treated mice had no detectable aneurysms (Fig. XI.11a right panel).

Simvastatin Decreased Neovascularization of Aneurysmal Tissue. Whole-mount CD31 staining of aortas demonstrated extensive proliferation of the vasa vasorum (VV) in areas of AAA dilation in the AngII treated mice. Simvastatin treatment significantly decreased the extent of neovascularization of the aneurysmal region (Fig. XI.11a right panel). In cross-sections of aortas from control mice, CD31 positive microvessels were essentially undetectable, while numerous capillary vessels could be observed in AngII treated aortas. In contrast, mice treated with simvastatin showed a marked decrease in capillary formation, $51.1 \pm 3.6\%$ ($n = 15$, $p < 0.05$) compared to mice treated with AngII alone (Fig. XI.11b right panel).

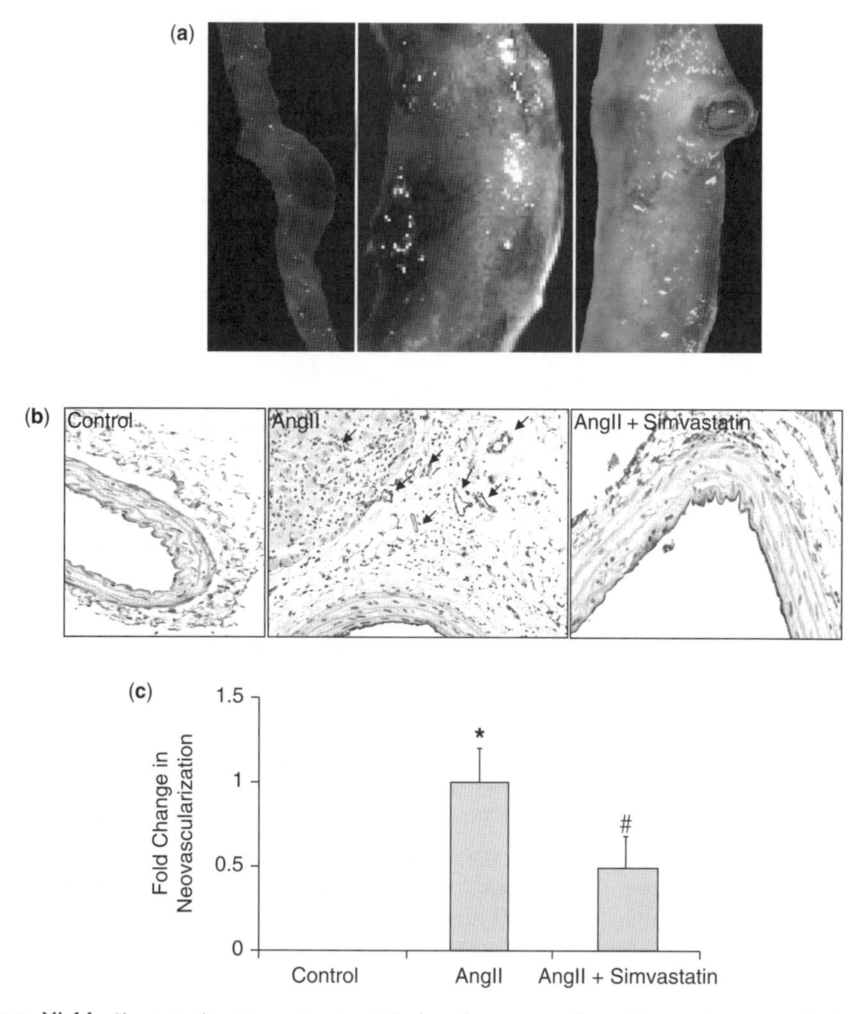

Figure XI.11. Simvastatin attenuates AngII-induced aneurysm formation and neovascularization in suprarenal aortic tissue. (a) Left panel: Representative whole-mount CD31 and Sudan staining of the aorta from an AngII treated mouse demonstrating multiple aneurysms (note the atheroma in the distal large aneurysm). Middle panels: high-power views of upper and lower regions shown in the left panel demonstrating extensive proliferation of VV. Right panel: high-power view of similar region from the aorta of a Simvastatin treated mouse. (b) CD31+ endothelial cells (brown color) in corresponding sections from aortas of control, AngII and AngII + Simvastatin treated mice. Arrows indicate new capillary formation. (c) Quantitative evaluation of CD31+ capillaries in media and adventitia normalized by area. $^{*}p < 0.05$ compared with control, $^{**}p < 0.05$ compared with AngII, data represent capillary counts in three fields in each of 15 sections. (See color insert.)

The finding that Simvastatin decreased both the formation of AAA and neovascularization of the aneurysms in *in vivo* models for angiogenesis suggested that new capillary blood vessel development might play a role in aneurysm formation in response to AngII and support the hypothesis that AngII stimulates neovascularization via the activation of Rho and that statins might inhibit angiogenesis at least in part via an antiangiogenic effect (Fig. XI.12).

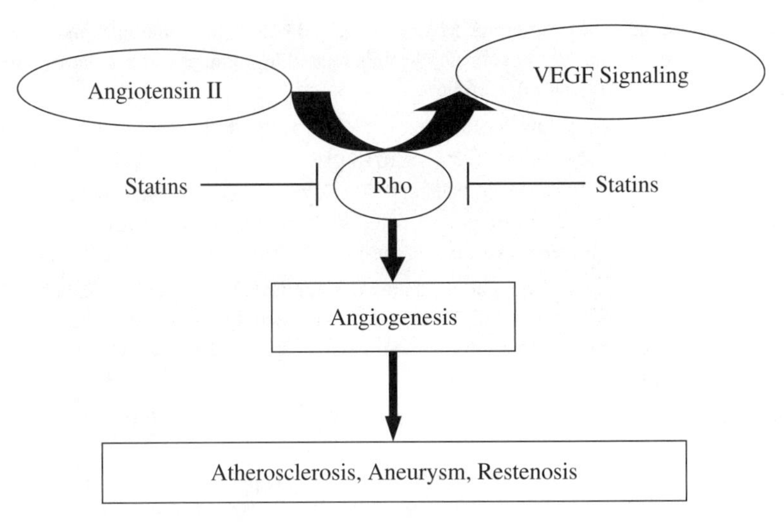

Figure XI.12. Proposed pathway for AngII stimulation of angiogenesis.

Role of Lipid Metabolism and Statins in Angiogenesis

These studies demonstrate a new relationship between angiogenesis, lipid metabolism, and RhoA dependent cell signaling. Both the corneal pocket assay and the chick chorioallantoic membrane assay demonstrate that FGF and VEGF stimulated neovascularization is reversed by Simvastatin in a dose-dependent manner. The use of cultures of the endothelial cells lining the human umbilical vein and the human dermal microvascular endothelial cells obtained from the foreskins of newborns has permitted the development of two powerful assays for *in vitro* studies of angiogenesis. These cell types not only develop structures that resemble microvessels, single-cell tubes or honeycombs that lack the smooth muscle cell outer coating of true capillaries, but may be used in monolayer cultures to study the molecular mechanisms involved in tube formation. Hence, not only have HUVECs proved a valuable tool in the study of the effects of stretch and stress on vascular biology [Bevilacqua, 1987; Davies, 1997; Libby, 2000; Yamada, 1992], but they have also contributed to our understanding of signaling pathways and growth factors that contribute to new blood vessel formation in the normal uterine cycle and in the study of diseases, such as atherosclerosis, rheumatoid arthritis, psoriasis, and diabetic retinopathy, whose pathogenesis involves new blood vessel formation. They also have contributed to an understanding of how neovascularization might play a role in the pathogenesis and treatment of vascular occlusive diseases and abdominal aortic aneurysm formation.

REFERENCES

Arenas IA, Xu Y, Lopez-Jaramillo P, Davidge ST. 2004. Angiotensin II-induced MMP-2 release from endothelial cells is mediated by TNF-alpha. Am J Physiol Cell Physiol. 286(4):C779–784.

Bellosta S, Via D, Canavesi M, Pfister P, Fumagalli R, Paoletti R, Bernini F. 1998. HMG-CoA reductase inhibitors reduce MMP-9 secretion by macrophages. Arterioscler Thromb Vasc Biol. 18(11):1671–1678.

Bevilacqua MP, Gimbrone MA, Jr. 1987. Inducible endothelial functions in inflammation and coagulation. Semin Thromb Hemost. 13(4):425–433.

Bevilacqua MP, Stengelin S, Gimbrone MA, Jr., Seed B. 1989. Endothelial leukocyte adhesion molecule 1: An inducible receptor for neutrophils related to complement regulatory proteins and lectins. Science. 243(4895):1160–1165.

Burns MP, DePaola N. 2005. Flow-conditioned HUVECs support clustered leukocyte adhesion by coexpressing ICAM-1 and E-selectin. Am J Physiol Heart Circ Physiol. 288(1):H194–204.

Choke E, Thompson MM, Dawson J, Wilson WR, Sayed S, Loftus IM, Cockerill GW. 2006. Abdominal aortic aneurysm rupture is associated with increased medial neovascularization and overexpression of proangiogenic cytokines. Arterioscler Thromb Vasc Biol. 26(9):2077–2082.

Dai G, Kaazempur-Mofrad MR, Natarajan S, Zhang Y, Vaughn S, Blackman BR, Kamm RD, Garcia-Cardena G, Gimbrone MA, Jr. 2004. Distinct endothelial phenotypes evoked by arterial waveforms derived from atherosclerosis-susceptible and -resistant regions of human vasculature. Proc Natl Acad Sci USA. 101(41):14871–14876.

Daugherty A, Cassis LA. 2004. Mouse models of abdominal aortic aneurysms. Arterioscler Thromb Vasc Biol. 24(3):429–434.

Daugherty A, Manning MW, Cassis LA. 2000. Angiotensin II promotes atherosclerotic lesions and aneurysms in apolipoprotein E-deficient mice. J Clin Invest. 105(11):1605–1612.

Davies PF. 1997. Overview: Temporal and spatial relationships in shear stress-mediated endothelial signalling. J Vasc Res. 34(3):208–211.

Dobrin PB, Baker WH, Gley WC. 1984. Elastolytic and collagenolytic studies of arteries. Implications for the mechanical properties of aneurysms. Arch Surg. 119(4):405–409.

Ernst JD. 2000. Bacterial inhibition of phagocytosis. Cell Microbiol. 2(5):379–386.

Freestone T, Turner RJ, Coady A, Higman DJ, Greenhalgh RM, Powell JT. 1995. Inflammation and matrix metalloproteinases in the enlarging abdominal aortic aneurysm. Arterioscler Thromb Vasc Biol. 15(8):1145–1151.

Gimbrone MA, Jr. 1976. Culture of vascular endothelium. Prog Hemost Thromb. 3:1–28.

Goldberger A, Middleton KA, Oliver JA, Paddock C, Yan HC, DeLisser HM, Albelda SM, Newman PJ. 1994. Biosynthesis and processing of the cell adhesion molecule PECAM-1 includes production of a soluble form. J Biol Chem. 269(25):17183–17191.

Goldstein JL, Brown MS. 1990. Regulation of the mevalonate pathway. Nature (London). 343(6257):425–430.

Gossen M, Bujard H. 1992. Tight control of gene expression in mammalian cells by tetracycline-responsive promoters. Proc Natl Acad Sci USA. 89(12):5547–5551.

Grundy SM. 1998. Statin trials and goals of cholesterol-lowering therapy [editorial; comment] [see comments]. Circulation. 97(15):1436–1439.

Herron GS, Unemori E, Wong M, Rapp JH, Hibbs MH, Stoney RJ. 1991. Connective tissue proteinases and inhibitors in abdominal aortic aneurysms. Involvement of the vasa vasorum in the pathogenesis of aortic aneurysms. Arterioscler Thromb. 11(6):1667–1677.

Herron GS, Werb Z, Dwyer K, Banda MJ. 1986. Secretion of metalloproteinases by stimulated capillary endothelial cells. I. Production of procollagenase and prostromelysin exceeds expression of proteolytic activity. J Biol Chem. 261(6):2810–2813.

Koh KK. 2000. Effects of statins on vascular wall: Vasomotor function, inflammation, and plaque stability. Cardiovasc Res. 47(4):648–657.

Kokura S, Wolf RE, Yoshikawa T, Granger DN, Aw TY. 1999. Molecular mechanisms of neutrophil-endothelial cell adhesion induced by redox imbalance. Circ Res. 84(5):516–524.

Kumar S, Li C. 2001. Targeting of vasculature in cancer and other angiogenic diseases. Trends Immunol. 22(3):129.

Laufs U, La Fata V, Plutzky J, Liao JK. 1998. Upregulation of endothelial nitric oxide synthase by HMG CoA reductase inhibitors. Circulation. 97(12):1129–1135.

Lerner EC, Qian Y, Blaskovich MA, Fossum RD, Vogt A, Sun J, Cox AD, Der CJ, Hamilton AD, Sebti SM. 1995. Ras CAAX peptidomimetic FTI-277 selectively blocks oncogenic Ras signaling by inducing cytoplasmic accumulation of inactive Ras-Raf complexes. J Biol Chem. 270(45): 26802–26806.

Libby P. 2000. Changing concepts of atherogenesis. J Intern Med. 247(3):349–358.

Massy ZA, Keane WF, Kasiske BL. 1996. Inhibition of the mevalonate pathway: Benefits beyond cholesterol reduction? [see comments]. Lancet. 347(8994):102–103.

Mehta PK, Griendling KK. 2007. Angiotensin II cell signaling: Physiological and pathological effects in the cardiovascular system. Am J Physiol Cell Physiol. 292(1):C82–97.

Muscella A, Marsigliante S, Carluccio MA, Vinson GP, Storelli C. 1997. Angiotensin II AT1 receptors and Na + /K+ ATPase in human umbilical vein endothelial cells. J Endocrinol. 155(3):587–593.

Nagata D, Mogi M, Walsh K. 2003. AMP-activated protein kinase (AMPK) signaling in endothelial cells is essential for angiogenesis in response to hypoxic stress. J Biol Chem. 278(33): 31000–31006.

Namiki A, Brogi E, Kearney M, Kim EA, Wu T, Couffinhal T, Varticovski L, Isner JM. 1995. Hypoxia induces vascular endothelial growth factor in cultured human endothelial cells. J Biol Chem. 270(52):31189–31195.

Nozawa F, Hirota M, Okabe A, Shibata M, Iwamura T, Haga Y, Ogawa M. 2000. Tumor necrosis factor alpha acts on cultured human vascular endothelial cells to increase the adhesion of pancreatic cancer cells. Pancreas. 21(4):392–398.

Ozawa Y, Kobori H. 2007. Crucial role of Rho-nuclear factor-kappaB axis in angiotensin II-induced renal injury. Am J Physiol Renal Physiol. 293(1):F100–109.

Ozawa Y, Kobori H, Suzaki Y, Navar LG. 2007. Sustained renal interstitial macrophage infiltration following chronic angiotensin II infusions. Am J Physiol Renal Physiol. 292(1):F330–339.

Park HJ, Kong D, Iruela-Arispe L, Begley U, Tang D, Galper JB. 2002. 3-hydroxy-3-methylglutaryl coenzyme A reductase inhibitors interfere with angiogenesis by inhibiting the geranylgeranylation of RhoA. Circ Res. 91(2):143–150.

Parmar KM, Larman HB, Dai G, Zhang Y, Wang ET, Moorthy SN, Kratz JR, Lin Z, Jain MK, Gimbrone MA, Garcia-Cardena G. 2006. Integration of flow-dependent endothelial phenotypes by Kruppel-like factor 2. J Clin Invest. 116(1):49–58.

Parmar KM, Nambudiri V, Dai G, Larman HB, Gimbrone MA, Jr., Garcia-Cardena G. 2005. Statins exert endothelial atheroprotective effects via the KLF2 transcription factor. J Biol Chem. 280(29):26714–26719.

Richard DE, Vouret-Craviari V, Pouyssegur J. 2001. Angiogenesis and G-protein-coupled receptors: Signals that bridge the gap. Oncogene. 20(13):1556–1562.

Sacks FM, Moye LA, Davis BR, Cole TG, Rouleau JL, Nash DT, Pfeffer MA, Braunwald E. 1998. Relationship between plasma LDL concentrations during treatment with pravastatin and recurrent coronary events in the Cholesterol and Recurrent Events trial. Circulation. 97(15):1446–1452.

Tsiara S, Elisaf M, Mikhailidis DP. 2003. Early vascular benefits of statin therapy. Curr Med Res Opin. 19(6):540–556.

Urbich C, Dernbach E, Zeiher AM, Dimmeler S. 2002. Double-edged role of statins in angiogenesis signaling. Circ Res. 90(6):737–744.

Vogt A, Qian Y, McGuire TF, Hamilton AD, Sebti SM. 1996. Protein geranylgeranylation, not farnesylation, is required for the G1 to S phase transition in mouse fibroblasts. Oncogene. 13(9):1991–1999.

Weis M, Heeschen C, Glassford AJ, Cooke JP. 2002. Statins have biphasic effects on angiogenesis. Circulation. 105(6):739–745.

Yamada T, Fan J, Shimokama T, Tokunaga O, Watanabe T. 1992. Induction of fatty streak-like lesions in vitro using a culture model system simulating arterial intima. Am J Pathol. 141(6): 1435–1444.

Yoon YS, Johnson IA, Park JS, Diaz L, Losordo DW. 2004. Therapeutic myocardial angiogenesis with vascular endothelial growth factors. Mol Cell Biochem. 264(1–2):63–74.

Zhang W, DeMattia JA, Song H, Couldwell WT. 2003. Communication between malignant glioma cells and vascular endothelial cells through gap junctions. J Neurosurg. 98(4):846–853.

12

HEMATOPOIETIC STEM CELL DEVELOPMENT IN THE PLACENTA

Katrin E. Rhodes and Hanna K.A. Mikkola

University of California Los Angeles, Los Angeles, CA 90095

INTRODUCTION

The placent and felties is an ephemeral fetal organ that is vital for the survival of the developing embryo and fetus. The placenta facilitates gas and nutrient exchange between the mother and the fetus via a specialized exchange system that involves an elaborate network of fetal vasculature and trophoblast lined maternal blood spaces [Cross, 2005]. Furthermore, the placenta has an important function in immune defense and is the primary source of hormones essential for a successful pregnancy. Recently, it was discovered that the placenta has another physiological role, as a hematopoietic organ, in which hematopoietic stem cells (HSCs) develop and expand. This discovery has opened new avenues of research focused on defining how this cytokine and growth factor rich organ also supports HSC development. Here, we review how the placenta was unveiled as a fetal hematopoietic organ and discuss its unique function as a microenvironment specialized for HSC development.

THE HEMATOPOIETIC SYSTEM

Hematopoietic stem cells are rare cells located within the bone marrow, and are responsible for maintaining blood homeostasis throughout an individual's lifetime. This goal is achieved through a tightly regulated balance of self-renewal and lineage differentiation events, wherein HSCs divide to replenish their pool size and produce lineage-committed progenitors destined for terminal differentiation into all of the different blood cell types, respectively. The regulation of HSC fate decisions is a compilation of extracellular cues from the local

Perinatal Stem Cells. Edited by C. L. Cetrulo, K. J. Cetrulo, and C. L. Cetrulo, Jr.

microenvironment, the niche, as well as intracellular signaling and transcriptional networks, which together coordinate gene expression programs that promote stem cell properties or induce differentiation. When the balance between self-renewal and differentiation is perturbed fatal blood diseases, such as leukemia or aplastic anemia, may be conceived. Due to the unique properties of HSCs, transplanted HSCs can regenerate the entire hematopoietic system of a recipient and thereby provide a cure for inherited and acquired blood diseases. The functional criteria for HSCs, self-renewal ability and multipotentiality, can be tested in experimental animals by transplantation into myeloablated recipients, which is the golden standard assay for adult HSC function.

The HSCs are generated during embryogenesis. As the nascent HSCs are very different from their adult counterparts by their functional properties, they have to migrate through a number of different microenvironments that expose them to distinct extracellular cues that promote their maturation toward adult HSCs. The HSCs emerge in multiple anatomical sites, and once generated, convene in the fetal liver to mature and expand their pool size in preparation for their final journey to the bone marrow. Until recently, the aorta–gonad mesonephros (AGM) region and adjacent major blood vessels (vitelline and umbilical arteries) were considered the main source of HSCs during embryogenesis (Table XII.1) [Cumano, 1996; de Bruijn, 2000; Godin, 1999; Medvinsky and Dzierzak, 1996; Muller, 1994]. However, because the number of transplantable HSCs found in the AGM is so low, it was questioned whether such a scarce population could be the sole contributor to the robust number of HSCs found in the fetal liver, which itself is incapable of *de novo* HSC generation. Therefore, it was hypothesized that other embryonic sites may contribute to the establishment of the fetal liver HSC pool. Indeed, recent studies have provided evidence that both the placenta and the yolk sac have the capacity to generate HSCs. However, the placenta is unique among the other fetal hematopoietic sites; it not only has the capacity for *de novo* hematopoiesis, but it also accrues a large reservoir of HSCs and protects them from signals that promote immediate differentiation.

HISTORICAL PERSPECTIVE ON PLACENTAL HEMATOPOIESIS

The placenta has not traditionally been regarded as a hematopoietic organ, however, the first reports suggesting that the placenta may function in blood development were already published in 1961. Interestingly, transplantations of mouse placental tissue into irradiated recipient mice demonstrated the ability to generate hematopoietic colonies in the spleen (CFU-S, colony forming unit spleen), and furthermore, these placenta-derived spleen colonies could also be successfully serially transplanted [Till and McCulloch, 1961]. Although it was later shown that CFU-S activity does not necessarily indicate the presence of HSCs, this study revealed that the placenta contained clonogenic hematopoietic cells with high proliferative potential. Subsequently, another early study suggested that the placenta harbors HSCs as placental tissue exhibited similar capacity to reconstitute the hematopoietic system of irradiated recipient mice to that of the adult bone marrow [Dancis, 1977]. Strikingly, HSC activity was not dependent on the continuous supply of blood to the placenta, as separation of placenta from the embryonic circulation did not abolish this activity. Furthermore, by performing a plaque-forming assay for placental tissue, another study showed that the midgestation placenta is a unique reservoir for B cell precursors, before any were found in the fetal liver [Melchers, 1979]. Inspite of these intriguing findings, further work on hematopoiesis in the placenta did not commence until decades later.

Interest in the placenta was restored after work in avians showed that the allantois, a tissue appendage derived from mesodermal precursors of the primitive streak, harbors

TABLE XII.1. Timeline of Placental Development: Developmental Placental Hematopoiesis

Timeline of Placental Development	Pre-Chorioallantoic Fusion (E7.5–8.25) <4 sp	Chorioallantoic Fusion (E8.5) 4–6 sp	Chorioallantoic/Trophoblast Interdigitation (E8.75–9.5) 8–28 sp	Labyrinth Vessel Branching (E10.0–11.0) 30–44 sp	Fully Vascularized Placenta (E11.5–E13.5) >44 sp	Late Gestational Placenta (>E13.5)
Hallmarks of Developmental Hematopoiesis	Onset of yolk sac primitive hematopoiesis Runx1–LacZ expression in yolk sac blood islands	Generation of definitive clonogenic progenitors in the yolk sac Runx1–LacZ expression in the pSP/AGM Initiation of heartbeat	Generation of definitive clonogenic progenitors in the pSP/AGM	Emergence of long-term reconstituting HSCs in the AGM	Beginning of HSC colonization into the fetal liver	Gross expansion of HSC pool in the fetal liver
Placental Hematopoiesis	Runx1–LacZ expression in the chorionic and part of the allantoic mesoderm Myeloerythroid potential of chorionic and allantoic mesoderm in OP-9 stroma coculture or explant culture		Generation of definitive clonogenic progenitors in the placenta	First long-term reconstituting HSCs in the placenta Runx1–LacZ+ cells within the chorioallantoic vessel walls and mesenchyme, and the lumen of labyrinth vessels *De novo* hematopoiesis within the placental vasculature in the absence of heartbeat	Major reservoir of HSCs in the placenta	Decline of placental HSC pool. Possible mobilization of HSCs from the placenta to the fetal liver

multipotential hematopoietic precursors [Caprioli, 1998, 2001]. Although birds do not develop a placenta, the allantois has both a similar developmental origin and function in nutrient and gas exchange as the placenta in eutherial mammals. Hematopoietic potential of the avian allantois was assessed by isolating the allantois of a quail embryo before the initiation of circulation, to exclude any nonallantoic cellular components, and grafting the quail allantois cells into a chick embryo. This avian xenograft model revealed that the cells derived from the quail allantois could contribute to the bone marrow of a postnatal chick, indicative of HSC activity. Both hematopoietic and endothelial cells were generated, suggesting that the allantois has the capacity to generate both blood cells and blood vessels [Caprioli, 1998].

DEVELOPMENT AND STRUCTURE OF THE MOUSE PLACENTA

The finding that the avian allantois possessed hematopoietic potential provoked the hypothesis that the placenta, which in part is comprised of the allantoic mesoderm, may also function as a hematopoietic organ. The placenta is an extraembryonic organ that develops from trophoectoderm and mesodermal tissues. During establishment of the mouse placenta, mesodermal precursors from the posterior primitive streak of the epiblast protrude into the exocoelomic cavity and form the allantoic bud (Fig. XII.1a). Of importance, the posterior primitive streak is also the source of the mesodermal precursors that establish hematopoiesis in the yolk sac and the embryo proper. The allantoic bud grows toward the ectoplacental cone to fuse with the chorionic mesoderm, which lines the exocoelomic cavity, and together

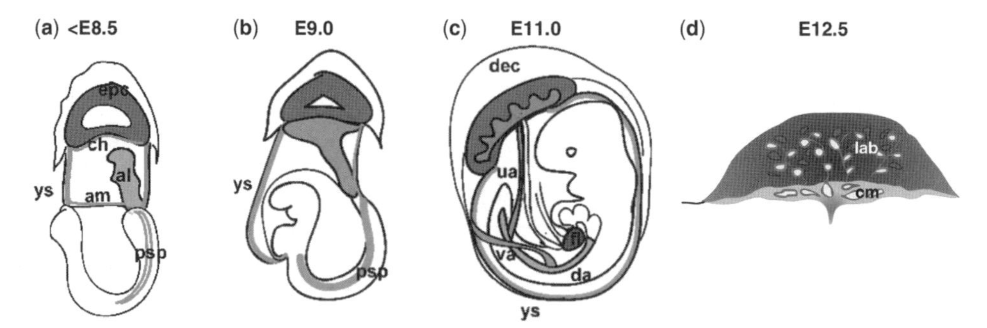

Figure XII.1. Development of the mouse placenta. (a) At E7.5–8.25 the allantois (red) has formed from mesodermal precursors from the primitive streak, and is growing toward the ectoplacental cone (brown). (b) Fusion of the allantois with the chorionic mesoderm occurs at E8.5, concomitant with the onset of heartbeat. Subsequently, chorioallantoic mesoderm interdigites with the trophoblasts and the placental vasculature starts to form. (c) By E10.5–11.0 large vessels that connect to the umbilical cord have formed in the chorioallantoic mesenchyme, and the feto-placental circulation is fully established. The placenta labyrinth is still developing and is therefore an active site of vasculogenesis–angiogenesis. (d) A E12.5 cross-section of the placenta displays the different regions of the placenta, namely, the chorioallantoic mesenchyme including the large vessels of the placenta (in red) and the placenta labyrinth, which is a unique region including trophoblast-lined maternal blood spaces (red spaces surrounded by brown trophoblasts) and fetal vessels lined by fetal endothelium (red vessels with lumens). al = allantois; ch = chorion; am = amnion; epc = ectoplacental cone; ys = yolk sac; psp = para-aortic splanchnopleura; dec = decidua; da = dorsal aorta; ua = umbilical artery; va = vitelline artery; fl = fetal liver; lab = labyrinth; cm = chorioallantoic mesenchyme. (See color insert.)

the chorioallantoic mesoderm coalesces with the trophoblasts to establish the fetal compartment of the placenta (Fig. XII.1b,d). The union of the allantoic and chorionic mesoderms initiates the development of two distinct regions of fetal vasculature within the placenta: the chorioallantoic vessels and the labyrinth vessels. The chorioallantoic vasculature, closest to the fetus, is comprised of large vessels that are surrounded by mesenchyme and directly connect to the umbilical cord vessels (Fig. XII.1c,d). This region of the placenta is primarily of mesodermal origin, although during midgestation it harbors tubular structures derived from ectoplacental endoderm that later form a cavity between chorioallantoic mesenchyme and the labyrinth called the Crypt of Duval [Duval, 1891; Ogura, 1998]. The labyrinth refers to the intricate fetal vascular network intertwined within the trophoblasts that line maternal blood spaces (Fig. XII.1d). This region directly mediates fetal-maternal exchange, and is positioned closer to the uterine wall. Chorioallantoic fusion occurs by E8.5, 5 sp (somite pair) stage, which coincides with the initiation of heartbeat and the onset of circulation in the embryo. The placenta starts to function in fetal-maternal exchange as maternal blood begins to flush the trophoblast lined maternal blood spaces (Fig. XII.1d). Adjacent to the labyrinth layer is the spongiotrophoblast layer, which provides structural support for the placenta. Giant cells, which are also a subtype of fetal trophoblasts, form a thin layer between the spongiotrophoblasts and the maternal decidua. Thereby, most of the placenta is of fetal origin, except for the maternal blood cells.

HEMATOPOIETIC ACTIVITY IN THE MOUSE PLACENTA

As the placenta is a chimeric organ composed of both fetal and maternal cells, deciphering whether the placenta has fetal hematopoietic potential requires tools to verify fetal origin of the hematopoietic cells. Alvarez-Silva et al. showed that the mouse placenta harbors a major pool of fetal-derived hematopoietic progenitors by using GFP+ reporter mice [Alvarez-Silva, 2003]. By breeding transgenic male mice with nontransgenic females, this model enables the distinction between maternal (GFP−) and fetal (GFP+) derived cells. By plating cells derived from the hematopoietic tissue rudiments into methylcellulose colony-forming assay, it was shown that the placenta harbors progenitors by 20 sp stage, shortly after the yolk sac (15 sp) and the caudal half of the embryo (18 sp), which comprises the para-aortic splanchopleura and eventually develops into the AGM (Table XII.1). In contrast, the fetal liver did not demonstrate colony-forming ability until much later in development. The frequency of the multipotential progenitors as compared to committed, unilineage progenitors within the placenta was much higher than in the fetal liver, implying that the hematopoietic programs in the placenta and the fetal liver are different from one another. Placental hematopoietic colonies were highly proliferative and could be replated over two months.

Paradoxically, clonogenic progenitors develop in the conceptus before definitive HSCs do, and in some anatomical sites, such as the early yolk sac, this progenitor activity is independent of HSC generation. Therefore, it was critical to assess HSC activity by stringent *in vivo* repopulation assays. Two studies subsequently verified that the placenta harbors true adult-reconstituting hematopoietic stem cells that can self-renew and generate multilineage progeny through serial transplantations [Gekas, 2005; Ottersbach and Dzierzak, 2005]. Ottersbach et al. used transgenic mice that expressed human-β-globin from a ubiquitous promoter or GFP from the Sca1 (Ly-6A) promoter to verify fetal origin of the cells. In the other study, Gekas et al. bred embryos that are heterozygous for the two different alleles of the pan-hematopoietic marker CD45 (CD45.1 and CD45.2), distinguishing them from maternal cells that only expressed CD45.2. The placenta was

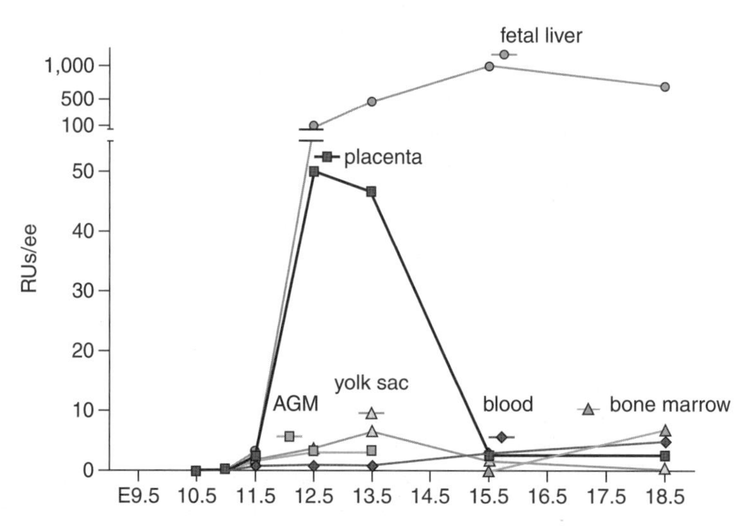

Figure XII.2. Kinetics of long-term reconstituting HSCs in the embryo and extraembryonic tissues. The graph depicts the number of HSCs with long-term reconstitution capability per tissue during development. The midgestation placenta harbors a large pool of HSCs, which diminishes toward the end of gestation while the fetal liver HSC pool is expanding. RU = reconstituting unit; ee = embryo equivalent. (See color insert.)

shown to harbor adult reconstituting HSCs as early as the AGM (E10.5–11.0), before any HSCs are circulating in the blood stream or have seeded the fetal liver (Fig. XII.2). Remarkably, in contrast to the AGM where the number of HSCs remains low, the HSC pool in the placenta increases drastically. At its peak, E12.5–13.5, the placenta harbors 15-fold more HSCs than the AGM, or the yolk sac (Fig. XII.2; Table XII.1). As the placental HSC pool decreases, there is a concomitant increase of HSCs in the fetal liver, suggesting that the placental HSCs may seed the fetal liver, which is directly downstream of the placenta in fetal circulation (Fig. XII.2). Furthermore, analysis of placental HSC and clonogenic progenitor frequencies by phenotype and functional assays provided evidence that the placental microenvironment supports HSC expansion and/or functional maturation without promoting their differentiation into myeloerythroid progenitors. In contrast, the fetal liver, in addition to harboring HSCs, produces a large number of progenitor cells that are committed for definitive erythroid differentiation. These data implicate inherent differences between fetal liver and placental HSC–progenitor populations and hematopoietic microenvironments, where the fetal liver, but not the placenta, supports active myeloerythroid differentiation during midgestation. These results are in accordance with the data from Alvarez-Silva et al., who showed that the hematopoietic developmental profile of placental and fetal liver cells in methylcellulose was very different, as the placenta harbored multipotential, high-proliferative potential progenitors, whereas the fetal liver was rich in unilineage progenitors [Alvarez-Silva, 2003]. Although the origin of placental HSCs remained yet undefined, these studies were the first to describe the placenta as a fetal hematopoietic organ that supports HSCs.

PHENOTYPE OF PLACENTAL HSCs

In order to investigate how the placental HSC pool is established, it is important to define markers that allow for isolation and visualization of placental HSCs. Since the surface markers change on developing HSCs during their progressive maturation, and because the functional properties that are required for engraftment of HSCs in the bone marrow in a standard adult transplantation assay are acquired only later during their development, purification of developing HSCs with accuracy is difficult [Godin and Cumano, 2007; Mikkola and Orkin, 2006]. So far, there are no markers that distinguish reliably a newly formed HSCs from a transient progenitor. Furthermore, it is important to distinguish fetal hematopoietic cells from maternal blood cells as both are circulating through the placenta. Based on the studies that have been performed, placental HSCs share the same surface markers as developing HSCs in other hematopoietic sites at the same stage of ontogeny. A marker combination that excludes circulating maternal cells and enriches for placental HSCs–multipotential progenitors throughout midgestation is CD34^{med}c-kit^{hi} [Gekas, 2005]. However, as CD34 and ckit are also expressed in endothelial progenitor cells, albeit at different levels (CD34^{hi}c-kit^{med}), the precise level of expression of these markers and their combination with other surface markers is important for isolating developing HSCs reliably. The first known blood specific marker that is expressed in nascent HSCs–progenitors in the yolk sac, AGM, and the placenta at the time of their emergence is CD41 (also expressed in megakaryocytes) [Bertrand, 2005; Corbel and Salaun, 2002; Ferkowicz, 2003; Matsubara, 2005; Mikkola, 2003; Rhodes, 2008]. However, by E12.5 CD41 is no longer expressed in LTR–HSCs, likely due to progressive downregulation during HSC maturation (unpublished observation). Although the first CD41+ HSC/progenitor cells do not express the panhematopoietic marker CD45, it becomes expressed in HSCs by E11.5 [Matsubara, 2005; Taoudi, 2005]. The adult LTR–HSC marker Sca1 is not expressed on the surface of nascent HSCs at their emergence in the AGM [de Bruijn, 2002], but becomes upregulated by E12.5. However, by using a transgenic Sca1–GFP mouse line, HSCs in the placenta, as well as the AGM, were enriched in the Sca1–GFP fraction even at an earlier stage when Sca1 was not yet expressed on the cell surface [de Bruijn, 2002; Ottersbach and Dzierzak, 2005]. Note, as Sca1 is also expressed in a subset of endothelial cells and in decidual cells of the placenta, combination with other markers is essential to purify placental HSCs. Other markers expressed in endothelial cells are also expressed in HSCs for varying developmental time periods. For example, CD31 (PECAM), the pan-endothelial marker remains expressed in HSCs throughout their ontogeny [Baumann, 2004; Taoudi and Medvinsky, 2007], while CD34 expression is downregulated during early postnatal life when the HSC pool shifts from actively cycling (fetal characteristic) to predominantly quiescent adult HSCs [Ogawa, 2001]. Interestingly, VE-cadherin, another vascular marker important in cell adherens junctions, is expressed in HSCs at their emergence, but is completely downregulated by E16.5, after fetal liver colonization [Kim, 2005; Taoudi, 2005]. The VE-cadherin expression pattern in placental HSCs has not been defined by transplantation, however, fluorescence activated cell sorting (FACS) analysis shows that placental CD34^{med}c-kit^{hi} HSC–progenitor population downregulates VE-cadherin between E11.5–12.5, as expected. Although the cellular origin of HSCs is still a matter of controversy, the abundant coexpression of endothelial markers on the emerging HSCs supports the hypothesis that HSCs may originate from hemogenic endothelium.

ORIGIN AND LOCALIZATION OF PLACENTAL HSCs

Tracing the origin of HSCs in the embryo is complicated because definitive HSCs develop a few days after transient blood progenitors, which start to circulate freely in the conceptus once heartbeat has been established, and there are no markers that distinguish reliably between a nascent HSC and a transient progenitor in the early embryo. As all hematopoietic cells develop from mesoderm, the chorionic and allantoic mesoderm would be the putative tissues of HSC origin if HSCs are generated in the placenta *de novo*. The localization of putative hemogenic precursors and HSCs in the placenta was mapped by a Runx1-LacZ reporter mouse. Runx1 is a transcription factor that is required for definitive hematopoiesis, and lack of Runx1 abolishes HSC formation leading to embryonic lethality by E12.5 [Wang, 1996]. Although not every Runx1 expressing cell is a hematopoitic stem–progenitor cell, Runx1 is expressed in all HSCs throughout ontogeny, therefore localizing its expression indicates possible sites of definitive hematopoiesis. The Runx1 locus is active also in the homozygous Runx1-LacZ null embryos, which are unable to generate HSCs, but display LacZ expression in the precursors that are attempting to generate HSCs. This model has been used to detect HSCs and their precursors in the AGM, where Runx1 expression is observed on the ventral side of the dorsal aorta, the site of HSC emergence (Table XII.1) [North, 1999]. Runx1 expression can first be observed in the conceptus at E7.5, when the blood islands in the yolk sac, the chorionic mesoderm, and parts of the allantoic mesoderm are labeled by β-galactosidase staining (Fig. XII.3a; Table XII.1) [Zeigler, 2006]. To define whether the allantoic and/or chorionic mesoderm have capacity for *de novo* hematopoiesis, these mesodermal tissues were isolated prior to placental fusion and circulation and assayed for hematopoietic potential. As it had been shown earleir that the precirculation allantois does not harbor hematopoietic progenitors that can proliferate on standard methylcellulose progenitor assays [Downs, 1998; Palis, 1999], hematopoietic potential was assessed by stimulating the cells from the tissues first by *in vitro* coculture on OP-9 stroma followed by methylcellulose cultures, or by a stroma free explant culture system where hemogenic precursors remain in contact with their neighboring cells [Corbel, 2007; Zeigler, 2006]. Strikingly, both the chorionic and allantoic mesoderm

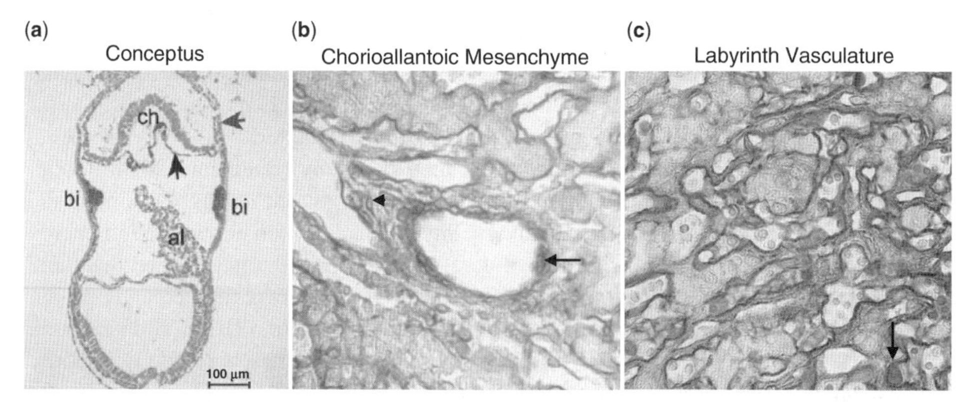

Figure XII.3. The Runx1–LacZ expression in the placenta. (a) A cross-section of a precirculation Runx1–LacZ conceptus documenting Runx1 expression in the chorionic mesoderm (black arrow) and the blood-islands of the yolk sac. The blue arrow denotes the ectoplacental cone. (b) The large vessels of the chorioallantoic mesenchyme harbor LacZ+ cells within the wall of the vessel (arrow) at E11.5, the time when HSCs emerge. The mesenchyme contains two distinct populations of LacZ+ cells, oblong-shaped cells that straddle within the stromal cells (arrowhead), and cuboidal LacZ+/cytokeratin+ cells derived from ectoplacental endoderm that have an organized structure (asterisk). (c) The LacZ+ definitive hematopoietic cells localize to the fetal labyrinth vasculature. (a–c) The Runx1–LacZ is in blue. (b,c) Laminin (pink) marks mesodermal tissues while cytokeratin (brown) marks trophoblasts and ectoplacental endoderm in the placenta. Bi = blood island; al = allantois. (See color insert.)

were capable of generating multipotential myelo-erythroid colonies (Table XII.1). These findings suggested that the mesodermal tissues destined to become the placenta have innate potential to generate multipotential hematopoietic precursors, similar to what had been shown earlier with the avian allantois. Furthermore, an earlier study in mice had shown that an ectopically transplanted murine allantois can contribute to the dorsal aorta and surrounding tissues [Downs and Harmann, 1997]. Although hematopoietic potential of the ectopic allantoic mesoderm was not addressed in this study, these findings imply a similar developmental potential of the allantoic mesoderm as the lateral plate mesoderm in the embryo proper.

The Runx1–LacZ mouse model was used, but at a later developmental age after chorioallantoic fusion, to define the anatomical location and timing of hematopoietic progenitor–stem cell emergence in the placenta [Ottersbach and Dzierzak, 2005; Rhodes, 2008]. Interestingly, the large vessels of the chorioallantoic mesenchyme also harbor LacZ+ cells integrated within the wall of the vessel lumen, highly reminiscent of the LacZ+ cells found within the ventral wall of the dorsal aorta in Runx1–LacZ embryos (Fig. XII.3b; Table XII.1). The presence of LacZ+ cells in the large vessels of the chorioallantoic mesenchyme in both the heterozygote Runx1$^{LacZ/+}$ and homozygote Runx1$^{LacZ/LacZ}$ embryos, which are unable to generate HSCs from their immediate precursors due to lack of Runx1, suggests that these vessels may be a site of HSC origin. Additionally, round LacZ+ definitive hematopoietic cells in Runx1$^{LacZ/+}$ placentas localized to the fetal labyrinth vessels. As the labyrinth in Runx1$^{LacZ/LacZ}$ placentas was devoid of these cells, it was suggested that the placental labyrinth vasculature may be a niche to which readily made HSCs colonize (Fig. XII.3c; Table XII.1). Furthermore, many of the round LacZ+ cells within the labyrinth vessels were mitotically active and formed small clusters, prompting the hypothesis that the placental labyrinth vasculature is a site for HSC expansion. In summary, these data suggested that the large vessels within the chorioallantoic mesenchyme may generate HSCs, similar to the large dorsal aorta and adjacent large vessels in the AGM region, while the small fetal labyrinth vessels of the placenta provide a niche in which HSCs may expand and mature in. In further support of this hypothesis, Runx1–LacZ+ cells in the chorioallantoic and labyrinthine blood vessels coexpress CD41, implying that nascent HSCs–progenitors emerge and reside in the vasculature [Rhodes, 2008].

It is important to note that in addition to being expressed in the nascent HSCs–progenitors, Runx1 is also expressed in some other cells that are not HSCs or even hematopoietic cells. For example, oblong-shaped LacZ+ cells fill the chorioallantoic mesenchyme and may be precursors for primitive macrophages that are found in the stroma (Table XII.1) [Rhodes, 2008]. The chorioallantoic mesenchyme also contains LacZ+ cells that form tubular structures that ultimately organize into a border between the chorioallantoic mesenchyme and the labyrinth (Crypt of Duval) [Duval, 1891; Ogura, 1998; Rhodes, 2008]. These cells are derived from ectoplacental endoderm. Although the specific function of these cells remains unclear it has been proposed that thay are later involved in calcium homeostasis [Kovacs, 2002]. Furthermore, as these cells secrete signaling molecules, such as Ihh (Indian Hedhehog), and are in close proximity to the large placental vessels where HSCs emerge during midgestation it has been postulated that they may provide local signals required for hematopoiesis in the placenta [Ogura, 1998; Rhodes, 2008].

The hypothesis that the placenta has a capacity for *de novo* hematopoiesis was confirmed using an Ncx1$^{-/-}$ knockout mouse model that lacks circulation [Rhodes, 2008]. These embryos are defective for the sodium–calcium exchange pump 1, which abolishes the initiation of heartbeat and thus prevents circulation of blood cells from their site of

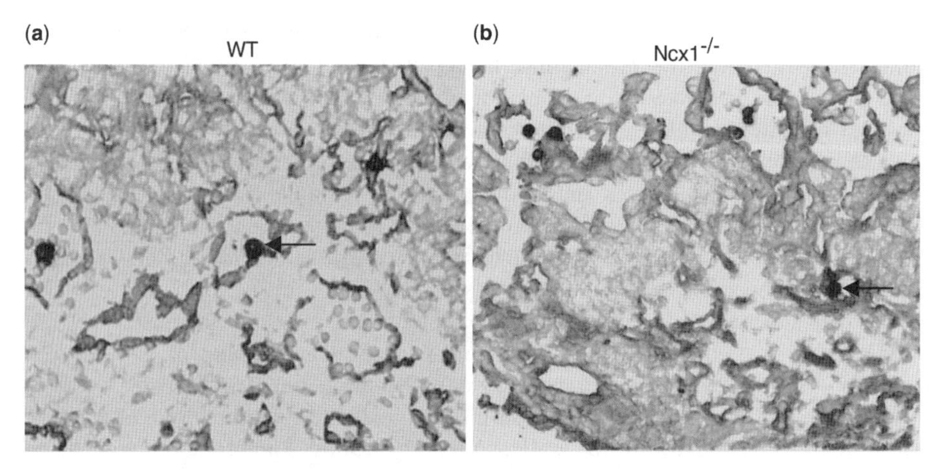

Figure XII.4. *De Novo* hematopoiesis in the placenta. (a) A wild-type placenta harboring CD41+ hematopoietic cells in the placental vasculature. (b) CD41+ hematopoietic cells emerge in the, vasculature in Ncx1$^{-/-}$ placenta in the absence of circulation. CD41 (blue) marks hematopoietic cells, whereas CD31 (red) identifies endothelium. Cytokeratin (brown) marks the trophoblasts. (See color insert.)

origin [Koushik, 2001; Lux, 2008]. Analysis of placentas from Ncx1$^{-/-}$ embryos revealed cells expressing CD41 attached to the luminal side of the large vessels within the chorioallantoic mesenchyme and developing labyrinth of the placenta (Fig. XII.4b), implying that *de novo* hematopoiesis occurs in these placental vessels (Table XII.1). As the Ncx1$^{-/-}$ embryos die by E10.5, before developing HSCs have acquired the ability to engraft and reconstitute adult bone marrow in transplantation assays, the developmental potential of placental hematopoietic cells could not be addressed *in vivo*. Importantly, placental tissues, as well as the AGM and yolk sac from Ncx1$^{-/-}$ embryos, demonstrated multilineage hematopoietic potential upon *in vitro* culture on OP9 and OP9-Dl1 stroma. Specifically, the ability to generate myeloerythroid, B- and T-lymphoid cells indicates that placental hematopoietic cells have the differentiation potential characteristic for HSCs, distinct from earlier, transient progenitor populations in the yolk sac that solely generate myeloerythroid cells. Taken together these studies implied that the placenta can initiate multilineage hematopoiesis de novo, and that this autonomous process is localized to the vasculature. However, further studies are required to discern which of the mesodermal components of the placenta give rise to the precursors that ultimately generate definitive HSCs. Although it is widely accepted that HSCs emerge from the vasculature further studies will be required to determine whether the HSCs are specified from a hemogenic endothelium or a precursor that originates from the subvascular mesenchyme.

SUMMARY AND FUTURE DIRECTIONS

Proper development and function of the placenta is essential for supporting healthy pregnancy. However, its vital functions are not only limited to its requirement in mediating fetal–maternal exchange, but also include a novel role as a fetal HSC niche. The uniqueness of the placenta as a hematopoietic organ relies on the findings that it is capable of *de novo* hematopoiesis, similar to the AGM and the yolk sac, but that it also establishes the first major

HSC pool in the conceptus. Although not formally proven, the establishment of the large pool of HSCs in the placenta is likely do to a combination of events: *de novo* HSC generation in the placenta, HSC seeding from the umbilical artery, the AGM and perhaps even the yolk sac, and HSC expansion in the placental vascular labyrinth. It is also possible that the drastic increase in functional HSCs is in part due to functional maturation of HSCs in the placenta, after which they can engraft in a standard transplantation assays. Remarkably, when HSCs reside in the placenta, they are not directed toward differentiation. These data imply that the placental microenvironment has unique properties that can support both HSC emergence and maintenance–expansion.

Despite the distinct evolutionary differences between vertebrate species, such as mouse, human, and avian, the establishment of the hematopoietic system during embryogenesis is a highly conserved process. In all organisms named above, HSCs emerge from the dorsal aorta in the embryo proper and other major vasculature, and the same transcription factors and signaling pathways appear to conduct this process [Martinez-Agosto, 2007]. Although the placenta is unique to eutherial mammals, structures with similar mesodermal origin are also found in other species. Note that, the avian allantois is a tissue rudiment with comparable developmental origin as the mammalian allantois, and it harbors multipotential hematopoietic cells with bone marrow reconstitution ability. It has even been suggested that in zebrafish, the vascular plexus termed the caudal hematopoietic tissue (CHT) and supports expansion and maturation of hematopoietic cells, may have similar functions in hematopoiesis as the mammalian placenta or the fetal liver [Burns and Zon, 2006; Murayama, 2006]. Furthermore, preliminary data suggests that the first trimester human placenta also harbors hematopoietic activity at comparable developmental time as the mouse placenta (unpublished data). The human placenta is known as a villous placenta, whereas the mouse placenta is labyrinthine type, and thereby the human and mouse placentas appear macroscopically different. However, the placentas are remarkably similar at the cellular and molecular level [Georgiades, 2002], and thereby it is likely that similar niche cells coordinate hematopoiesis in the placenta in both species. Further mechanistic studies will be required to define the key niche cells and molecular cues that support placental hematopoiesis in the chorioallantoic and labyrinth vasculature. Indeed, the placenta is rich in growth factors and cytokines that are secreted by various placental cell types, such as the endothelium, stromal cells, ectoplacental endoderm–Crypt of Duval, trophoblasts. Isolating these niche cells and defining the molecular cues that dictate distinct stages of HSC specification, emergence, maturation, and expansion will be of great value, and may have a major impact for clinical applications that aim to recreate fetal hematopoietic niches *in vitro* for expansion of HSCs from cord blood, or generation from pluripotent cells, such as human ES cells or induced pluripotent stem (iPS) cells.

REFERENCES

Alvarez-Silva M, Belo-Diabangouaya P, Salaun J, Dieterlen-Lievre F. 2003. Mouse placenta is a major hematopoietic organ. Development. 130(22):5437–5444.

Baumann CI, Bailey AS, Li W, Ferkowicz MJ, Yoder MC, Fleming WH. 2004. PECAM-1 is expressed on hematopoietic stem cells throughout ontogeny and identifies a population of erythroid progenitors. Blood. 104(4):1010–1016.

Bertrand JY, Giroux S, Golub R, Klaine M, Jalil A, Boucontet L, Godin I, Cumano A. 2005. Characterization of purified intraembryonic hematopoietic stem cells as a tool to define their site of origin. Proc Natl Acad Sci USA. 102(1):134–139.

Burns CE, Zon LI. 2006. Homing sweet homing: Odyssey of hematopoietic stem cells. Immunity. 25(6):859–862.

Caprioli A, Jaffredo T, Gautier R, Dubourg C, Dieterlen-Lievre F. 1998. Blood-borne seeding by hematopoietic and endothelial precursors from the allantois. Proc Natl Acad Sci USA. 95(4):1641–1646.

Caprioli A, Minko K, Drevon C, Eichmann A, Dieterlen-Lievre F, Jaffredo T. 2001. Hemangioblast commitment in the avian allantois: Cellular and molecular aspects. Dev Biol. 238(1):64–78.

Corbel C, Salaun J. 2002. AlphaIIb integrin expression during development of the murine hemopoietic system. Dev Biol. 243(2):301–311.

Corbel C, Salaun J, Belo-Diabangouaya P, Dieterlen-Lievre F. 2007. Hematopoietic potential of the pre-fusion allantois. Dev Biol. 301(2):478–488.

Cross JC. 2005. How to make a placenta: Mechanisms of trophoblast cell differentiation in mice—a review. Placenta. 26(Suppl A):S3–S9.

Cumano A, Dieterlen-Lievre F, Godin I. 1996. Lymphoid potential, probed before circulation in mouse, is restricted to caudal intraembryonic splanchnopleura. Cell. 86(6):907–916.

Dancis J, Jansen V, Brown GF, Gorstein F, Balis ME. 1977. Treatment of hypoplastic anemia in mice with placental transplants. Blood. 50(4):663–670.

de Bruijn MF, Ma X, Robin C, Ottersbach K, Sanchez MJ, Dzierzak E. 2002. Hematopoietic stem cells localize to the endothelial cell layer in the midgestation mouse aorta. Immunity. 16(5):673–683.

de Bruijn MF, Speck NA, Peeters MC, Dzierzak E. 2000. Definitive hematopoietic stem cells first develop within the major arterial regions of the mouse embryo. Embo J. 19(11):2465–2474.

Downs KM, Gifford S, Blahnik M, Gardner RL. 1998. Vascularization in the murine allantois occurs by vasculogenesis without accompanying erythropoiesis. Development. 125(22):4507–4520.

Downs KM, Harmann C. 1997. Developmental potency of the murine allantois. Development. 124(14):2769–2780.

Duval M. 1891. Le placenta des rongeurs. J Anat Physiol. Paris 27(1891):24–73, 344–395, 515–612.

Ferkowicz MJ, Starr M, Xie X, Li W, Johnson SA, Shelley WC, Morrison PR, Yoder MC. 2003. CD41 expression defines the onset of primitive and definitive hematopoiesis in the murine embryo. Development. 130(18):4393–4403.

Gekas C, Dieterlen-Lievre F, Orkin SH, Mikkola HK. 2005. The placenta is a niche for hematopoietic stem cells. Dev Cell. 8(3):365–375.

Georgiades P, Ferguson-Smith AC, Burton GJ. 2002. Comparative developmental anatomy of the murine and human definitive placentae. Placenta. 23(1):3–19.

Godin I, Cumano A. 2007. [Hematopoietic stem cells: Where do they come from at last?]. Med Sci (Paris). 23(8–9):681–684.

Godin I, Garcia-Porrero JA, Dieterlen-Lievre F, Cumano A. 1999. Stem cell emergence and hemopoietic activity are incompatible in mouse intraembryonic sites. J Exp Med. 190(1):43–52.

Kim I, Yilmaz OH, Morrison SJ. 2005. CD144 (VE-cadherin) is transiently expressed by fetal liver hematopoietic stem cells. Blood. 106(3):903–905.

Koushik SV, Wang J, Rogers R, Moskophidis D, Lambert NA, Creazzo TL, Conway SJ. 2001. Targeted inactivation of the sodium-calcium exchanger (Ncx1) results in the lack of a heartbeat and abnormal myofibrillar organization. Faseb J. 15(7):1209–1211.

Kovacs CS, Chafe LL, Woodland ML, McDonald KR, Fudge NJ, Wookey PJ. 2002. Calcitropic gene expression suggests a role for the intraplacental yolk sac in maternal-fetal calcium exchange. Am J Physiol Endocrinol Metab. 282(3):E721–732.

Lux CT, Yoshimoto M, McGrath K, Conway SJ, Palis J, Yoder MC. 2008. All primitive and definitive hematopoietic progenitor cells emerging prior to E10 in the mouse embryo are products of the yolk sac. Blood III(7):3435–3438. Prepublished online as a Blood First Edition Paper on Oct 11, 2007; DOI: 10.1182/blood-2007-08-107086.

Martinez-Agosto JA, Mikkola HK, Hartenstein V, Banerjee U. 2007. The hematopoietic stem cell and its niche: A comparative view. Genes Dev. 21(23):3044–3060.

Matsubara A, Iwama A, Yamazaki S, Furuta C, Hirasawa R, Morita Y, Osawa M, Motohashi T, Eto K, Ema H, Kitamura T, Vestweber D, Nakauchi H. 2005. Endomucin, a CD34-like sialomucin, marks hematopoietic stem cells throughout development. J Exp Med. 202(11):1483–1492.

Medvinsky A, Dzierzak E. 1996. Definitive hematopoiesis is autonomously initiated by the AGM region. Cell. 86(6):897–906.

Melchers F. 1979. Murine embryonic B lymphocyte development in the placenta. Nature (London). 277(5693):219–221.

Mikkola HK, Fujiwara Y, Schlaeger TM, Traver D, Orkin SH. 2003. Expression of CD41 marks the initiation of definitive hematopoiesis in the mouse embryo. Blood. 101(2):508–516.

Mikkola HK, Orkin SH. 2006. The journey of developing hematopoietic stem cells. Development. 133(19):3733–3744.

Muller AM, Medvinsky A, Strouboulis J, Grosveld F, Dzierzak E. 1994. Development of hematopoietic stem cell activity in the mouse embryo. Immunity. 1(4):291–301.

Murayama E, Kissa K, Zapata A, Mordelet E, Briolat V, Lin HF, Handin RI, Herbomel P. 2006. Tracing hematopoietic precursor migration to successive hematopoietic organs during zebrafish development. Immunity. 25(6):963–975.

North T, Gu TL, Stacy T, Wang Q, Howard L, Binder M, Marin-Padilla M, Speck NA. 1999. Cbfa2 is required for the formation of intra-aortic hematopoietic clusters. Development. 126(11): 2563–2575.

Ogawa M, Tajima F, Ito T, Sato T, Laver JH, Deguchi T. 2001. CD34 expression by murine hematopoietic stem cells. Developmental changes and kinetic alterations. Ann NY Acad Sci. 938: 139–145.

Ogura Y, Takakura N, Yoshida H, Nishikawa SI. 1998. Essential role of platelet-derived growth factor receptor alpha in the development of the intraplacental yolk sac/sinus of Duval in mouse placenta. Biol Reprod. 58(1):65–72.

Ottersbach K, Dzierzak E. 2005. The murine placenta contains hematopoietic stem cells within the vascular labyrinth region. Dev Cell. 8(3):377–387.

Palis J, Robertson S, Kennedy M, Wall C, Keller G. 1999. Development of erythroid and myeloid progenitors in the yolk sac and embryo proper of the mouse. Development. 126(22):5073–5084.

Rhodes KE, Gekas C, Wang Y, Lux CT, Francis CS, Chan DN, Conway S, Orkin SH, Yoder MC, Mikkola HK. 2008. The emergence of hematopoietic stem cells is initiated in the placental vasculature in the absence of circulation. Cell Stem Cell. 2(3):252–263.

Taoudi S, Medvinsky A. 2007. Functional identification of the hematopoietic stem cell niche in the ventral domain of the embryonic dorsal aorta. Proc Natl Acad Sci USA. 104(22):9399–9403.

Taoudi S, Morrison AM, Inoue H, Gribi R, Ure J, Medvinsky A. 2005. Progressive divergence of definitive haematopoietic stem cells from the endothelial compartment does not depend on contact with the foetal liver. Development. 132(18):4179–4191.

Till JE, McCulloch EA. 1961. A direct measurement of the radiation sensitivity of normal mouse bone marrow cells. Radiat Res. 14:213–222.

Wang Q, Stacy T, Binder M, Marin-Padilla M, Sharpe AH, Speck NA. 1996. Disruption of the Cbfa2 gene causes necrosis and hemorrhaging in the central nervous system and blocks definitive hematopoiesis. Proc Natl Acad Sci USA. 93(8):3444–3449.

Zeigler BM, Sugiyama D, Chen M, Guo Y, Downs KM, Speck NA. 2006. The allantois and chorion, when isolated before circulation or chorio-allantoic fusion, have hematopoietic potential. Development. 133(21):4183–4192.

13

FETAL CELL MICROCHIMERISM, A LOW-GRADE NATURALLY OCCURRING CELL THERAPY

Michèle Leduc, Selim Aractingi, and Kiarash Khosrotehrani

Developmental Physiopathology Laboratory, UPMC University, Paris, France

INTRODUCTION

Over the last decade, new areas of research have emerged describing physiological events during fetal development that may have long-lasting effects on maternal health. Among them, fetal cell microchimerism has been one of the most studied.

This phenomenon is defined by the presence and persistence of a low number of cells from a donor in a genetically different recipient. During the last decade, it has become clear that cells traffic between mother and fetus. In this chapter, we will detail recent finding on the nature of these cells as well as the implications of this phenomenon.

THE PLACENTA IS A SOURCE OF STEM CELLS

For years, the yolk sac and the aorta–gonad–mesonephros (AGM) have been considered as two hematopoietic organs in the embryo. Recently, a number of studies have established the placenta as yet another hematopoietic organ that plays an important role in hematopoietic stem cells. The data supporting this role of placenta will be discussed more thoroughly elsewhere. Alvarez-Silva et al. showed that the mouse placenta contains a high frequency of multipotential clonogenic progenitors [Alvarez-Silva, 2003]. Further analying the repopulation capacity of placental cells, Gekas et al. reported that the number of HSCs from 12.5–13.5-days murine placenta is 15-fold higher than in the AGM [Gekas, 2005]. Most of these placental HSCs were found within the c-kit^{hi}CD34+ population, which represent up to 1–2% of nucleated cells obtained from collagenase-treated placentas. The majority

Perinatal Stem Cells. Edited by C. L. Cetrulo, K. J. Cetrulo, and C. L. Cetrulo, Jr.

(87%) of colony forming (CFU−) cells in the placenta were also c-kit+ CD34+. Interestingly, Sca-1 expression, another marker for hematopietic progenitors, has also been described on 78% of placental CD34+ cells [Ottersbach, 2005]. In humans, Challier et al. also described the presence of fetal erythroblasts in first trimester placentas. These erythroblasts were labeled with GATA-2, which is considered as a marker of primitive hematopoiesis and c-kit labeling of endothelial cells at similar gestational age was suggestive of hematopoiesis [Challier, 2005]. Placenta also contains mesenchymal stem cells (MSCs). Many investigators demonstrated that cells isolated and expanded from different parts of human second-trimester and term placentas (e.g., AF amniotic fluid, amnion, or deciduas) express typical markers of MSCs, such as CD90, CD105, CD166, CD49e, SH3, SH4, HLA-ABC, and are devoid of CD31, CD34, CD45, CD49d, CD123, HLA-DR [Bailo, 2004; Fukuchi, 2004; In 't Anker, 2004]. De coppi et al. described a population of multipotent mesenchymal cells isolated from amniotic fluid with expression of Oct-4 or Nanog characteristic of embryonic stem cells. This cell population could be maintained *in vitro* and could produce *in vitro* multiple cell types including adipocytes, osteoblasts, chondroblasts, neuroblasts, endothelial cells, myocytes, and hepatocytes. Yen et al. also isolated a population of multipotent cells from human term placentas (PDMCs). Furthermore, these PDMCs displayed the ability to differentiate into mesodermal-lineage cells, such as osteoblasts or adipocytes, as well as ectodermal neuron-like cells [Yen, 2005, 2007, 2008].

TRANSFER OF FETAL PROGENITOR CELLS DURING GESTATION

Overall these studies demonstrate the presence of different populations of stem cells with a broad plasticity residing within the placenta at different gestational ages in humans as well as in mice. Consequently, a seeding of these progenitors occurs into the maternal circulation probably due to breaches in the trophoblast villi.

Fetal cells enter the maternal circulation during all human and murine pregnancies [Ariga, 2001; Khosrotehrani, 2005a] and persist in some women for decades after delivery [Bianchi, 1996]. Fetal cells are indeed found in the circulation of 30–70% of healthy women with a prior history of pregnancy [Artlett, 2002; Lambert, 2002]. Fetal cells in maternal circulation have various phenotypes. Nucleated erythroblasts, lymphocytes, and trophoblast cells are mostly identified during gestation. Interestingly, among the chimeric fetal cells in maternal tissues, the progenitors that have been described in the placenta can be retrieved. Guetta et al. demonstrated the presence of fetal hematopoietic progenitor cells in maternal blood during the second trimester (15–25 weeks of getational age) [Guetta, 2003]. Using *in situ* hybridization, they reported male-presumably fetal cells in the CD34 sorted population in all (22/22) blood samples obtained from women carrying a male fetus. For most women, these results were confirmed using a Y chromosome specific nested polymerase chain reaction (PCR) amplification technique. The average number of fetal CD34+ cells was 13 in 20-mL maternal blood. In agreement with these results, fetal CD34+ cells were detected in 95% of pregnancies using a rhesus monkey model of fetal cell microchimerism [Jimenez, 2005]. The cells were detected during the second and third trimester. The estimated frequency of the fetal CD34+ cells was 4.5/50,000 maternal CD34+ cells. The long-term persistence of fetal microchimeric hematopoietic progenitor cells was reported for the first time by Bianchi et al. [Bianchi, 1996]. Fetal male cells could be detected in the flow-cytometry sorted CD34+CD38+ population in six out of eight women with a history of male pregnancy. In addition, in one patient fetal CD34+CD38− cells could be detected. This study suggested the long-term engraftment of fetal lymphoid progenitors (CD34+CD38+) or hematopoietic stem cells (CD34+CD38−) transferred during pregnancy [Bianchi, 1996]. More recently, Adams et al. confirmed these results by studying women who underwent CD34-enrichment apheresis after chemotherapy and/or G-CSF mobilization. Male cells, presumably fetal in origin, were detected in 48% of CD34-enriched samples [Adams, 2003]. The estimation of the number of fetal CD34+ cells was from <1–357/million of maternal cells. Finally, in their study, Guetta et al. reported fetal CD34+ cells in the circulation of 2 out of 10 women with a prior history of male pregnancy [Guetta, 2003]. The quantification of the number of fetal cells by *in situ* hybridization among the maternal CD34+ population was 18–24-cells/20 mL maternal blood. In this study, the time interval since the last male delivery was between 6 and 24 years. However, these studies were based on the expression of the CD34 antigen and do not conclusively demonstrate the "stemness" of these cells.

More functional studies accordingly demonstrated that, in semisolid cultured maternal blood obtained at delivery, fetal hematopoietic progenitor cells formed erythroid and granulocyte-monocyte colonies [Osada, 2001]. Progenitor cells of fetal origin could not be detected 1 year after delivery. Interestingly, in a similar study performed on second trimester maternal blood, hematopoietic progenitors of fetal origin were found to include more primitive progenitors that could give rise to granulocyte, erythrocyte, macrophage, megakaryocyte (CFU-GEMM), and also CFU-blasts, probably reflecting the presence of more immature progenitors at this gestational age [Valerio, 1997].

In addition to the hematopoietic progenitors, fetal mesenchymal stem cells are detected in maternal blood, although at a much lower frequency [O'Donoghue, 2003]. The presence of fetal mesenchymal stem cells in 1 out of 20 tested women after first trimester termination was demonstrated based on their proliferation and differentiation capacity [O'Donoghue, 2003]. More recently, the long-term persistence of mesenchymal stem cells has been demonstrated as well. O'Donoghue et al. studied bone marrow from women with prior male pregnancies [O'Donoghue, 2004]. Using Y chromosome specific *in situ* hybridization, male-presumably fetal-vimentin or laminin expressing cells were found in the lamellae of cortical bone or in connective tissue adjacent to trabecular bone up to 51 years after delivery. In this study, mesenchymal stem cells were isolated by cell culture of flushed rib bone marrow. In 9/9 women with a prior male pregnancy versus 0/5 women without any male child, male fetal mesenchymal stem cells were isolated. These fetal mesenchymal stem cells persisting long term in maternal bone marrow had the following markers: CD45−, CD14−, CD11a−, CD49b−, CD49d^{low}, SH2+, SH3+, Vimentin+, CD29+, CD49e+, CD106+, HLA-Class II- [O'Donoghue, 2004]. The adipogenic and osteogenic capacities of these cells of fetal origin was as well proven in this study as a further demonstration of the persistence of fetal mesenchymal stem cells in maternal bone marrow during decades.

FETAL PROGENITOR CELLS IN MATERNAL TISSUES

Fetal chimeric progenitor cells home to the maternal damaged areas as part of the repair process. This hypothesis is now supported by a large number of human and animal studies as detailed below. Recently, the capacity of fetal lymphoid progenitor cells acquired during pregnancy to develop in maternal thymus and bone marrow into functional T and B cells was demonstrated for the first time in wild-type, as well as immunodeficient mothers [Khosrotehrani, 2008]. The cells of fetal origin consistently harbored a paternal genetic marker that was absent in the mother in a variety of study models. Thymic development of fetal T cells was demonstrated by the expression of cell surface markers of "immature double positive" thymocytes, such as low T cell Receptor, presence of both CD4 and CD8 and an absence of the IL7 receptor [Akashi, 1998; Ellmeier, 1999]. In addition, single positive CD4 or CD8+ lymphocytes of fetal origin with high IL7R and TCR expression patterns were also described. Altogether, the presence of fetal thymocytes at various T cell differentiation stages strongly implied that fetal T cell progenitors enter and differentiate in the maternal thymus. The resulting fetal lymphocytes were functional and could restore allogenic and antigen specific T cell responses in immunodeficient mothers. These results are supported by a strain of previous findings in humans showing the presence of lymphocytes of fetal origin in maternal circulation from healthy women. Fetal cells could indeed be identified in the CD3, CD4, or CD8+ sorted populations of, respectively, 70, 31, and 64% of tested patients [Evans, 1999; Lambert, 2002]. In addition, other hematopoietic cell types of fetal origin also have been reported. The multilineage phenotype of fetal cells described long term after gestation clearly suggests the presence of fetal progenitors in maternal bone marrow.

Apart from the hematopoietic system, fetal cells have been described in a variety of maternal tissues in both human and mice [Khosrotehrani, 2005b]. These cells can adopt various phenotypes in maternal injured tissues and express markers of hepatocytes or epithelial cells in addition to leukocytes [Khosrotehrani, 2004]. Fetal cells have been identified in maternal liver affected with a variety of disorders ranging from hepatitis C to primary biliary cirrhosis. These cells had the morphology of liver cells and expressed hepatocyte markers [Guettier, 2005; Stevens, 2004]. Similar results were obtained in animal models. Maternal liver after chemical injury or alcohol abuse was invaded by fetal cells that expressed albumin [Khosrotehrani, 2007; Wang, 2004]. Note, this recruitment of fetal cells occurred over time during the regeneration process. In addition, the type of injury leading to fetal cell recruitment seemed important since other types of injury, such as partial hepatectomy, did not result in similar homing of fetal cells [Khosrotehrani, 2007]. Similarly, fetal derived cells have been shown to adopt the phenotype of thyroid cells [Srivatsa, 2001], cardiomyocytes [Bayes-Genis, 2005], kidney tubular cells [Wang, 2004], and even various cells associated with the central nervous system, such as neurons, astrocytes, or glial cells [Tan, 2005]. All these reports emphasized the need for maternal injury to recruit fetal cells. In the absence of injury, fetal cells were either undetectable or at a very low level. This result is further demonstrated in recent experiments during gestation. Skin inflammation alone was able to recruit fetal endothelial progenitors that expressed CD31 and VEGFR2 and that could form entire vessels derived from fetal cells [Nguyen Huu, 2007].

FUTURE THERAPIES

The results reported above show that the plasticity of the fetal progenitor cells found in the placenta can be observed in maternal injured tissues *in vivo*. Therefore, we believe that the study of fetal cell microchimerism can be informative on the putative capacity of placental stem cells acquired naturally. Isolation and identification of the cell populations that explain the observed plasticity described above could be the starting point of effective cellular therapies. In the current debate over the use of embryonic stem cells for treatment of disease, the use of fetal stem cell populations that can be acquired without major ethical controversy could prove significant.

REFERENCES

Adams KM, Lambert NC, Heimfeld S, Tylee TS, Pang JM, Erickson TD, Nelson JL. 2003. Male DNA in female donor apheresis and CD34-enriched products. Blood. 102(10):3845–3847.

Akashi K, Kondo M, Weissman IL. 1998. Role of interleukin-7 in T-cell development from hematopoietic stem cells. Immunol Rev. 165:13–28.

Alvarez-Silva M, Belo-Diabangouaya P, Salaun J, Dieterlen-Lievre F. 2003. Mouse placenta is a major hematopoietic organ. Development. 130:5437–5444.

Ariga H, Ohto H, Busch MP, Imamura S, Watson R, Reed W, Lee TH. 2001. Kinetics of fetal cellular and cell-free DNA in the maternal circulation during and after pregnancy: Implications for non-invasive prenatal diagnosis. Transfusion. 41:1524–1530.

Artlett CM, Cox LA, Ramos RC, Dennis TN, Fortunato RA, Hummers LK, Jimenez SA, Smith JB. 2002. Increased microchimeric CD4+ T lymphocytes in peripheral blood from women with systemic sclerosis. Clin Immunol. 103:303–308.

Bailo M, Soncini M, Vertua E, Signoroni PB, Sanzone S, Lombardi G, Arienti D, Calamani F, Zatti D, Paul P, Albertini A, Zorzi F, Cavagnini A, Candotti F, Wengler GS, Parolini O. 2004. Engraftment potential of human amnion and chorion cells derived from term placenta. Transplantation. 78(10):1439–1448.

Bayes-Genis A, Bellosillo B, de la CO, Salido M, Roura S, Ristol FS, Soler C, Martinez M, Espinet B, Serrano S, Bayes DL, Cinca J. 2005. Identification of male cardiomyocytes of extracardiac origin in the hearths of women with male progeny: Male fetal cell microchimerism of the hearth. J Hearth Lung Transplant. 24:2179–2183.

Bianchi DW, Zickwolf GK, Weil GJ, Sylvester S, DeMaria MA. 1996. Male fetal progenitor cells persist in maternal blood for as long as 27 years postpartum. Proc Natl Acad Sci USA. 93:705–708.

Challier JC, Galtier M, Cortez A, Bintein T, Rabreau M, Uzan S. 2005. Immunocytological evidence for hematopoiesis in the early human placenta. Placenta. 26:282–288.

Ellmeier W, Sawada S, Littman DR. 1999. The regulation of CD4 and CD8 coreceptor gene expression during T cell development. Annu Rev Immunol. 17:523–554.

Evans PC, Lambert N, Maloney S, Furst DE, Moore JM, Nelson JL. 1999. Longterm fetal microchimerism in peripheral blood mononuclear cell substets in healthy women and women with scleroderma. Blood. 93:2033–2037.

Fukuchi Y, Nakajima H, Sugiyama D, Hirose I, Kitamura T, Tsuji K. 2004. Human placenta-derived cells have mesenchymal stem/progenitor cell potential. Stem Cells. 22(5):649–658.

Gekas C, Dieterlen-Lievre F, Orkin SH, Mikkola HK. 2005. The placenta is a niche for hematopoietic stem cells. Dev Cell. 8:365–375.

Guetta E, Gordon D, Simchen MJ, Goldman B, Barkai G. 2003. Hematopoietic progenitor cells as targets for non-invasive prenatal diagnosis: Detection of fetal CD34+ cells and assessment of post-delivery persistence in the maternal circulation. Blood Cells Mol Dis. 30(1):13–21.

Guettier C, Sebagh M, Buard J, Feneux D, Ortin-Serrano M, Gigou M, Tricottet V, Reynes M, Samuel D, Feray C. 2005. Male cell microchimerism in normal and diseased female livers from fetal life to alduthood. Hepathology. 42:35–43.

In 't Anker PS, Scherjon SA, Kleijburg-van der Keur C, de Groot-Swings GM, Claas FH, Fibbe WE, Kanhai HH. 2004. Isolation of mesenchymal stem cells of fetal or maternal origin from human placenta. Stem Cells. 22(7):1338–1345.

Jimenez DF, Leapley AC, Lee CI, Ultsch MN, Tarantal AF. 2005. Fetal CD34+ cells in the maternal circulation and long-term microchimerism in rhesus monkeys (Macaca mulatta). Transplantation. 79(2):142–146.

Khosrotehrani K, Jonhson KL, Cha DH, Salomon N, Bianchi DW. 2004. Transfer of fetal cells with multilineage potential to maternal tissue. JAMA. 292:75–80.

Khosrotehrani K, Johnson KL, Guégan S, Stroh H, Bianchi DW. 2005a. Natural history of fetal cell microchimerism during and following murine pregnancy. J Reprod Immunol. 66:1–12.

Khosrotehrani K, Bianchi DW. 2005b. Multi-lineage potential of fetal cells in maternal tissue: A legacy reverse. J Cell Sci. 118:1559–1563

Khosrotehrani K, Reyes RR, Johnson KL, Freeman RB, Salomon RN, Peter I, Stroh H, Guégan S, Bianchi DW. 2007. Fetal cells participate over time in the response to specific types of murine maternal hepatic injury. Hum Reprod. 22:654–661.

Khosrotehrani K, Leduc M, Bachy V, Nguyen Huu S, Oster M, Abbas A, Uzan S, Aractingi S. 2008. Pregnancy allows the transfer and differentiation of fetal lymphoid progenitors into functional T and B cells in mothers. J Immunol. 180:889–897.

Lambert NC, Lo YM, Erickson TD, Tylee TS, Guthrie KA, Furst DE, Nelson JL. 2002. Male microchimerism in healthy women and women with scleroderma: Cells or circulating DNA? A quantitative answer. Blood. 100:2845–2851.

Nguyen Huu S, Oster M, Uzan S, Chareyre F, Aractingi S, Khosrotehrani K. 2007. Maternal neoangiogenesis during pregnancy partly derives from fetal endothelial progenitor cells. Proc Natl Acad Sci USA. 104:1871–1876.

O'Donoghue K, Choolani M, Chan J, de la Fuente J, Kumar S, Campagnoli C, Bennett PR, Roberts IA, Fisk NM. 2003. Identification of fetal mesenchymal stem cells in maternal blood: Implications for non-invasive prenatal diagnosis. Mol Hum Reprod. 9(8):497–502.

O'Donoghue K, Chan J, de la Fuente J, Kennea N, Sandison A, Anderson JR, Roberts IA, Fisk NM. 2004. Microchimerism in female bone marrow and bone decades after fetal mesenchymal stem-cell trafficking in pregnancy. Lancet. 364(9429):179–182.

Osada H, Doi S, Fukushima T, Nakauchi H, Seki K, Sekiya S. 2001. Detection of fetal HPCs in maternal circulation after delivery. Transfusion. 41(4):499–503.

Ottersbach K, Dzierzak E. 2005. The murine placenta contains hematopoietic stem cells within the vascular labyrinth region. Dev Cell. 8(3):377–387.

Srivatsa B, Srivatsa S, Johnson KL, Samura O, Lee SL, Bianchi DW. 2001. Microchimerism of presumed fetal origin in thyroid specimens from women: A case-control study. Lancet. 358:2034–2038.

Stevens AM, McDonnell WM, Mullarkey ME, Pang JM, Leisenring W, Nelson JL. 2004. Liver biopsies from human females contain male hepatocytes in the absence of transplantation. Lab Invest. 84:1603–1609.

Tan XW, Liao H, Sun L, Okabe M, Xiao ZC, Dawe GS. 2005. Fetal microchimerism in the maternal mouse brain: A novel population of fetal progenitor or stem cells able to cross the blood-brain barrier? Stem Cells. 23(10):1443–1452.

Valerio D, Altieri V, Antonucci FR, Aiello R. 1997. Characterization of fetal hematopoietic progenitors circulating in maternal blood of seven aneuploid pregnancies. Prenat Diagn. 17(12): 1159–1169.

Wang Y, Iwatani H, Ito T, Horimoto N, Yamato M, Matsui I, Imai E, Hori M. 2004. Fetal ells in mother rats contribute to the remodeling of liver and kidney after injury. Biochem Biophys Res Commun. 325:961–967.

Yen BL, Huang HI, Chien CC, Jui HY, Ko BS, Yao M, Shun CT, Yen ML, Lee MC, Chen C. 2005. Isolation of multipotent cells from human term placenta. Stem Cells. 23(1):3–9.

Yen ML, Chien CC, Chiu IM, Huang HI, Chen YC, Hu HI, Yen BL. 2007. Multilineage differentiation and characterization of the human fetal osteoblastic 1.19 cell line: A possible in vitro model of human mesenchymal progenitors. Stem Cells. 25(1):125–131.

Yen BL, Chien CC, Chen YC, Chen JT, Huang JS, Lee FK, Huang HI. 2008. Placenta-derived multipotent cells differentiate into neuronal and glial cells in vitro. Tissue Eng Part A. 14(1):9–17.

INDEX

Perinatal Stem Cells. Edited by C. L. Cetrulo, K. J. Cetrulo, and C. L. Cetrulo, Jr.